The
Shikimate
Pathway

Biosynthesis of Natural Products Series

Consultant Editor
Sir Derek Barton, F.R.S.
Professor of Organic Chemistry
Imperial College of Science and Technology

Series Editor
Gordon Kirby
Regius Professor of Chemistry
Glasgow University

Endpaper. Metabolites of the Shikimate Pathway

The Shikimate Pathway

Edwin Haslam
Reader in Chemistry, University of Sheffield

London Butterworths

THE BUTTERWORTH GROUP

ENGLAND
Butterworth & Co (Publishers) Ltd
London: 88 Kingsway, WC2B 6AB

AUSTRALIA
Butterworths Pty Ltd
Sydney: 586 Pacific Highway, NSW 2067
Melbourne: 343 Little Collins Street, 3000
Brisbane: 240 Queen Street, 4000

CANADA
Butterworth & Co (Canada) Ltd
Toronto: 14 Curity Avenue, 374

NEW ZEALAND
Butterworths of New Zealand Ltd
Wellington: 26–28 Waring Taylor Street, 1

SOUTH AFRICA
Butterworth & Co (South Africa) (Pty) Ltd
Durban: 152–154 Gale Street

First published 1974

© Butterworth & Co (Publishers) Ltd, 1974

ISBN 0 408 70569 8

Filmset and printed in England by
Cox & Wyman Ltd, London, Fakenham and Reading

Preface

Many of the organic chemist's traditional natural products exhibit life cycles with strikingly similar characteristics. Encapsulated within the early part of the cycle are a few brief moments of fame and glory—the possession of some novel structural feature or the display of a unique chemical reactivity—but once these have been assimilated into the mainstream of organic chemistry, the life of the natural product almost invariably decays once more into gentle obscurity. A few are spared this fate. For some man finds a profitable use and for a very few rescue from oblivion occurs when it is discovered that they occupy a key position in one of life's vital processes. Such a natural product is shikimic acid which was discovered almost ninety years ago by J. F. Eykmann in *Illicium* species. Its biochemical significance was not realised, however, until the 1950s when Bernard Davis recognised it as an important intermediate in one of the major pathways of metabolism of aromatic compounds in nature. Characteristically this pathway now bears its name—the shikimate pathway.

Studies of metabolic pathways may be carried out at different levels and from differing viewpoints which pose questions and problems to many scientific disciplines. The present text attempts to give a bird's eye view of the shikimate pathway and an appreciation of its implications for the life of a range of organisms. As such it is hoped that it will be of general value to practitioners in the many and varied areas of biochemical research associated with metabolism. The selection and manner of presentation of the material inevitably reflects, however, the author's own interests and commitments in this field and whilst this has led to an abbreviated discussion, and occasional omission, of some important topics it has hopefully permitted a more detailed and balanced presentation of others.

It is a pleasure to record my gratitude to a number of colleagues for their advice and comments and in particular to Dr John Guest of the Department of Microbiology, University of Sheffield, for his detailed and constructive criticisms of some of the more biochemical material in the text. My thanks are also due to my wife who prepared

Preface

the manuscript for the publishers and who, in doing so, eliminated many of my more extravagant outrages against the Queen's English. Finally it is appropriate to record my indebtedness to three organic chemists: Professor R. D. Haworth and Lord Todd, under whom I first studied and who sowed the seeds of my interest (albeit slowly to bear fruit) in this area of research, and latterly Professor W. D. Ollis who has provided the much needed stimulus and the environment for its practice.

Chemistry Department   E. Haslam
University of Sheffield

Foreword

The diverse pathways by which complex organic molecules are synthesised in Nature have long excited the curiosity of natural product chemists and of biochemists.

Early biosynthetic theories had two main foundations. First, common building-units were discerned within a series of related natural product structures. Second, the mechanisms involved in the linkage of units and the transformation of intermediates were assumed to resemble, more or less closely, those of processes familiar in the chemical laboratory. With the advent of labelling techniques, employing both radioactive and stable isotopes, biosynthetic theories have been subjected to intensive experimental investigation and many shown, in fact, to be well-founded.

The pathways leading to many important classes of natural products are now known in impressive detail. Furthermore, the dynamic relationships between the primary metabolic processes common to most organisms and the secondary pathways leading to more complex, specialised metabolites are now becoming much clearer. Notably, in recent years, the use of precursors labelled stereoselectively with tritium and deuterium has provided fascinating insight into the subtle stereochemical control characteristic of most enzymically catalysed reactions.

The publication of a series of monographs on natural product biosynthesis is timely. Each volume contains a comprehensive and authoritative survey of selected areas of the subject providing the specialist with a perspective view of his field of research and the newcomer with the information and stimulus to plan his own investigations.

It is most fortunate that the Series Editor, who is so well qualified for the task, has accepted the onerous responsibilities of Editorial Office, and equally fortunate that he has been able to assemble so able and authoritative a team of authors.

Professor Sir Derek Barton, F.R.S.

For Margaret

Contents

Introduction

Knowledge concerning the structure of naturally occurring compounds has grown at a tremendous rate over the last ten to fifteen
years and this dramatic progress has been made possible largely by
the advent of a whole array of new physical methods of structural
analysis, many of which were unheard of thirty years ago. Knowledge of both the chemical and physiological properties of this
multitude of new substances has increased at a far more modest rate
and not surprisingly this discrepancy has begun to profoundly influence the whole philosophy of natural products research. Formerly
structure determination itself was the major goal but increasingly
this object is seen to be a restricted one and only the beginning rather
than the end of the story.

One facet of this continuing drama concerns the biogenesis of
natural products since from the biological point of view, knowledge
of the biogenesis of a substance is just as important as knowledge of
its structure. Biochemical studies of fundamental importance have
revealed the existence of several general biosynthetic pathways to
natural products, such as the acetate–malonate pathway, the mevalonate route to isoprenoids and the shikimate pathway of aromatic
amino acid metabolism. The purpose of this text is to broadly
illustrate the nature and the importance of the shikimate pathway
and to demonstrate its involvement in the biosynthesis of many
natural products, particularly those containing aromatic structures.

The book comprises six chapters and the first of these is devoted
to a description of the steps in the pathway which lead from carbohydrate to the individual aromatic amino acids and the manner in
which the differing strategies of control of the flow of metabolites
are exercised. One chapter discusses the chemistry of important
intermediates on the pathway and draws attention to the ways in
which specifically isotopically labelled precursors may be prepared.
The remaining four sections describe the numerous and structurally
diverse metabolites which are derived from the aromatic amino
acids or from intermediates in the pathway. Identical compounds in
the chemical sense are often synthesised in different ways in different

organisms and the shikimate pathway provides many examples of metabolites of this type. Wherever possible therefore the distinction between organisms has been maintained and the final two chapters are thus concerned exclusively with those metabolites obtained from higher plants.

Several metabolites of the shikimate pathway such as the various isoprenoid quinones and the catecholamines have clearly defined biochemical roles in the life of the organism. However, for a great many aromatic metabolites, particularly those from higher plants, their physiological function, if any, is poorly understood. This represents a problem of considerable magnitude for future work in this area and it is one to which attention should now be given.

NOMENCLATURE AND ABBREVIATIONS

Standard abbreviations used in both the biochemical and chemical literature have been employed in the text and are listed in the Biochemical Journal publication, 'Policy of the Journal and Instructions to Authors' issued by the Biochemical Society, London, 1972. Wherever possible enzymes have been named at the first mention by use of the EC (Enzyme Commission) numbers (Recommendations of the International Union of Biochemistry on the nomenclature and classification of Enzymes, Elsevier, Amsterdam, 1965).

Structural formulae have been given for most of the compounds discussed in the text and the major change in nomenclature which has been utilised concerns the numbering of the ring systems of both $(-)$-shikimic acid and $(-)$-quinic acid. Revised systems of numeration have been employed for these compounds to accord respectively with normal IUPAC procedures and with the tentative rules for cyclitol nomenclature (IUPAC information bulletin, No. 32, 1968; *European J. Biochem.*, 1968, **5**, 1).

Organic acids have been named both as the anionic form (e.g. shikimate) and as the free acid (e.g. shikimic acid) to accord with the context of the discussion. Where the emphasis in the text is largely chemical, they are named as acids but where the emphasis is biochemical and the transformations most probably involve the anion of the acid, they are named as anions.

1

The Shikimate Pathway: Biosynthesis of the Aromatic Amino Acids

1.1 INTRODUCTION

Organisms differ markedly in their ability to carry out the chemical reactions involved in the biosynthesis of the amino acids which are constituents of proteins. Most micro-organisms and plants are competent in all such syntheses whilst most animals lack about half of these capabilities. Amongst the amino acids which cannot be produced by *de novo* synthesis in animals are the three aromatic amino acids L-phenylalanine (1), L-tyrosine (2) and L-tryptophan (3). In plants and micro-organisms they are formed by a metabolic process known as the shikimate pathway[1, 2].

(−)-Shikimic acid was first described as a natural product from the plant *Illicium religiosum* Sieb. by Eykmann in 1885 and it was from the Japanese name of this plant, shikimi-no-ki, that the name shikimic acid was derived. Eykmann's own investigations coupled with the detailed structural and stereochemical studies of Fischer and Dangschat[5–8], Freudenberg[9] and Karrer[10] led to the formulation of both (−)-shikimic acid and the closely related natural product (−)-quinic acid as (4) and (5) respectively*. Fischer and Dangschat[5, 7] also remarked on the close structural similarity of known aromatic

---

* The numeration of the ring carbon atoms in both (−)-shikimic acid and (−)-quinic acid has been revised to accord with IUPAC nomenclature rules. Thus (−)-shikimic acid is numbered through the double bond and (−)-quinic acid is numbered according to the tentative rules for the nomenclature of cyclitols (*European J. Biochem.*, 1968, **5**, 1–12). Both of these forms of numeration conflict with earlier forms established by usage in the chemical and biochemical literature.

4 Biosynthesis of the Aromatic Amino Acids

secondary plant metabolites such as gallic acid (6) to both (−)-shikimic acid and (−)-quinic acid and they suggested a tentative biogenetic relationship. However, the full significance of this observation and the realisation that (−)-shikimic acid was an obligatory intermediate in an important biochemical pathway was not recognised until the early 1950s when the elegant work of Davis[1], Sprinson[2] and later Gibson[11] and their collaborators revealed the nature of the metabolic pathway which leads from carbohydrate to the aromatic amino acids (1–3). (−)-Shikimic acid was the first of the intermediates in this pathway to be established[12] and this early association led predictably to the pathway's familiar name.

1.2 THE COMMON PATHWAY

An outline of the shikimate pathway from carbohydrate through chorismate to the aromatic amino acids and other metabolically important compounds is shown in *Figure 1.1*. The major branch point occurs at chorismate and that part of the metabolic sequence from carbohydrate to chorismate is generally referred to as the common pathway.

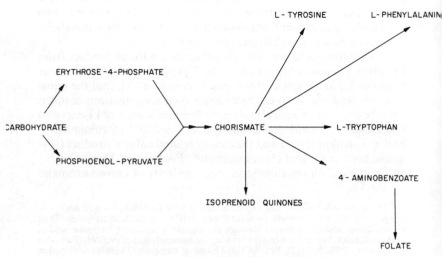

Figure 1.1. An outline of the Shikimate Pathway

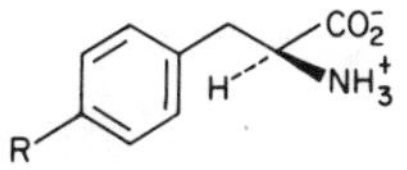

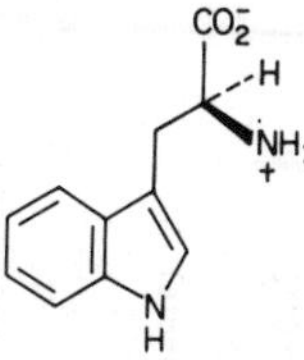

(1, R=H, L-phenylalanine)
(2, R=OH, L-tyrosine)

(3, L-tryptophan)

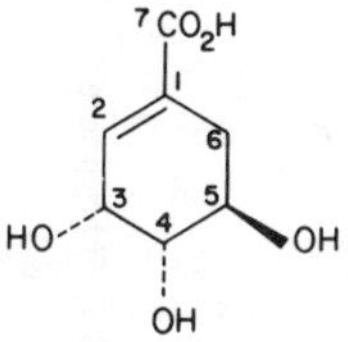

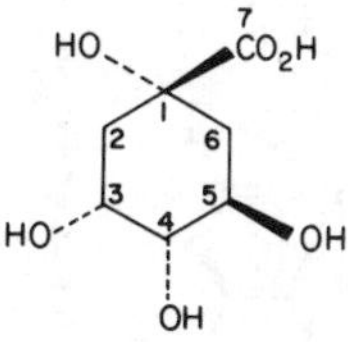

(4, shikimic acid)

(5, quinic acid)

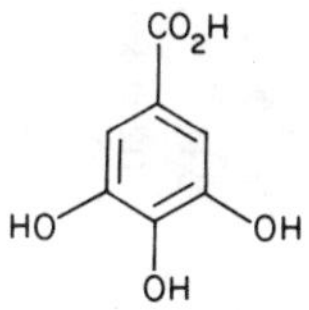

(6, gallic acid)

1.2.1   METHODOLOGY AND IDENTIFICATION OF INTERMEDIATES

The elucidation of the steps in the common part of the pathway from carbohydrate precursors to the formation of chorismate remains a classic example of the combined use of auxotrophic mutants, isotopically labelled precursors and finally enzyme studies to define a metabolic sequence. An auxotrophic mutant of an organism is a nutritional mutant. The genetic change in such mutant strains causes an inability to carry out one of the reactions in a biosynthetic pathway leading to one or more essential metabolites. Analysis of the sequence of steps in the biosynthetic pathway is facilitated by the fact that each mutant may accumulate in its culture fluid the substrate for the enzymic reaction which is blocked and may utilise later members of the sequence to replace the required metabolites[13, 14].

Davis[15, 16] produced auxotrophic mutants of *Escherichia coli*, *Aerobacter aerogenes*, *Bacillus subtilis* and *Salmonella typhimurium* by ultra-violet irradiation of a bacterial culture in a minimal medium of salts plus D-glucose. Selection of the mutant strains was dependent on the use of the observation[17] that penicillin kills bacterial cells only

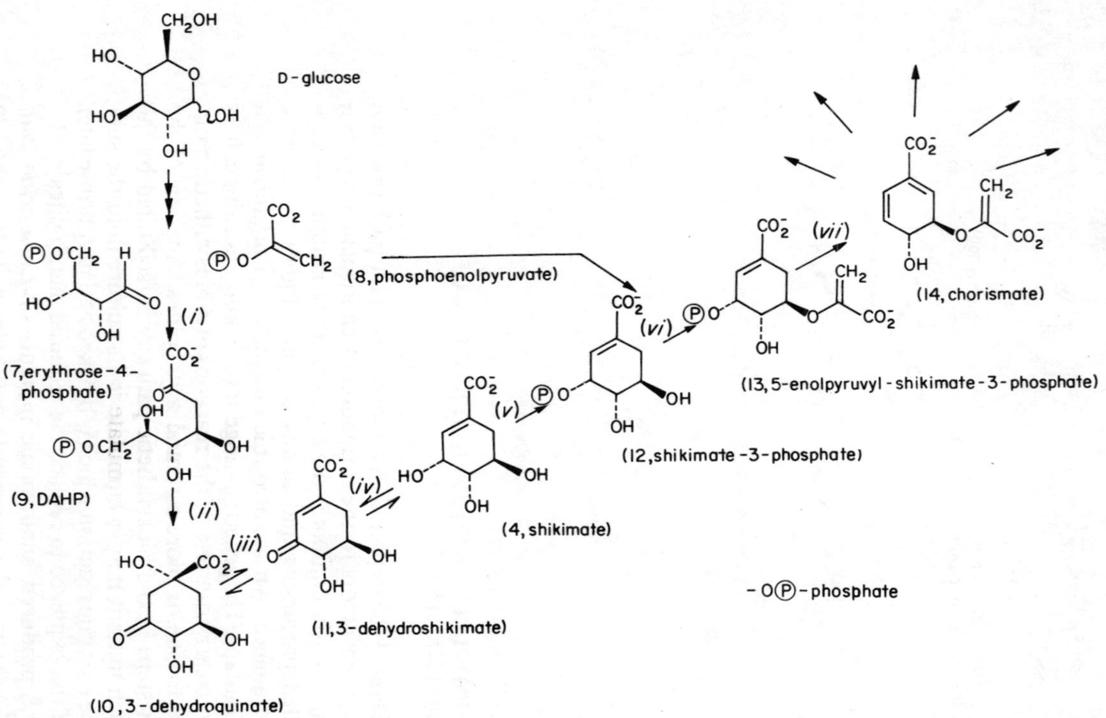

Figure 1.2. The biosynthesis of chorismate; the common pathway in micro-organisms. All acids are formulated in their anionic forms:

(i) DAHP synthetase
(ii) 3-Dehydroquinate synthetase (NAD$^+$, Co^{2+})
(iii) 3-Dehydroquinate dehydratase
(iv) 3-Dehydroshikimate reductase (NADPH)

(v) Shikimate kinase (ATP)
(vi) 5-Enolpyruvylshikimate-3-phosphate synthetase (PEP)
(vii) Chorismate synthetase

under conditions of growth. The irradiated culture was first grown on a medium supplemented with yeast extract and casein hydrolysate to produce a mixture of wild type and mutant strains and then finally on the minimal medium plus penicillin. Since only the wild type strains were able to grow under these latter conditions these were the only forms of the bacterium killed by the penicillin. The resting cells of the mutant forms were unaffected and were subsequently isolated. Essentially the same method was developed simultaneously and independently by Lederberg and Zinder[18].

Mutant strains of *Escherichia coli* and *Aerobacter aerogenes* were described[15, 16] which had a quintuple requirement of aromatic substrates (L-phenylalanine, L-tyrosine, L-tryptophan, 4-amino-benzoate and 4-hydroxybenzoate) for growth. Certain of these mutants were found to accumulate (−)-shikimic acid (4) in their culture filtrates and other mutants, blocked in earlier reactions in the pathway, were able to utilise (−)-shikimic acid (4) to replace the aromatic substrates. These observations[12, 13] established with great probability that (−)-shikimic acid was a common precursor for each of these aromatic compounds. Experiments of this type permitted each of the intermediates in the common pathway, 3-dehydroquinic acid[19] (10), 3-dehydroshikimic acid[20] (11), (−)-shikimic acid[12, 13] (4), shikimic acid-3-phosphate[21, 22] (12), 5-enolpyruvylshikimic acid-3-phosphate[23, 24] (13) and chorismic acid[25, 26] (14), to be isolated and characterised and for the pathway between 3-dehydroquinic acid (10) and chorismic acid (14) to be formulated (*Figure 1.2*). It may be noted, however, that, apart from an aromatic mutant of *Bacillus subtilis*, bacterial mutants blocked between shikimate and shikimate-3-phosphate (*Figure 1.2*) have not been isolated. Failure to obtain mutants of this class was first noted by Davis and Mingioli[14] and the identification of shikimate-3-phosphate (12) as an intermediate was based primarily on the isolation of this compound in nearly all of the aromatic auxotrophs which accumulated (−)-shikimic acid.

1.2.2 ENZYMOLOGY

Enzymes involved as catalysts in each of the steps from 3-dehydro-quinate (10) to chorismate (14) in the common pathway have all been subsequently isolated and characterised from bacterial mutants. Methods of assay for each form of activity have been described[27]. Mitsuhashi and Davis[28] first isolated 3-dehydroquinate dehydratase (E.C. 4.1.2.10) the enzyme which is responsible for the dehydration step (10→11). With a partially purified extract they showed that the equilibrium ratio of 3-dehydroshikimate (11) to 3-dehydroquinate (10) was 15 at pH 7.4 and is thus favourable to the forward reaction in the direction of biosynthesis. 3-Dehydroshikimate reductase[29] (E.C. 1.1.1.25) has a requirement for NADPH as cofactor and a pH optimum of 8.5. The equilibrium constant (2.8×10^8) determined at neutral pH again strongly favours the forward reaction.

The ability to phosphorylate shikimic acid in the presence of ATP has been demonstrated for many micro-organisms and the enzyme shikimate kinase which is responsible for this activity has been shown to operate at a pH optimum of 7.0 and to have a requirement for magnesium or manganese ions[30, 31]. Morell and Sprinson[32] observed the presence of two shikimate kinase iso-enzymes in *Salmonella typhimurium* and in *Bacillus subtilis* a similar observation has been reported[33]. One of the enzyme activities in *Bacillus subtilis* was part of a multi-enzyme aggregate and the other was in a minor protein fragment which was present outside the aggregate. The existence of two shikimate kinase iso-enzymes, Morell and Sprinson suggested[32], might account for the absence in many bacteria of single gene mutants lacking shikimate kinase activity.

The enzyme 5-enolpyruvylshikimate-3-phosphate synthetase was obtained from *Escherichia coli* K-12 mutant 58-278 by Levin and Sprinson[24] and it mediates the transfer of the enolpyruvyl side chain to the hydroxyl group at position 5 of shikimate-3-phosphate (12). At its pH optimum of 5.4–6.2 the enzyme showed a ready reversibility and an equilibrium constant of 12 was estimated. The final enzyme in the common pathway is chorismate synthetase, which catalyses the 1,4- conjugate elimination of phosphate from 5-enolpyruvylshikimate-3-phosphate (13) to give chorismate (14). It was isolated from cell free extracts of *Escherichia coli* B-37 and had a pH optimum of 7.8–8.5. Under aerobic conditions the enzyme was inactive and was activated most effectively by a reduced FAD regenerating system in an atmosphere of hydrogen or nitrogen. The enzyme was also slowly activated by NADH and it was suggested that ferrous ion composed a labile part of the enzyme system which was responsible for its sensitivity towards oxygen.

1.2.3 THE INITIAL STEPS

A condensation between phosphoenolpyruvate (8) and D-erythrose-4-phosphate (7) initiates the common part of the shikimate pathway[2]. Nutritional studies with bacterial mutants failed initially to reveal any intermediates earlier in the pathway than the first cyclic compound 3-dehydroquinic acid (10) and elucidation of the first two steps in the pathway relied on isotopic tracer studies and the isolation of the appropriate enzymes. Variously labelled ^{14}C derivatives of D-glucose (15) were converted to (−)-shikimic acid (4) by *Escherichia coli* mutants and the distribution of the label in the (−)-shikimic acid was determined by chemical degradation[35] (*Figure 1.3*).

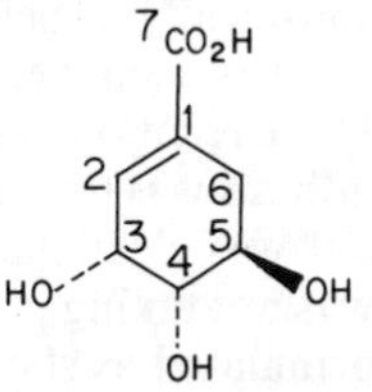

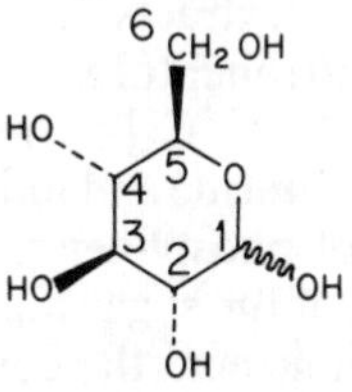

C-atom in (−)-shikimic acid	C-atom in D-glucose	fraction
1	2	0·4
	5	0·5
2	1	0·25
	6	0·6
3	2	0·22
	5	0·6
4	3 or 4	0·9
5	2	0·24
	3 or 4	0·59
6	1	0·4
	6	0·5
7	3 or 4	0·86

Figure 1.3. Distribution of isotopic label in (−)-shikimic acid from D-glucose-[^{14}C]

This pattern of labelling, when it was analysed, prompted Sprinson[2] to propose that in *Escherichia coli* (−)-shikimic acid was biosynthesised from a three carbon fragment of glycolysis (8) and a four carbon sugar (7) which was derived from the pentose phosphate pathway of carbohydrate metabolism.

The identification of these two precursors was greatly facilitated by experiments at the enzymic level and by the observation that cell free extracts of *Escherichia coli* 83-24 were able to convert various sugar phosphates and in particular D-altroheptulose-7-phosphate to (−)-shikimic acid, albeit in low yields (∼ 5 per cent). When D-altroheptulose-7-phosphate and D-fructose-1,6-diphosphate were incubated together the yield of (−)-shikimic acid was almost doubled and it was implied, on the basis of earlier work[37], that these two substrates underwent a series of reactions leading to D-altroheptu-lose-1,7-diphosphate (16) before conversion to (−)-shikimic acid. In support of this idea enzymically prepared D-altroheptulose-1,7-diphosphate was transformed almost quantitatively to (−)-shikimic acid by bacterial extracts[38]. Furthermore isotopically labelled 4,5,6,7-[$^{14}C_4$]-D-altroheptulose-1,7-diphosphate was converted exclusively to 2,3,4,5-[$^{14}C_4$]-(−)-shikimic acid. These reactions were inhibited by both fluoride and iodoacetate but synthesis was restored by the addition of phosphoenolpyruvate. Since it was clear that a direct cyclisation of the sugar diphosphate (16) was not taking place to give (−)-shikimic acid the conversion was formulated as shown in *Figure 1.4*. A sequence of reactions was proposed which involved cleavage of the substrate to 1,3-dihydroxy-2-propanone phosphate (17) and D-erythrose-4-phosphate (7), oxidation of the keto phosphate (17) to phosphoenolpyruvate (8) and finally condensation of (7) and (8) to give (−)-shikimic acid.

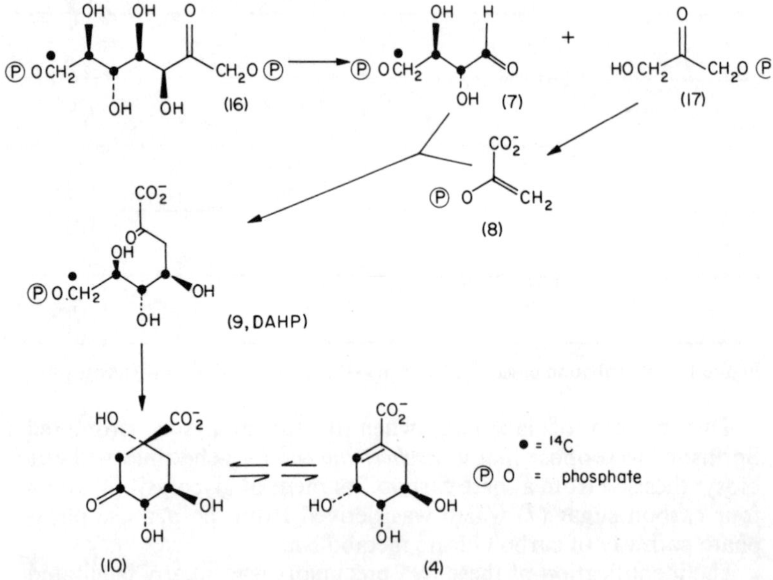

Figure 1.4. Conversion of D-altroheptulose-1,7-diphosphate (16) to (−)-shikimic acid[36, 37]

Sprinson formulated[2] the condensation reaction between the two phosphate esters (7 and 8) as an aldol condensation and predicted that the products of this reaction would be orthophosphate and 3-deoxy-D-arabinoheptulosonic acid-7-phosphate, DAHP, (9). The D-arabino configuration was predicted for (9) since it was assumed that cyclisation of DAHP gave 3-dehydroquinate (10) directly and hence that the absolute configuration at the chiral centres at C-4 and C-5 of (9) corresponded to that at C-4 and C-5 in 3-dehydroquinate (10). This view was supported by chemical synthesis of 3-deoxy-D-arabinoheptulosonic acid-7-phosphate (9) and the synthetic product was converted to 3-dehydroquinate (10) quantitatively by bacterial extracts[39]. The synthetic sample of DAHP (9) was also shown to be identical with the product formed enzymically from phosphoenolpyruvate (8) and D-erythrose-4-phosphate (7) by enzyme preparations which were no longer capable of carrying out the further conversion of DAHP (9) to 3-dehydroquinate (10). This enzyme DAHP synthetase (EC 4.1.2.15) was found to be specific for the substrates (7 and 8), showed no requirement for metal ions and other cofactors and was essentially irreversible[40].

3-Dehydroquinate synthetase, the enzyme responsible for the cyclisation of DAHP (9) to give 3-dehydroquinate (10), the first cyclic intermediate in the shikimate pathway, was obtained in partially purified form[41] from *Escherichia coli*. The enzyme required Co^{2+} and NAD^+ (but not $NADP^+$) for full activity. No intermediates were isolable when these cofactors were removed but it was observed in a kinetic analysis of the enzymic transformation that the release of orthophosphate, the disappearance of DAHP and the formation of 3-dehydroquinate all proceeded at the same rate. These observations indicated, it was suggested, that one enzyme was responsible for the whole sequence of reactions necessary for the conversion. On the basis of these observations, Sprinson and his collaborators formulated a working hypothesis for the steps involved in the cyclisation of the substrate DAHP and this is discussed in more detail later.

1.2.4 THE ROLE OF QUINIC ACID

(−)-Quinic acid (5), which like (−)-shikimic acid (4) is a widely distributed natural product in the plant kingdom[42], is an effective growth factor for *Aerobacter aerogenes* mutants which can respond to 3-dehydroquinate (10)[43]. An NAD^+ dependent quinate dehydrogenase (EC 1.1.1.24) was observed in extracts of such mutants and of wild type strains of *Aerobacter aerogenes*[44]. This enzyme catalysed the reversible dehydrogenation of quinate to 3-dehydroquinate (10) and thus made quinate (5) itself available as a carbon

12 Biosynthesis of the Aromatic Amino Acids

source for aromatic amino acid metabolism. In addition it was possible to select *Neurospora crassa* strains able to grow on quinic acid (5) and enzymic analyses provided evidence for an inducible quinate dehydrogenase in the organism[45, 46]. However, this enzyme was shown to be part of an inducible degradative enzyme system which metabolised 3-dehydroquinate (10) via protocatechuate and the β-ketoadipate pathway and was quite distinct from the synthetic apparatus of the shikimate pathway. All the available evidence favours the conclusion that although some micro-organisms can utilise quinic acid as a metabolite[47] this acid is not a normal intermediate on the shikimate pathway in these systems.

1.3 MECHANISTIC STUDIES OF THE COMMON PATHWAY

The reactions of the shikimate pathway pose a number of interesting questions of both a mechanistic and stereochemical nature. Detailed studies have been made of the DAHP synthetase, 3-dehydroquinate dehydratase, 5-enolpyruvylshikimate-3-phosphate synthetase and chorismate synthetase reactions which have added further important knowledge to this area of molecular biology. Whilst these investigations have done nothing to detract from the important dictum[48] that 'cells obey the laws of chemistry', they have nevertheless revealed some of the distinctive facets of enzyme chemistry and have highlighted some of the important differences between enzyme and related chemically catalysed reactions.

1.3.1 DAHP SYNTHETASE

The mechanism of the condensation of phosphoenolpyruvate (8) and D-erythrose-4-phosphate (7), the DAHP synthetase reaction,

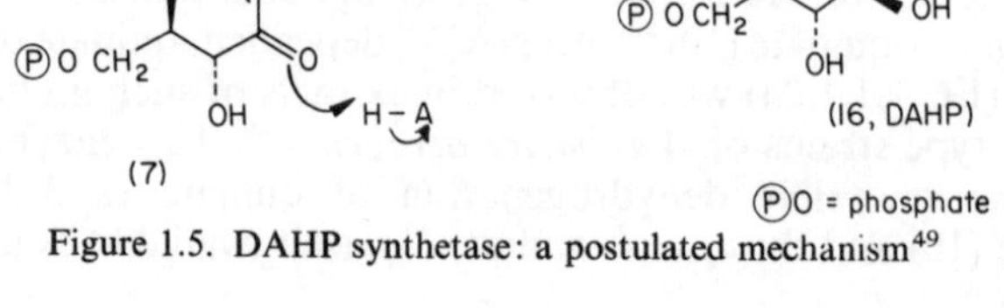

Figure 1.5. DAHP synthetase: a postulated mechanism[49]

has been the subject of some speculation. In 1955 Sprinson and his collaborators postulated[49] a concerted process (*Figure 1.5*). Later kinetic observations[50-52], however, suggested that the DAHP synthetase reaction operated by a 'ping-pong' mechanism[53] and because of its propensity to protect the enzyme against denaturation phosphoenolpyruvate (8) was assumed to be the substrate bound first to the enzyme. DeLeo and Sprinson[54] and Nagano and Zalkin[52] carried out experiments with ^{18}O labelled substrates and showed that the reaction involved C–O as opposed to P–O bond cleavage in the phosphoenolpyruvate (8). On the basis of these results DeLeo and Sprinson[54] proposed the alternative mechanism shown in *Figure 1.6* in which the substrate—phosphoenolpyruvate— was first transferred to a nucleophilic group on the enzyme (for example a carboxylate group) and then reacted with the second substrate D-erythrose-4-phosphate. In support of this mechanism DeLeo and Sprinson showed that DAHP (16) formed in a medium containing tritiated water was extensively labelled, as would be predicted by a mechanism involving the intermediate formation of a methyl group from the methylene of phosphoenolpyruvate (8).

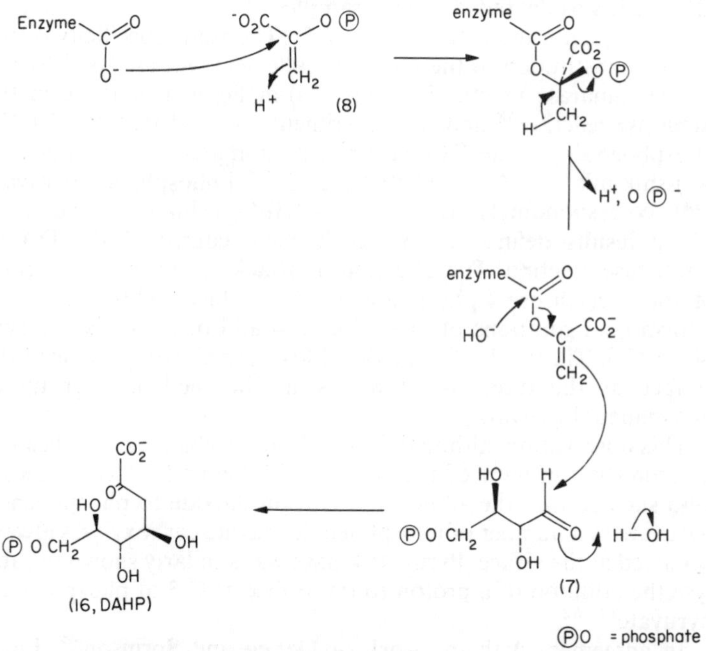

Figure 1.6. DAHP synthetase: a postulated 'ping-pong' mechanism[54]

In contrast the investigations of Floss and his collaborators[55, 56] have shown that in the conversion to DAHP (16) the methylene protons of the majority of molecules of phosphoenolpyruvate (8) cannot have become part of a freely rotating methyl group since the reaction at this carbon atom was demonstrated to occur in a predominantly stereospecific manner. Onderka, Floss and Carroll[56] prepared 3(R)-3-[^{3}H]-phosphoglycerate (19) and 3(S)-3-[^{3}H]-phosphoglycerate (23) from 1-[^{3}H]-D-mannose (18) and 1-[^{3}H]-D-glucose (22) respectively using the glycolytic enzymes, including phosphomannose isomerase or phosphoglucose isomerase. These phosphoglycerates were incubated with phosphoglyceromutase and enolase and gave two specimens of phosphoenolpyruvate tritiated at C-3. Since the enolase reaction has been shown[57] to catalyse the *anti* elimination of the elements of water from 2-phosphoglycerate it was deduced that the tritiated phosphoenolpyruvate from 1-[^{3}H]-D-mannose had the E configuration (20) and that from 1-[^{3}H]-D-glucose had the Z configuration (24). Each sample of phosphoenolpyruvate was converted to shikimate by incubation with D-erythrose-4-phosphate (7) and an enzyme extract from *Escherichia coli* 83-24. Each sample of 6-[^{3}H]-(−)-shikimic acid (21 or 25) was degraded to give samples of 2(R S)-3(^{3}H)-malic acid by the sequence indicated (*Figure 1.7*). The stereospecificity of the tritium substitution in the racemic malic acid was analysed by use of the fumarase reaction[58]. Only 2(S)-malic acid is active in the fumarase reaction[58] and the experiments showed that 3(E)-3-[^{3}H]-phosphoenolpyruvate (20) gave rise predominantly to 6(S)-6-[^{3}H]-(−)-shikimic acid (21) and that 3(Z)-3-[^{3}H]-phosphoenolpyruvate (24) correspondingly gave 6(R)-6-[^{3}H]-(−)-shikimic acid (25). These results defined the stereochemical course of the DAHP synthetase reaction. The direction of attack at the carbonyl group of the D-erythrose-4-phosphate on the *re* face follows from the known configurations of DAHP at C-4 and of (−)-shikimic acid (4) at C-5. The work of Onderka, Floss and Carroll[56] defined the *si* face as the direction of attack at the methylene group of phosphoenolpyruvate.

This observation, although it is without mechanistic significance, is interesting in view of the studies of Rose and his colleagues[59] who showed that the addition of carbon dioxide to phosphoenolpyruvate in a number of phosphoenolpyruvate carboxylases always occurred at the *si* face. Pyruvate kinase was similarly shown to catalyse the addition of a proton to the *si* face at C-3 of phosphoenolpyruvate[60–62].

In agreement with the work of DeLeo and Sprinson[54], Floss, Onderka and Carroll[56] observed some incorporation of tritium from tritiated water at C-3 of DAHP and hence C-6 of shikimate in the enzymic synthesis from phosphoenolpyruvate. Several explanations were considered for this experimental result and the related observation of partial scrambling of the stereospecific tritium label in phosphoenolpyruvate in the conversion to shikimate. The authors favoured the view that an intermediate carbanion is formed at C-3 of phosphoenolpyruvate during the condensation and that this normally attacks the aldehyde carbon of D-erythrose-4-phosphate to give DAHP. The carbanion, in their view, can undergo reversible protonation as a side reaction and hence incorporate tritium from the tritiated water medium.

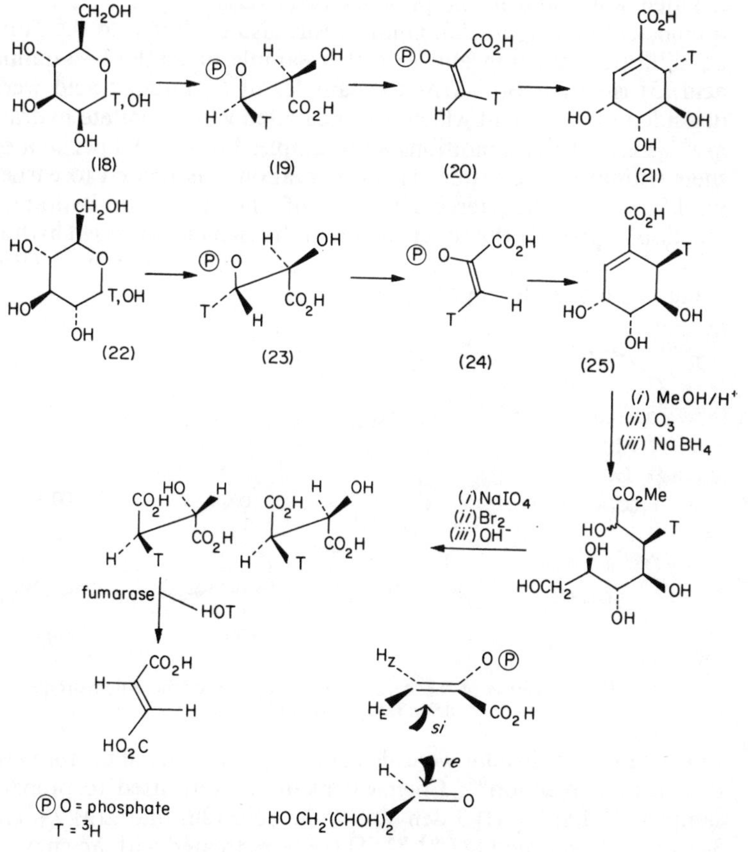

Figure 1.7. Stereochemical features of the DAHP synthetase reaction[55, 56]

1.3.2   3-DEHYDROQUINATE DEHYDRATASE

Hanson and Rose[63] demonstrated that the addition of water to 3-dehydroshikimate (11) to form 3-dehydroquinate (10) catalysed by ·3-dehydroquinate dehydratase follows the unusual stereochemical course of a *cis* addition. Smith, Turner and Haslam[64] showed similarly that the forward biosynthetic reaction is a *cis* elimination reaction. (−)-Quinic acid (5) labelled with tritium in the 2 position was obtained by Hanson and Rose[63] using an extract of *Aerobacter aerogenes* A 170-143 which contained the enzymes quinate dehydrogenase and 3-dehydroquinate dehydratase. 3-Dehydroshikimate was equilibrated with this enzyme extract in tritiated water and in the presence of NADH to give 2-[^{3}H]-(−)-quinic acid (5), *Figure 1.8*. Quinate was also subjected to the same equilibration reaction and a further sample of 2-[^{3}H]-(−)-quinic acid (5) isolated. Both tritiated samples of (−)-quinic acid were degraded to citric acid which was incubated with aconitate hydratase[65]. Under these conditions both samples lost over 95 per cent of their tritium to the water. This observation was related to earlier work[66, 67] on the stereochemistry of the aconitate hydratase reaction to provide the conclusion that the enzyme catalysed hydration of 3-dehydroshikimate (11) to give 3-dehydroquinate (10), occurs with overall *cis* geometry.

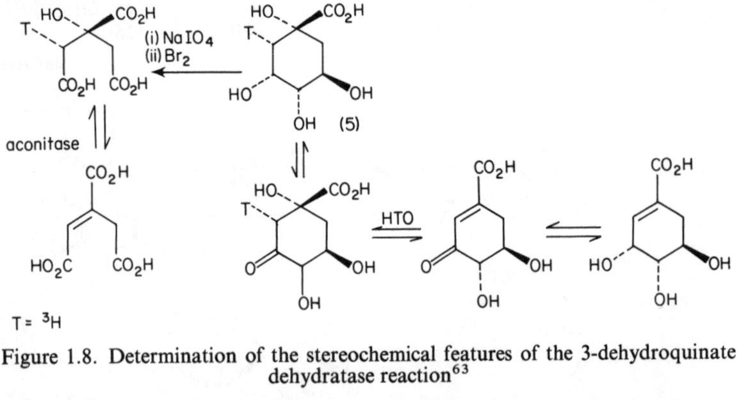

Figure 1.8. Determination of the stereochemical features of the 3-dehydroquinate dehydratase reaction[63]

The same conclusions were also obtained in a study of the forward dehydration reaction[64]. Chemical methods were used to prepare samples of 2(s)-2-[^{2}H]-3-dehydroquinic acid (26) and 2(R)-2-[^{2}H]-3-dehydroquinic acid (27)[64, 68]. These were treated with an enzyme preparation from *Escherichia coli* 83-24 and an enzyme system for the cyclic generation of NADPH to give samples of (−)-shikimic acid (4) which were examined by proton magnetic resonance and mass spectroscopy. Loss of the deuterium label was observed with 2(R)-2-[^{2}H]-3-dehydroquinic acid (27) as substrate and (−)-shikimic acid labelled with deuterium at the 2 position was formed from 2(S)-2-[^{2}H]-3-dehydroquinic acid (26).

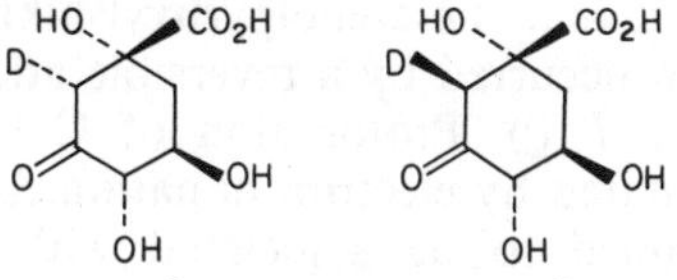

(27, D = ^{2}H)   (26, D = ^{2}H)

The *cis* elimination and addition of water observed in the reversible 3-dehydroquinate dehydratase reaction represents the first example of this type recorded in enzyme chemistry. It contrasts with the known *trans* addition of water to other similar $\alpha\beta$ unsaturated compounds such as *cis*-aconitate and fumarate[66, 67]. Hanson and Rose[63] formulated the addition as taking place in two distinct steps. First addition of water (or hydroxide anion) to give the enol or enolate anion (28) followed by protonation of the enol at the α

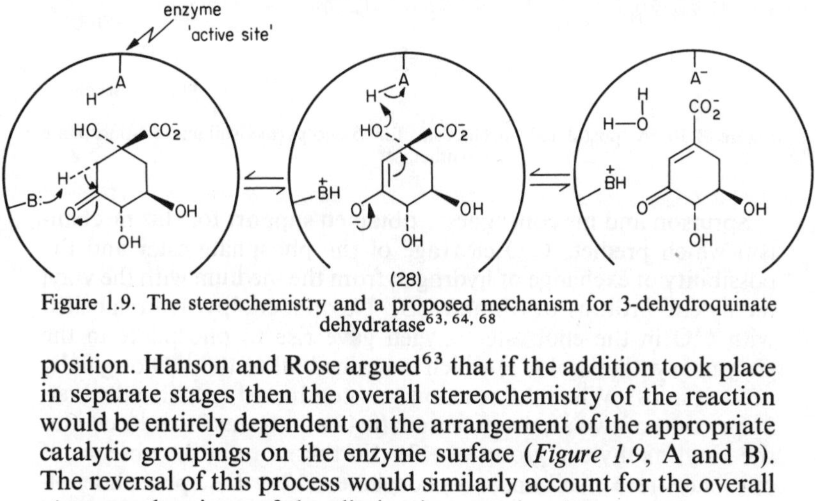

Figure 1.9. The stereochemistry and a proposed mechanism for 3-dehydroquinate dehydratase[63, 64, 68]

position. Hanson and Rose argued[63] that if the addition took place in separate stages then the overall stereochemistry of the reaction would be entirely dependent on the arrangement of the appropriate catalytic groupings on the enzyme surface (*Figure 1.9*, A and B). The reversal of this process would similarly account for the overall *cis* stereochemistry of the elimination reaction (*Figure 1.9*).

1.3.3   5-ENOLPYRUVYLSHIKIMATE-3-PHOSPHATE SYNTHETASE

The formation of 5-enolpyruvylshikimate-3-phosphate (13) represents a rare type of reaction in which the enolpyruvyl fragment of phosphoenolpyruvate (8) is transferred, apparently unchanged, to the molecule of (−)-shikimate-3-phosphate (12). The only other reaction of this class which is known occurs in the formation of UDP-N-acetylenolpyruvylglucosamine, an intermediate in the synthesis of UDP-N-acetylmuramic acid[69, 70]. Levin and Sprinson[23, 24] considered that the 5-enolpyruvylshikimate-3-phosphate synthetase reaction occurred by a reversible addition–elimination mechanism (*Figure 1.10*). Protonation of C-3 of phosphoenolpyruvate (8), facilitated by electron donation from the enol ester oxygen, was assumed to be associated with the synchronous nucleophilic attack at C-2 of the substrate by the 5-hydroxyl group of shikimate-3-phosphate (12). In the second stage of the reaction elimination of orthophosphate from the intermediate (29) yielded the product (13).

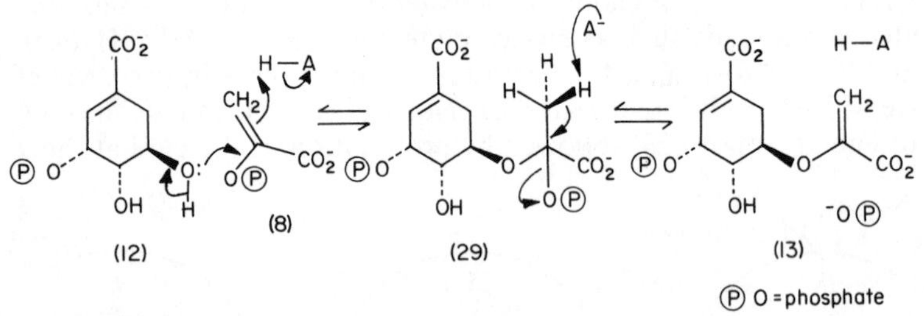

Figure 1.10. A postulated mechanism for 5-enolpyruvylshikimate-3-phosphate synthetase[23, 24]

Sprinson and his colleagues[71] obtained support for this mechanism which predicts C-O cleavage of the phosphate ester and the possibility of exchange of hydrogen from the medium with the vinyl methylene protons of (8) and (13). Phosphoenolpyruvate labelled with ^{18}O in the enol ester oxygen gave rise to phosphate in the enzymic synthesis of (13) which contained all of the ^{18}O originally present in (8). When the reaction was conducted in tritiated water, 21 per cent of two atoms of tritium (0.42 atom) was introduced into the vinyl methylene group of (13). Furthermore in the absence of one substrate, shikimate-3-phosphate (12), the phosphoenolpyruvate (8) was not labelled with tritium when it was equilibrated with the enzyme in tritiated water. 5-Enolpyruvylshikimate-3-phosphate (13) and phosphoenolpyruvate (8) isolated from an enzymic reaction in deuteriated water were both shown to contain deuterium. More than one atom of deuterium was introduced into (13) in this reaction and mass spectroscopy showed it to be distributed in d_0 (8 per cent), d_1 (53 per cent) and d_2 (39 per cent) species.

Analysis of the proton magnetic resonance spectrum indicated that the deuterium in (13) was confined entirely to the vinyl methylene group and was distributed equally between each of the two positions. Sprinson and his group[71] interpreted these results in terms of the general mechanism shown in *Figure 1.10*. They also suggested that a methyl group with unrestricted rotation was formed in the intermediate (29) in order to explain the results of the deuteriation experiment.

1.3.4 CHORISMATE SYNTHETASE

A 1,4-conjugate elimination of phosphoric acid transforms 5-enolpyruvylshikimate-3-phosphate (13) to chorismate (14). Two groups[55, 56, 72], working independently, both arrived at the conclusion that the 6 pro-R hydrogen atom was labile in the chorismate synthetase reaction and hence that the reaction occurred with an overall *trans*-geometry of the eliminated groups. Hill and Newkome[72] prepared synthetically two samples of ($\pm$)-shikimic acid in which the 6 pro-R and 6 pro-S protons were respectively replaced by deuterium in the natural form of the acid. Each sample was converted to L-phenylalanine and L-tyrosine using the organism *Escherichia coli* 156-53M31 and the aromatic amino acids were analysed for their deuterium content by mass spectroscopy. Onderka, Carroll and Floss[56] used specimens of 6(R) and 6(S)-6-[^{3}H]-($-$)-shikimic acid (25, 21) prepared enzymically as above. After admixture with 7-[^{14}C]-($-$)-shikimic acid they were converted separately to chorismic acid by cell free extracts of *Aerobacter*

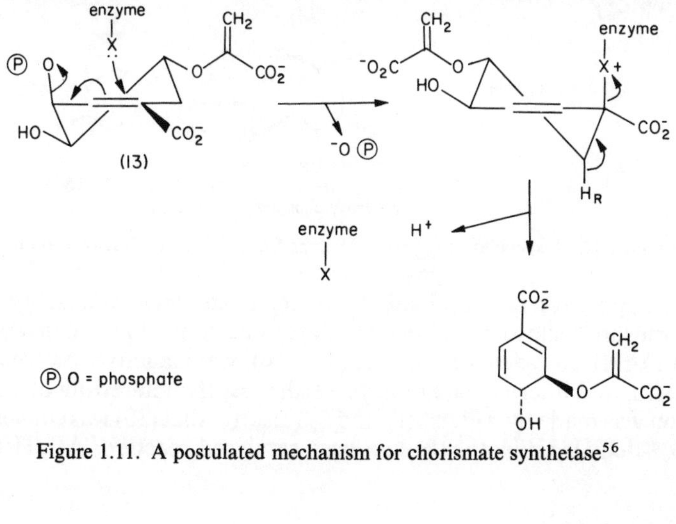

Figure 1.11. A postulated mechanism for chorismate synthetase[56]

aerogenes 62-1. The changes in $^3H/^{14}C$ ratio observed in these reactions showed that only the 6(R)-6-[^{3}H]-shikimic acid suffered a substantial loss of tritium in the conversion to chorismic acid. A two stage X group mechanism[73] was postulated in order to account for the *trans* stereochemistry of the process (*Figure 1.11*).

1.3.5   MISCELLANEOUS PROBLEMS

Major unsolved problems still surround the mechanism by which DAHP (9) is transformed enzymically to 3-dehydroquinate (10) in the early stages of the shikimate pathway (*Figure 1.2*). No intermediates have been detected in this reaction and Sprinson, Rothschild and Srinivasan[39] proposed a mechanism for the enzyme which is shown in *Figure 1.12*. In this sequence loss of orthophosphate from DAHP (9) was facilitated by oxidation at C-5. Selective reduction at C-5 then gave 3,7-dideoxy-D-threohepto-2,6-diulosonic acid (30) and finally an internal aldol condensation was postulated to give 3-dehydroquinate (10). The oxidation-reduction sequence was assumed to involve the cofactor NAD^+ which was required in the enzymic conversion.

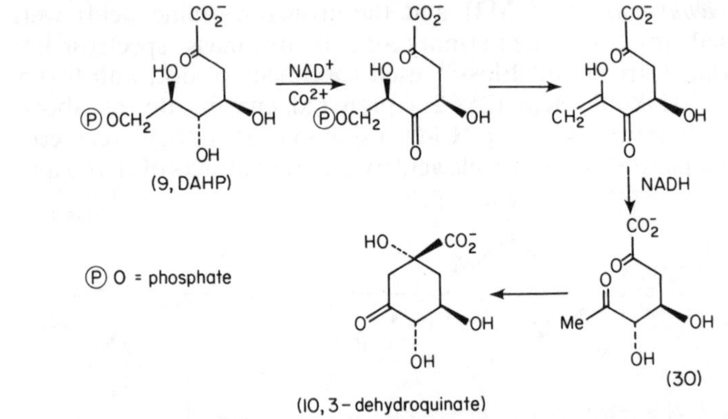

Figure 1.12. A speculative sequence of steps from DAHP to 3-dehydroquinate[39]

Support for this mechanism was removed when Adlersberg and Sprinson[74] showed that bacterial extracts, active in the conversion of DAHP to 3-dehydroquinate (9 → 10), were inactive towards the postulated intermediate (30). Nevertheless, the interesting observation was made by Adlersberg and Sprinson[74] that (30) was converted to 3-dehydroquinate (10) in a base catalysed reaction. At pH 11 a

72 per cent conversion was recorded and at neutral pH and 37° a 50 per cent yield was obtained in an imidazole buffer.

1.4 BIOSYNTHESIS OF THE AROMATIC AMINO ACIDS FROM CHORISMATE

The main stem of the shikimate pathway terminates at chorismic acid (14), *Figure 1.2*. From this point five synthetic routes to essential metabolites—the three aromatic amino acids, *p*-amino benzoic acid and the folate group of co-enzymes and the isoprenoid quinones—branch out (*Figure 1.1*).

1.4.1 THE L-PHENYLALANINE AND L-TYROSINE PATHWAYS

Two separate routes of biosynthesis diverge from chorismate (14) and lead to the amino acids L-phenylalanine (1) and L-tyrosine (2), *Figure 1.13*. Both pathways pass through the same intermediate prephenate (31) which may be envisaged to be derived from (14) by a Claisen rearrangement. Rather unusual circumstances surrounded the discovery of prephenic acid. L-Phenylalanine auxotrophs of *Escherichia coli* were found[75] to accumulate a growth factor for themselves and this effect was traced[76, 77] to the preceding accumulation of an unstable precursor; initially this was named prephenylalanine and later prephenic acid. It was observed that this compound under mildly acidic conditions—such as were developed in the culture medium during growth—was converted into its own normal metabolic product phenylpyruvic acid (32). This substance was an active growth factor for L-phenylalanine auxotrophs and transamination then gave L-phenylalanine (1). Acid in the culture medium was thus able to replace the enzyme, prephenate dehydratase[78], missing from the mutant cell.

In bacteria and other micro-organisms, unlike higher organisms, L-phenylalanine (1) is not normally a precursor of L-tyrosine (2). Davis[79] postulated that in these organisms prephenic acid is the precursor of L-tyrosine (2) and subsequent work at the enzymic level by Schwink and Adams[80] established this proposal. Thus in *Escherichia coli* prephenic acid (31) is oxidatively aromatised to *p*-hydroxyphenylpyruvic acid (33) by a soluble, NAD^+ dependent, enzyme—prephenate dehydrogenase. With appropriate fortification —addition of L-glutamate and pyridoxal phosphate as cofactors —extracts of *Escherichia coli* then convert *p*-hydroxyphenylpyruvic acid quantitatively to L-tyrosine by transamination.

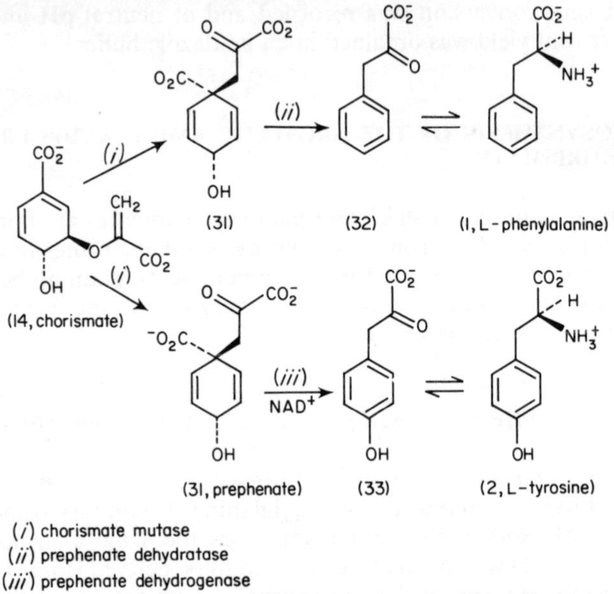

(i) chorismate mutase
(ii) prephenate dehydratase
(iii) prephenate dehydrogenase

Figure 1.13.   The biosynthesis of L-phenylalanine and L-tyrosine from chorismate in *Escherichia coli*

In *Escherichia coli*, *Salmonella typhimurium* and *Aerobacter aerogenes* two soluble multi-activity enzymes or enzyme complexes function in the utilisation of chorismate (14) for L-phenyl-alanine and L-tyrosine synthesis[81-87]. An enzyme or enzyme complex (P-protein) containing chorismate mutase and prephenate dehydratase activities has been isolated and partially purified from *Escherichia coli*, *Salmonella typhimurium* and *Aerobacter aerogenes*. The enzyme complex catalyses the transformation of chorismate (14) to phenylpyruvate (32) and both enzymic activities are retained in physical association after chromatography on DEAE cellulose. Kinetic analysis indicated that in isolated enzyme systems direct synthesis of phenylpyruvate (32) from chorismate (14) does not occur. Prephenate (31) once formed dissociates from the enzyme surface and accumulates in the reaction medium. After a lag period it is converted to phenylpyruvate (32). Schmit, Artz and Zalkin[86] also obtained evidence to show that functionally distinct sites (catalytic and regulatory) exist on the P-protein from *Salmonella typhimurium* for chorismate mutase and prephenate dehydratase activities. The P-protein was obtained from *Escherichia coli* K-12 by Davidson, Blackburn and Dopheide[87] who showed that it existed in solution mainly as a dimer of similar (and probably identical) sub-units of molecular weight 40000 and thus bore a strong resemblance to the native T-protein from the same organism.

The T-protein which catalyses the conversion of chorismate (14) to *p*-hydroxyphenylpyruvate (33) was isolated and extensively purified from *Escherichia coli* and *Aerobacter aerogenes* by Gibson and his group[81-84] and from *Salmonella typhimurium* by Dayan and Sprinson[88]. During the purification of the T-protein to near homogeneity by gel filtration, the ratio of the two enzyme activities (chorismate mutase and prephenate dehydrogenase) remained constant and this supported the suggestion that the protein (molecular weight 76000) was an aggregate which catalysed two steps in the metabolic sequence. Gibson later demonstrated[84] that the enzyme complex was reversibly dissociable into two sub-units of approximately equal size. Once separated the individual sub-units were inactive and this was interpreted in terms of the idea that the active sites in the complex are created by the association of sub-units. This behaviour of the T-protein is therefore a typical example of the frequently observed requirement for interaction between separate polypeptide chains for the expression of enzyme activity.

In other organisms which have been examined the situation is different from that observed in *Escherichia coli* and *Aerobacter aerogenes*. Thus in *Neurospora crassa* a single chorismate mutase was found, uncomplexed with either of the prephenate utilising enzymes[89, 90]. On the other hand two chorismate mutases were found in yeast[91] but both of these were readily separable from the corresponding prephenate dehydratase and prephenate dehydrogenase activities.

1.4.2   THE L-TRYPTOPHAN BRANCH

Early studies of the biosynthesis of L-tryptophan (3) established that both anthranilic acid[92] and indole[93] were able to replace L-tryptophan as a growth factor for some bacteria. The suggestion that these compounds were normal precursors of the aromatic amino acid was confirmed by Tatum, Bonner and Beadle[94] who showed that some L-tryptophan requiring mutants of *Neurospora* sp. responded to anthranilic acid or indole whilst others responded only to indole and accumulated anthranilic acid. Somewhat later other mutants were found which were blocked between indole and L-tryptophan and accumulated indole in the culture medium.

A combination of mutant and enzyme studies subsequently permitted the identification of the three intermediates (36, 37 and 38,

Figure 1.14) in the L-tryptophan biosynthetic pathway between anthranilate (34) and indole. Isotopic tracer work[95, 96] established that the carboxyl group of anthranilate (34) was lost in the conversion to L-tryptophan and that the two carbon atoms which go to form the pyrole ring of the indole nucleus were most probably derived from C-1 and C-2 of a ribose derivative. Enzymatic studies[97, 98] led to the isolation of indole glycerol-3-phosphate (38) and to the recognition of 5-phosphoribosyl-1-pyrophosphate (35) as the most efficient source of the five carbon atoms which complete the pyrrole ring and the glycerol phosphate side chain of (38)[96, 99]. Subsequently two sets of *Escherichia coli* and *Salmonella typhimurium* mutants were described by Smith and Yanofsky[100] which were found incapable of converting anthranilate to indoleglycerol-3-phosphate (38) in the presence of the ribose derivative (35) or ATP and ribose-5-phosphate. Mixtures of the two types were however able to bring about the reaction. Extracts of one set of mutants were found to catalyse the reaction of anthranilate (34) and 5-phosphoribosyl-1-pyrophosphate (35) to give the deoxyribulose phosphate (37). Extracts of mutants of the second type converted (37) to indoleglycerol-3-phosphate (38) and thus established the status of the deoxyribulose phosphate (37) as an intermediate on the pathway of L-tryptophan biosynthesis. The initial product of reaction, formed between the anthranilate (34) and 5-phosphoribosyl-pyrophosphate, is 1-*o*-carboxyphenyl-D-ribosylamine-5-phosphate (36). This compound, although not isolated, was formulated for a long time as an intermediate and evidence for its formation has recently been obtained in *Escherichia coli*, *Salmonella typhimurium* and *Aerobacter aerogenes*[101-103]. A rearrangement of the Amadori type leads to the formation of the deoxyribulose derivative (37) from (36).

Clarification of both the first and final stages of L-tryptophan biosynthesis has been obtained principally from enzyme studies (*Figure 1.14*). The early work of Tatum and Bonner[104] showed that the mechanism of L-tryptophan synthesis from indole (40) involved a condensation with L-serine (39) and used pyridoxal phosphate as cofactor. More recently the tryptophan synthetase of *Escherichia coli* (L-serine hydrolase, adding indole, EC 4.21.20) has been isolated and shown to be composed of non-identical and readily separable protein sub-units[105-107]. The A or α sub-unit (molecular weight 29 500) was shown to be a single polypeptide chain, containing paradoxically not a single L-tryptophan residue[108, 109], and the B or β_2 sub-unit (molecular weight 108 000) a dimer composed of identical polypeptide chains[110]. Highly purified preparations of the separated α and β_2 sub-units were shown to catalyse distinct reactions (*i* and *ii*).

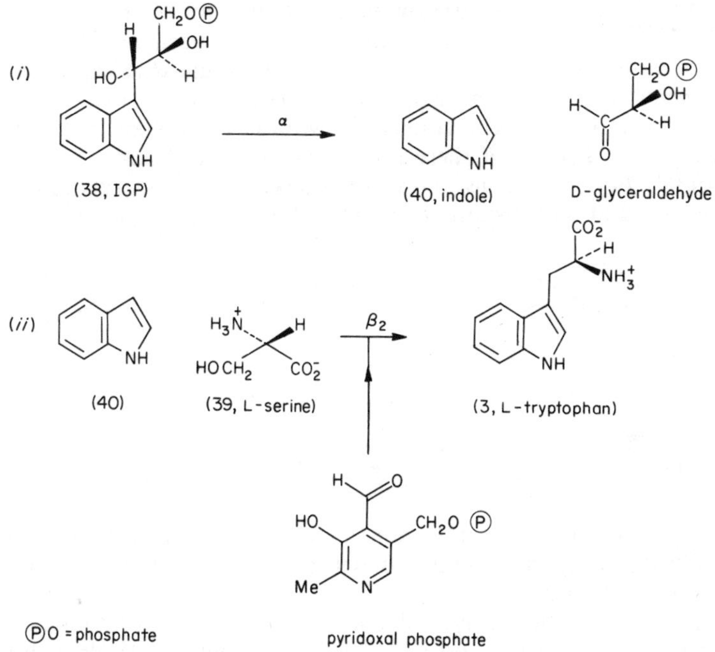

(i)   (38, IGP)   →α   (40, indole)   D-glyceraldehyde

(ii)   (40)   (39, L-serine)   →β$_2$ (3, L-tryptophan)

Ⓟ O = phosphate   pyridoxal phosphate

However, the α and β$_2$ sub-units were shown to combine when mixed and in combination as the fully associated L-tryptophan synthetase αβ$_2$α complex they catalysed a reaction which is the sum of (i) and (ii) but in which indole (40)—a product of (i) and a substrate for (ii)—was not detectable as a free intermediate. Thus indole (40), although it can enter the pathway between indoleglycerol-3-phosphate (38) and L-tryptophan (3) either as a growth supplement or an accumulated product, may not be a true physiological intermediate. Yanofsky and his collaborators[106] have observed that the full αβ$_2$α complex has an enhanced catalytic activity relative to the component α and β$_2$ sub-units and that there is a concomitant gain in specificity for the reaction (ii). Thus the αβ$_2$α complex was 100 times more active than the α sub-unit in reaction (i) and 30 times more active than the β$_2$ sub-unit in reaction (ii). In addition these workers observed that the intermediate αβ$_2$ complex was about one half as active per β$_2$ sub-unit as the tetrameric complex and they concluded that each α unit contributed equally to each of the two identical active sites in the full αβ$_2$α enzyme. During the process of synthesis of the complex it has been assumed that the β chains dimerise rapidly and then become associated with the α chains to give the tetrameric species.

26 Biosynthesis of the Aromatic Amino Acids

In addition to the striking effects of sub-unit interaction upon catalytic activity and specificity which have been noted for the L-tryptophan synthetase from *Escherichia coli* the $\alpha\beta_2\alpha$ enzyme and its α and β_2 sub-units have been shown to catalyse a variety of other reactions not entirely connected with L-tryptophan biosynthesis[111-115]. Typical of these are the reactions (*iii*) and (*iv*) shown below. Both reactions represent alternative transformations of L-serine and both require the pyridoxal phosphate cofactor[112, 113].

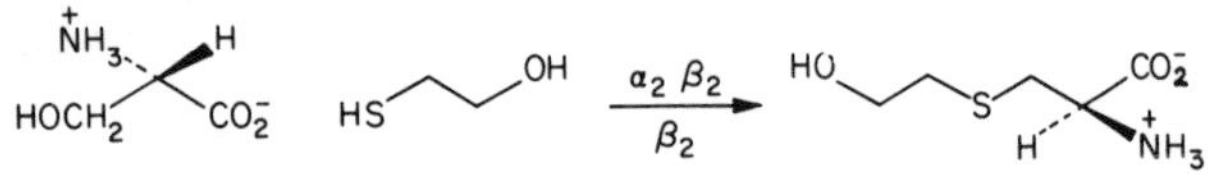

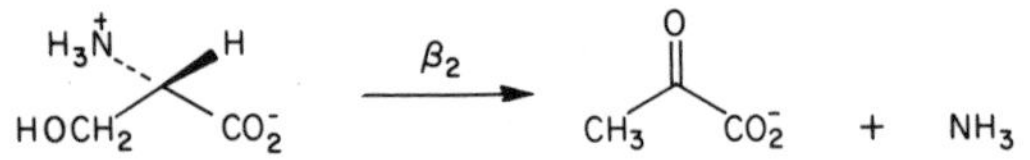

The L-tryptophan synthetases of fungal origin which have been studied appear to be quite different from the bacterial enzyme. A single undissociable protein species catalyses the biosynthetic reaction from indoleglycerol phosphate (38) to L-tryptophan (3) and the reactions (*i*) and (*ii*)[116-120] and a mutant of *Neurospora crassa* has been obtained which is active only in reaction (*ii*) above[120].

The formation of anthranilate from chorismate[25] is the first step on the biosynthetic route leading specifically to L-tryptophan (*Figure 1.14*). The reaction is a complex one and its mechanism is not yet fully understood[11, 121, 122]. Several experimental observations have been made at the enzymic level. Srinivasan and Rivera[123] showed that the amide nitrogen of L-glutamine was the amino donor in the formation of anthranilate (34). Isotopic tracer studies[124] with *Escherichia coli* B-37 established that the carbon skeleton of anthranilate (34) was identical to that of shikimate and that the amino group was added at C-2 of chorismate (14), (and hence C-6 of shikimate). Tamir and Srinivasan[125] also showed that in this reaction the enolpyruvyl group of chorismate (14) was eliminated as pyruvate after initial protonation of the enol methylene group.

Anthranilate synthetase the enzyme which catalyses the formation of anthranilate (34) from chorismate (14) and L-glutamine (*Figure 1.14*) has been isolated and characterised from a number of microorganisms. Two main types have been identified in bacteria. Type I has been obtained uncontaminated with other enzymes of the

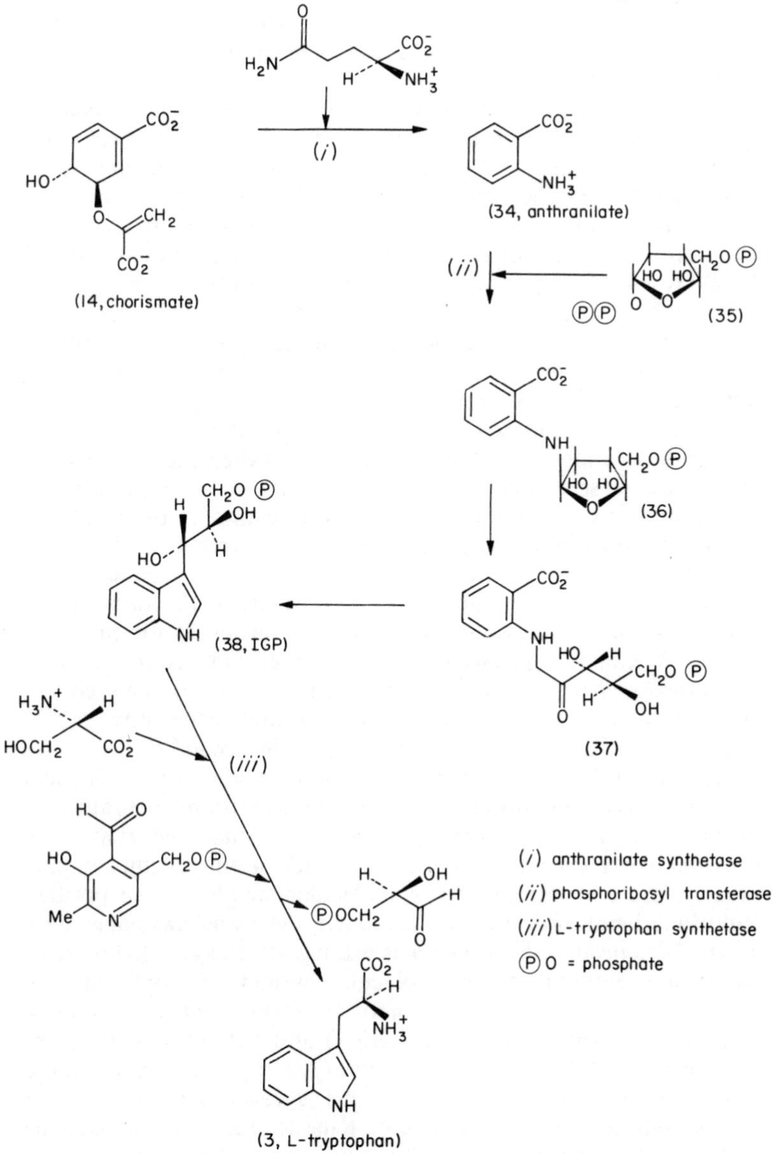

Figure 1.14. The biosynthesis of L-tryptophan from chorismate

pathway from *Bacillus subtilis*[126], *Pseudomonas* sp.[127] and *Serratia marcescens*[128, 129]. On the other hand, type II is always found associated with the second enzyme of the L-tryptophan pathway—anthranilate 5-phosphoribosyl-pyrophosphate transferase (PR transferase)—and this complex has been isolated from *Salmonella typhimurium*[130–132], *Escherichia coli*[133–135] and *Aerobacter aerogenes*[11]. Estimates of the molecular weight of the aggregate[131] have varied from 260 000 to 290 000 dependent on the source but anthranilate synthetase (component I) when separated from the PR transferase (component II) has a molecular weight in the region of 60 000 to 64 000. Separation of the two components is rather difficult but they can be conveniently prepared as separate entities using mutant strains which are deficient in one component or the other. The two components associate spontaneously to re-form the enzyme aggregate when they are mixed.

The enzyme complex (type II) is capable of utilising either L-glutamine or ammonia as amino donor but when the anthranilate synthetase is separated from the PR transferase it cannot use L-glutamine as the amino donor substrate and is only able to utilise ammonia in this step. The anthranilate synthetase of *Escherichia coli* is activated by the PR transferase and it has been suggested that the latter activates a glutamine binding site on the anthranilate synthetase or provides a separate binding site itself for this substrate. A kinetic analysis of the native enzyme complex from *Escherichia coli* has shown[133] the reaction mechanism to be a sequential one in which the presence of one substrate at the active site of the enzyme does not affect the binding of the second.

Zalkin and his colleagues[129–131] have made some pertinent observations on the sub-unit structure of the anthranilate synthetase from *Salmonella typhimurium* (type II enzyme) and that from *Serratia marcescens* (type I) and have made an interesting comparison of the two enzymes. The molecular weight of the purified multifunctional anthranilate synthetase–PR transferase aggregate from *Salmonella typhimurium* was estimated[131] as 280 000 by sedimentation equilibrium. A molecular weight of approximately 62 000 was established for each of the non identical sub-units, anthranilate synthetase (component I) and PR transferase (component II), and from a consideration of the approximate size of the native enzyme it was suggested that this aggregate was composed of two sub-units of each component. Kinetic studies of the enzyme indicated that the binding of chorismate and L-tryptophan was to component I of the aggregate and that end-product inhibition of anthranilate synthetase activity by L-tryptophan was due to antagonism of chorismate binding. Hwang and Zalkin[129] also purified the single enzyme anthranilate synthetase from *Serratia marcescens* and they showed it to have similar kinetic properties to the anthranilate synthetase (component I) from *Salmonella typhimurium*. The enzyme (molecular weight $\sim$ 141 000) was shown to be a tetrameric complex composed of non identical sub-units of molecular weight $\sim$ 60 000 and 21 000, in which the smaller sub-unit functions as a glutamine binding protein. Very similar observations on the enzyme from *Serratia marcescens* were also made by Robb, Hutchinson and Belser[128].

Treatment of the anthranilate synthetase–PR transferase from *Salmonella typhimurium* with trypsin resulted in the formation of a modified enzyme with properties very similar to those of the type I (e.g. *Serratia marcescens*). The action of trypsin removed a major proportion of the PR–transferase sub-unit to give an enzyme complex containing two unaltered sub-units of component I (anthranilate synthetase) and two sub-units of a modified PR–transferase (molecular weight $\sim$ 15 000–19 000). The properties of this altered enzyme were remarkably similar to those of the native enzyme from *Serratia marcescens* and retained all the properties of the L-glutamine dependent reaction of the original enzyme from *Salmonella typhimurium*.

In the fungus *Neurospora crassa*, four unlinked genes specify the proteins which catalyse the five enzymic steps involved in the conversion of chorismate (14) to L-tryptophan (3). DeMoss and Gaertner[136] have isolated from this organism an enzyme complex (molecular weight $\sim$ 240 000) which contains not only the anthranilate synthetase but also the N-5′-phosphoribosylanthranilate isomerase and indoleglycerol phosphate synthetase enzymes. The last two enzymes are present, as in *Escherichia coli*, in a bifunctional enzyme formed from a single polypeptide chain and this component may be partially dissociated from the *Neurospora crassa* complex by treatment with *p*-mercuribenzoate. Similar patterns of enzyme aggregation in the L-tryptophan pathway have been found in other fungi and it is significant to note that the three reactions catalysed by the complex do not occur sequentially along the metabolic pathway.

1.5   CONTROL OF AROMATIC AMINO ACID METABOLISM IN MICRO-ORGANISMS

Striking advances have been made in recent years towards an understanding of the molecular processes involved in metabolic regulation[137, 138]. At the enzymic level metabolic control may be exercised broadly in two ways; by an alteration in the number of enzyme molecules (control of enzyme synthesis) and by a regulation of enzyme activity. In a branched metabolic pathway, such as the shikimate pathway, an end-product usually controls enzymes specific to its own synthesis. Thus repression of synthesis and end-product inhibition of activity are frequently encountered for enzymes after the branch point. In the common part of the pathway key enzymes often occur in multiple forms and an end-product then regulates this part of the metabolic sequence through its action on the synthesis or activity of one of these proteins. Amongst alternative patterns of control of a branched pathway to which reference should also be made, is that of concerted or multivalent control in which all the end-products act in concert to cause a lowering in the activity or synthesis of a particular enzyme.

Several excellent examples of this regulatory diversity and of the differing strategies of control exerted by different organisms have been observed in studies of the shikimate pathway in micro-organisms (*Figures 1.15* and *1.16*).

1.5.1 THE COMMON PATHWAY

Studies on the regulation of the common pathway of aromatic biosynthesis in several micro-organisms have shown that control of the first reaction (*Figure 1.2*), the conversion of D-erythrose-4-phosphate (7) and phosphoenolpyruvate (8) to 3-deoxy-D-arabino-heptulosonic acid-7-phosphate (9, DAHP), catalysed by the enzyme DAHP synthetase (EC 4.1.2.15) is an important factor in the overall control of the pathway[139]. In a number of enteric bacteria this enzyme exists in multiple molecular forms each of which is under the feedback control of a specific end-product. Thus in *Escherichia coli* there are three DAHP synthetases (iso-enzymes), the activity and formation of which are controlled by the three aromatic amino acids[50, 140–145]. The formation and activity of DAHP synthetase (tyr) is controlled by L-tyrosine, that of DAHP synthetase (phe) by L-phenylalanine and that of DAHP synthetase (trp) by L-tryptophan. Several workers have reported a failure to observe any significant inhibition of DAHP synthetase (trp) activity by L-tryptophan[140, 141, 143, 144, 146, 147]. However, it has been pointed out that in wild type *Escherichia coli* DAHP synthetase (trp) is produced in much smaller amounts than either of the other DAHP synthetases and the low levels of activity may have been the cause of the difficulties in identifying inhibition by L-tryptophan[148]. With *Escherichia coli* K-12 recombinant strains were isolated which contained only one of the three DAHP synthetase iso-enzymes and Pittard, Camakaris and Wallace[148] were able to show by growth of a strain which produced only the single DAHP synthetase (trp) that this was inhibited by L-tryptophan. Similar results were reported by Gibson and Pittard[11].

The same pattern of three iso-enzymes each sensitive to regulation by one of the aromatic amino acid end-products was also observed in *Salmonella typhimurium*[27], *Neurospora crassa*[149-151] and *Claviceps paspali*[152] but in *Saccharomyces cerevisiae*[91, 153] only two product sensitive DAHP synthetases were found. In *Bacillus subtilis* Jensen and Nester demonstrated[154,155] the existence of a single DAHP synthetase which was strongly inhibited by chorismate or prephenate and a similar situation was also noted for *Staphylococcus epidermidis*[33] and Bacillus *licheniformis*[156]. In *Bacillus subtilis* this mechanism of feedback control is augmented by the elaboration of multiple forms of the enzyme shikimate kinase (*Figure 1.2*, 4 and 12). Both prephenate and chorismate exercise feedback control over the shikimate kinase enzymes, one form of which is located in an enzyme complex with DAHP synthetase and chorismate mutase and the other of which is not part of this high molecular weight aggregate[33]. In later work, Nakatsukasa and Nester[157] obtained from *Bacillus subtilis* a bifunctional enzyme complex which retained the DAHP synthetase and chorismate mutase activities. When this complex was mixed with fractions which showed only traces of shikimate kinase activity, full catalytic activity of shikimate kinase was restored. Nakatsukasa and Nester[157] suggested that in *Bacillus subtilis* chorismate mutase was the key enzyme in a trifunctional multi-enzyme complex which functions as a unit of feedback control and that the active site region of the chorismate mutase was crucial in order to maintain the DAHP synthetase and shikimate kinase in a catalytically functional state. This situation was not, however, found in *Bacillus licheniformis* in which the regulation of aromatic amino acid biosynthesis is similar to that in *Bacillus subtilis*. In *Bacillus licheniformis* the three enzymes DAHP synthetase, shikimate kinase and chorismate mutase were found uncomplexed and were readily separable[156].

Enzyme aggregation is of widespread occurrence. In the majority of cases which have been studied, the enzymes found in an aggregate operate sequentially on a given substrate and it is thought that the assembly of enzymes channels the substrate and leads to an increase in the overall efficiency of the process. In the context of this hypothesis, the function of the complex in *Bacillus subtilis* is uncertain since the reactions catalysed by the three enzymes in the aggregate do not follow one another directly on the shikimate pathway. Nester, Lorence and Nasser suggested[33] that in this instance the complex may provide more efficient control over entrance into the metabolic sequence leading to the individual aromatic amino acids than would result from structurally independent enzymes.

In contrast, Giles and his colleagues have suggested that the multi-enzyme complex, which they isolated from *Neurospora crassa* and other fungi[46,158] and which contains the complete sequence of biosynthetic enzymes from DAHP (9) to 5-enolpyruvyl-shikimate-3-phosphate (13), has developed in response to a need to channel a particular metabolite—3-dehydroquinate (10). They found[45] that *Neurospora crassa* produced an inducible 3-dehydro-quinate dehydratase (*Figure 1.2*, 10 → 11), in addition to that which forms part of the biosynthetic complex. The inducible enzyme serves a catabolic function which leads from 3-dehydroquinate (10) to protocatechuate. Giles and his collaborators suggested that in this case complex formation offers a way of segregating the biosynthetic and catabolic enzymes that would otherwise compete for the same intermediate (10). In bacteria which do not possess the catabolic enzymes, this biosynthetic enzyme aggregate is not found and is presumably unnecessary.

1.5.2   THE L-PHENYLALANINE AND L-TYROSINE PATHWAYS

In bacteria such as *Escherichia coli*, *Aerobacter aerogenes* and *Salmonella typhimurium* the first two reactions of both the L-phenylalanine and L-tyrosine pathways are carried out by enzyme complexes, respectively the P and T-proteins, which contain chorismate mutase and the appropriate prephenate metabolising enzyme. L-Phenylalanine and L-tyrosine, in experiments with cell free extracts, generally show their typical end-product inhibition effects much more strongly on the prephenate metabolising enzymes than on the corresponding chorismate mutase. Thus L-phenylalanine causes in all cases an inhibition of prephenate dehydratase activity and in *Aerobacter aerogenes* and *Salmonella typhimurium* it also causes a diminution in the activity of the associated chorismate mutase P-enzymes[81-86]. Repression of synthesis of the P-protein is also caused by L-phenylalanine in *Aerobacter aerogenes* and *Escherichia coli* and in these organisms L-tyrosine similarly affects the synthesis of the two enzymes of the T-protein complex. L-Tyrosine is a feedback inhibitor of the second enzyme prephenate dehydrogenase but in *Escherichia coli* and *Aerobacter aerogenes* it has little effect on the chorismate mutase T activity.

33

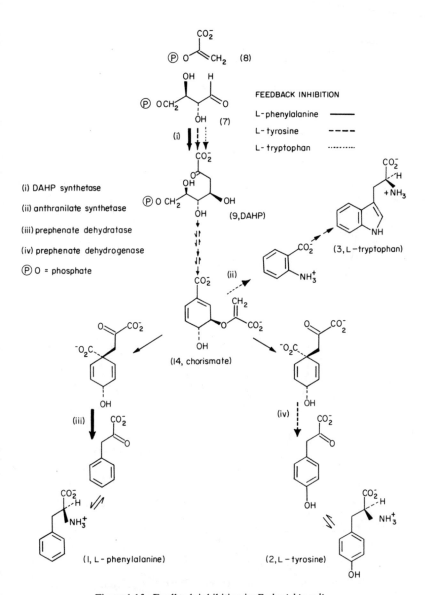

Figure 1.15. Feedback inhibition in *Escherichia coli*

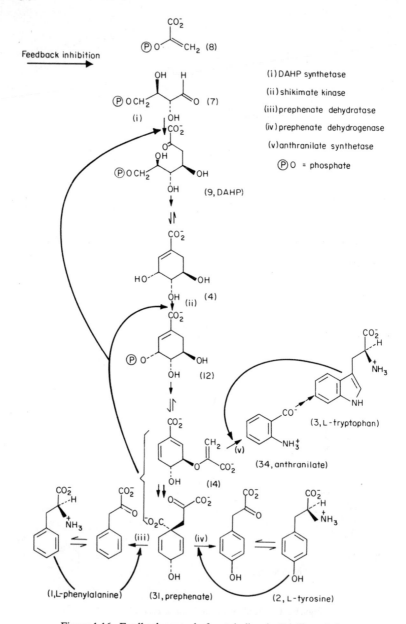

Figure 1.16. Feedback control of metabolism in *Bacillus subtilis*

Slightly different patterns of control have been adopted by other organisms. Thus *Neurospora crassa*[89] and *Claviceps paspali*[152] contain a single chorismate mutase whose activity is modulated by both L-phenylalanine and L-tyrosine. The end-product inhibition of activity may be reversed and the chorismate mutase activity enhanced by L-tryptophan. In *Claviceps paspali*, L-phenylalanine and L-tyrosine also regulate the activity of the corresponding prephenate metabolising enzymes. The chorismate mutases and prephenate enzymes are also separable in *Saccharomyces cerevisiae*[91] and in this organism chorismate mutase activity is induced and activated by L-tryptophan. *Bacillus subtilis*[33, 155, 159, 160] contains two chorismate mutases both of which are insensitive to feedback inhibition but are susceptible to repression and in this organism the enzymes which utilise prephenate are subject to feedback control by the respective amino acid end-products.

1.5.3 THE L-TRYPTOPHAN PATHWAY

The regulation of L-tryptophan biosynthesis has been most closely studied in *Escherichia coli*[132–139, 161]. In other micro-organisms the sequence of chemical reactions is identical with that observed in *Escherichia coli* although differences in enzyme organisation have been noted[118–120, 136]. The activity of the first enzyme on the pathway from chorismate (14), anthranilate synthetase, is, in all cases which have been examined, subject to feedback inhibition by L-tryptophan[25, 89, 91, 152, 162–164]. In several organisms anthranilate synthetase is an oligomer containing non-identical sub-units and Zalkin and Chen[165], in a study of the enzyme from *Serratia marcescens*, showed that one sub-unit catalyses the reaction of chorismate (14) and ammonia and contains a binding site for L-tryptophan. The other sub-unit contains a binding site for L-glutamine and functions in the transfer of the amide amino group of L-glutamine to the catalytic site. Zalkin and Chen[165], in their study, probed by physicochemical means the conformational changes which occurred on the binding of substrates and the inhibitor L-tryptophan to the enzyme. They proposed a model for the inhibition of anthranilate synthetase in which L-tryptophan binds to the regulatory site and maintains the enzyme in a conformation having a poor affinity for both substrates. Alternatively it was suggested that the binding of the substrate chorismate causes a conformational change of the enzyme which probably facilitates the binding of L-glutamine the second substrate.

The conversion of chorimate (14) to L-tryptophan (3), *Figure*

1.14, involves five enzyme controlled reactions. These enzymes are delineated by a group of five contiguous genes—the tryptophan operon—in *Escherichia coli*[166]. Each gene is believed to control the formation of a different polypeptide and those specified by genes trp *A, B, C* and *E* (*Figure 1.17*) have been obtained in a pure form. The proteins derived from the *A* and *B* genes form the tetrameric tryptophan synthetase complex, that from the *C* gene, indoleglycerol

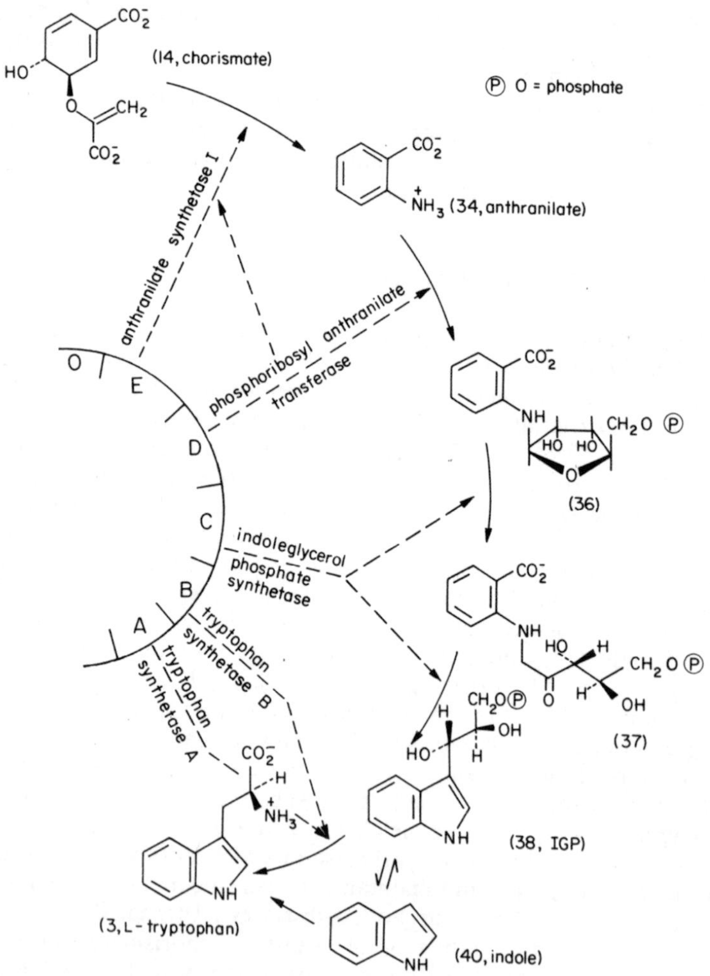

Figure 1.17. L-Tryptophan biosynthesis in *Escherichia coli*: the L-tryptophan operon[166]

phosphate synthetase, catalyses two sequential reactions in trypto-phan biosynthesis (*Figure 1.17*, 36 → 38). The D and E genes specify the anthranilate synthetase complex and the D gene poly-peptide, the second component of the complex, also catalyses the phosphoribosyl anthranilate transferase reaction. The expression of all the genes is controlled by a single operator locus (O) and it has been shown that L-tryptophan co-ordinately represses formation of all the enzymes of its own biosynthetic pathway from chorismate[134, 167]. L-Tryptophan is also an allosteric inhibitor of anthranilate synthetase and of the associated phosphoribosyl anthranilate transferase reaction in *Escherichia coli* and this property probably derives from the fact that these two enzymes share a common protein specified by the D gene. The tryptophan operon of *Salmonella typhimurium* bears many similarities to that found in *Escherichia coli*. The presence in both operons has also been noted of an internal promoter element which is responsible for non-co-ordinate synthesis of the five L-tryptophan synthetic enzymes which is observed under repression conditions.

1.6 THE SHIKIMATE PATHWAY IN HIGHER PLANTS

Evidence for the operation of the shikimate pathway in higher plants has been obtained mainly from tracer experiments and from a limited range of enzyme studies. These have shown that not only do the pathways of biosynthesis of L-phenylalanine, L-tyrosine and L-tryptophan involve the same intermediates as in bacteria and fungi (*Figures 1.2, 1.13, 1.14*), but that higher plants can also con-vert the aromatic amino acids into a plethora of characteristic natural products or secondary metabolites[168, 169, 170].

1.6.1 METHODOLOGY AND IDENTIFICATION OF INTERMEDIATES

Several tracer studies of the biosynthesis of (−)-shikimic acid in higher plants have been made and these all accord with the view that, as in micro-organisms, the seven-carbon skeleton of the alicyclic acid is derived by an initial condensation of D-erythrose-4-phosphate (7) and phosphoenolpyruvate (8), *Figure 1.2*. Uniform labelling of (−)-shikimic acid was obtained by exposure of living plant tissues to ^{14}C carbon dioxide for extended periods[171–174]. Short-term exposure gave (−)-shikimic acid with 25 per cent of the isotope in the carboxyl group[174]. This was rationalised on the basis of the known pathway of carbon in the photosynthetic cycle[175] which predicts a predominance of the carbon isotope in the carboxyl group of phosphoenolpyruvate (8), and hence the carboxyl group of (−)-shikimic acid (4), during the initial stages of photosynthesis. Towers and Yoshida[176] similarly obtained (−)-[^{14}C]-shikimic acid after feeding U-[^{14}C]-D-ribose to *Pinus resinosa*. Degradation showed the (−)-shikimic acid to contain almost a third of its radioactivity in the carboxyl group and this was interpreted in terms of an initial breakdown of the pentose sugar to phosphoenolpyruvate (8) prior to incorporation into the acid (4). Dewick and Haslam[174] in an analogous series of experiments to those conducted by Sprinson and his collaborators[35, 36] with *Escherichia coli* mutants, showed that D-glucose, labelled with ^{14}C at C-1 or C-6, administered to *Ginkgo biloba* and *Rhus typhina* gave (−)-shikimic acid and (−)-quinic acid isotopically labelled predominantly (∼ 80 per cent) at C-2 and C-6. Moreover in *Ginkgo biloba* it was observed, in agreement with the work of Sprinson[35, 36], that labelling at C-2 was rather greater than that at C-6.

Isotopic tracer studies have also shown that (−)-shikimic acid is an excellent precursor of the three aromatic amino acids in plant tissues[177–180]. The conversion of L-phenylalanine to L-tyrosine has also been observed in plants[181–185] and this is a biosynthetic reaction which is common in higher organisms but not in micro-organisms.

1.6.2   ENZYMOLOGY

Further support for the operation of the shikimate pathway in higher plants has been provided by studies on enzymes isolated and characterised from plant sources. DAHP synthetase activity has thus been demonstrated in extracts of several plant tissues[186]. An enzyme preparation was also obtained[187] from sweet potato which catalysed the formation of DAHP from D-erythrose-4-phosphate (7) and phosphoenolpyruvate (8) and had properties very similar to those enzymes isolated from bacterial sources. Nandy and Ganguli[188] have similarly demonstrated the presence of DAHP synthetase activity in mung bean (*Phaseolus aureus*) by showing that extracts of mung bean seedlings converted a mixture of the two substrates (7 and 8) to 3-dehydroshikimate (11).

Such extracts presumably also contained the enzyme 3-dehydro-quinate dehydratase (EC 4.2.1.10) and Gamborg[189] has shown that this enzyme along with the next enzyme in the pathway, shiki-mate dehydrogenase (EC 1.1.1.25), is present in extracts from a number of plant cell tissue cultures. Both of these enzymes have been isolated and partially purified from plant sources[190-193] and several studies have shown how their activities vary within different plant tissues and under different physiological conditions of growth[194, 195]. The results of this work corroborate the hypothesis that the highest enzyme activities are located at sites in the plant where rapid protein synthesis or the formation of aromatic secondary metabolites occurs[170].

Enzymes which catalyse the steps from shikimate (4) to the aromatic amino acids have been less widely studied in plants. A single chorismate mutase has been obtained from pea seedlings (*Pisum sativum*) by Cotton and Gibson[196]. The enzyme showed similarities to the analogous enzymes from *Saccharomyces cerevisiae*[91] and *Neurospora crassa*[89] and was inhibited by L-phenylalanine and L-tyrosine but activated by L-tryptophan. The end-product inhibition of this enzyme was not additive and was reversed by L-tryptophan. The presence of the prephenate metabolising enzymes and the transaminase which catalyses the formation of L-phenylalanine and L-tyrosine has been demonstrated in *Phaseolus* sp.[197-200] and Gamborg and Keely[200] have made a study of the prephenate dehydrogenase from *Phaseolus vulgaris*.

Of the enzymes specific to L-tryptophan biosynthesis only tryptophan synthetase (L-serine hydro-lyase, adding indole, EC 4.2.1.20) has been subject to a systematic study in plants. Delmer and Mills[201] obtained the enzyme from callus tissue of *Nicotiana tabacum* and demonstrated that it closely resembled the protein aggregate from *Escherichia coli*. The enzyme catalysed the terminal reactions of L-tryptophan synthesis and was separable into two protein sub-units (A and B) by differential precipitation with ammonium sulphate. Both components were essential for the conversion of indoleglycerolphosphate (38) to the aromatic amino acid but component B alone catalysed the conversion of indole to L-tryptophan. Crude extracts of the callus tissue of *Nicotiana tabacum* have also been obtained[202, 203] which catalyse the first step unique to L-tryptophan biosynthesis, the conversion of chorismate (14) to anthranilate (34), *Figure 1.14*. The catalytic activity was destroyed by heat and no reaction occurred in the absence of Mg^{2+} or L-glutamine. In cell extracts this anthranilate synthetase activity was inhibited by L-tryptophan and this inhibition was competitive with chorismate. During a discussion of these results Belser, Murphy, Delmer and Mills[202] raised the question of the relevance of feedback control of enzyme activity in the cells of a higher plant growing under conditions which are not subject to appreciable nutritional variation. These authors concluded that regulation would be important not only in relation to biochemical efficiency but also to differentiation. They also pointed out that control of L-tryptophan synthesis offered in addition a mode of control of the level of the hormone indole acetic acid (IAA) in the cell.

1.6.3   (−)-SHIKIMIC AND (−)-QUINIC ACID IN PLANTS

A distinctive feature of the shikimate pathway and its operation in higher plants is the widespread occurrence, often in substantial quantities, of both (−)-shikimic acid and (−)-quinic acid in the free state in plant tissues[204]. Various studies have concluded that both (−)-shikimic acid and (−)-quinic acid are actively metabolised during the growth of the plant[171, 172, 205–207]. Occasionally the question of the existence of a more direct route from (−)-quinic acid to the aromatic amino acids has been raised[171, 172, 207] but at this stage it is more reasonable to assume that quinate has access to the common part of the shikimate pathway via 3-dehydroquinate (10) and the enzyme quinate dehydrogenase (EC 1.1.1.24). This enzyme has been well documented in *Aerobacter aerogenes*[44] and tissue cultures of the mung bean[189] but attempts to detect it in most higher plants have usually failed[170, 189]. It is therefore interesting to note that a number of micro-organisms and particularly fungi can utilise quinate or shikimate as carbon sources for growth[46, 47, 208], probably by conversion to protocatechuate and then degradation by the β-ketoadipate pathway[209]. The oxidation of quinate (and shikimate) is mediated by one, or more, dehydrogenases whose synthesis is directly inducible. In a survey of sixty-six strains of *Acetomonas oxydans*, Whiting and Coggins[210] isolated a particulate enzyme system which oxidised quinate (5) to 3-dehydroquinate (10) and shikimate (4) to 3-dehydro-shikimate (11). The enzyme had properties quite distinct from the NAD^+ dependent quinate oxidoreductase from *Aerobacter aerogenes* and evidence was obtained to suggest that cytochrome 555 was involved in the electron transport system. It is possible that an analogous enzyme system may also be involved in the incorporation of quinate into the shikimate pathway in plants.

1.6.4   META-CARBOXY AROMATIC AMINO ACIDS AND L-DOPA

Recent investigations of intermediary nitrogen metabolism in higher plants have been notable for the discovery of a large group of amino acids that are not normally encountered as components of proteins[211, 212]. These acids are usually found free or occasionally as malonyl or γ-glutamyl derivatives and a great many of them have carbon skeletons identical or very similar to those of the protein α-amino acids. Four m-carboxy substituted aromatic amino acids (45–48) have been isolated from plants[213–218] and their distribution appears to be associated closely with the plant families Cruciferae, Resedaceae, Iridaceae and Curcurbitaceae. Morris and Thompson[217, 218] and Larsen[173, 219, 220] have investigated the biosynthesis of the phenylglycine derivatives (46 and 48) and have concluded that they are most probably derived by a chain shortening pathway involving loss of the carboxyl carbon atom from the corresponding m-carboxyphenylalanine (45) or tyrosine (47) derivatives. Thus 2-[^{14}C]-m-carboxyphenylalanine administered to Iris gave[217, 218] m-carboxyphenylglycine (46) with complete retention of the label in the aliphatic carboxyl group. The work of Larsen[219, 220] and later Larsen, Onderka and Floss[221] has revealed in greater detail the pathways of biosynthesis of (45) and (47). Larsen[173, 219, 220] showed that U-[^{14}C]-(−)-shikimic acid was incorporated into the amino acids (45) and (47) with retention of the complete carbon skeleton of shikimate including its carboxyl group. In subsequent work 6-[^{3}H]-6R-(−)-shikimic acid (41) and 6-[^{3}H]-6s-(−)-shikimic acid (42) were fed to *Reseda lutea* and L-phenylalanine (1), L-tyrosine (2) and the m-carboxyl aromatic amino acids (45 and 47) were isolated. Degradation of the L-phenylalanine (1) and L-tyrosine (2) showed that, as in bacteria, the pro 6s-hydrogen of (−)-shikimic acid is retained and the pro 6R-hydrogen is lost in the conversion to chorismate (14) and hence (1) and (2). Tritium incorporation into the m-carboxyaromatic amino acid (45) also only resulted when the precursor was 6-[^{3}H]-6R-(−)-shikimic acid (41). Degradation of the tritiated sample of (45) via 2-phenylethylamine and benzoic acid to acetanilide and p-bromoacetanilide showed that the tritium label was located exclusively *para* to the C_3-side chain. Larsen[219, 220] originally rationalised these observations in terms of an alternative rearrangement of chorismate (14) to 3-(3-carboxy-6-hydroxy-cyclohexa-2,4-dienyl)-pyruvate. This proposal has, however, since been modified and a pathway (*Figure 1.18*) postulated in which *iso*chorismate (43) is formed from chorismate (14) and then rearranges to (44) in a process analogous to the chorismate–prephenate conversion. The cyclohexadienyl derivative (44) then yields the amino acids (45) and (47) by dehydration or dehydrogenation followed by transamination.

A further important non-protein aromatic amino acid found in a restricted group of plants and derived from shikimate is 3,4-dihydroxy-L-phenylalanine, L-dopa, (49)[222]. The amino acid has been implicated quite closely in alkaloid biosynthesis and occurs in relatively high concentration in some plants such as *Vicia faba*[223, 224] where it gives rise during senescence or when the cellular structure is damaged to melanin formation and the characteristic blackening of the tissue. Evans and Raper[225] showed that L-tyrosine was directly hydroxylated by plant preparations with

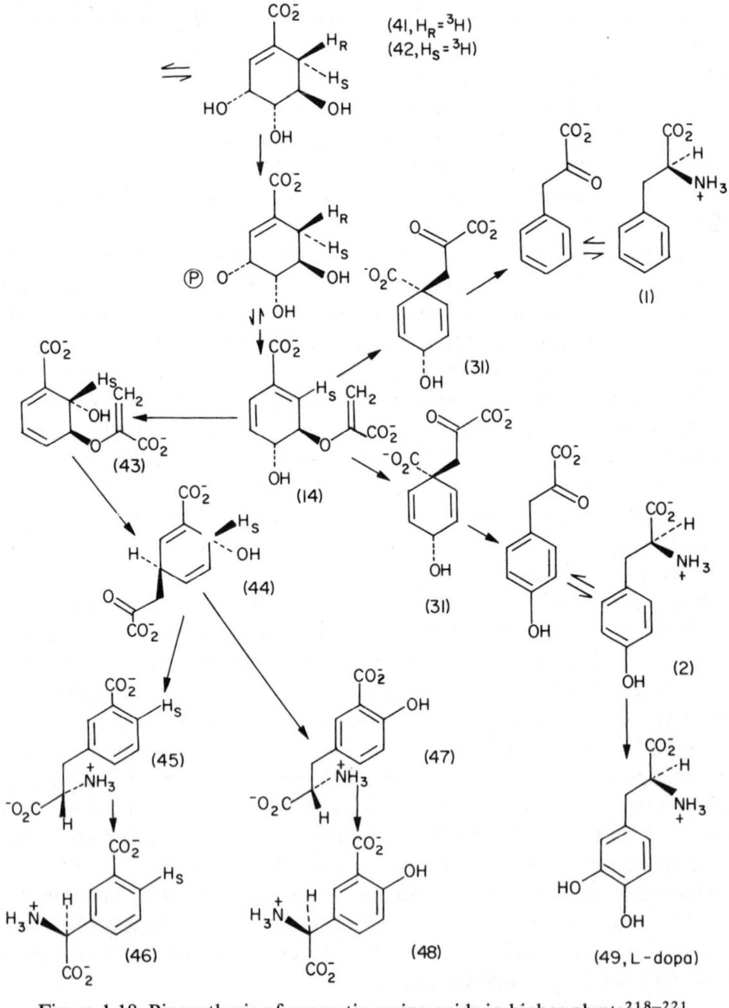

Figure 1.18 Biosynthesis of aromatic amino acids in higher plants[218-221]

tyrosinase activity to give L-dopa (49). Subsequently Liss[226] has demonstrated the direct transformation of [^{14}C]-L-tyrosine to L-dopa (49) in *Euphorbia lathyrus* but little detailed work has otherwise been carried out on this reaction in plants

REFERENCES

1. Davis, B. D. (1955). *Adv. Enzymol.*, **16**, 287
2. Sprinson, D. B. (1961). *Adv. Carbohydrate Chem.*, **15**, 235
3. Eykmann, J. F. (1885). *Rec. Trav. Chim.*, **4**, 32
4. Eykmann, J. F. (1891). *Chem. Ber.*, **24**, 1278
5. Fischer, H. O. L. and Dangschat, G. (1934). *Helv. Chim. Acta*, **17**, 1196. 1200
6. Fischer, H. O. L. and Dangschat, G. (1935). *Helv. Chim. Acta*, **18**, 1204, 1206
7. Fischer, H. O. L. and Dangschat, G. (1937). *Helv. Chim. Acta*, **20**, 705
8. Fischer, H. O. L. and Dangschat, G. (1932). *Chem. Ber.*, **65**, 1009
9. Freudenberg, K., Meisenheimer, H., Lane, J. T. and Plankenhorn, E. (1940). *Annalen*, **543**, 162
10. Karrer, P. and Link, K. P. (1927). *Helv. Chim. Acta*, **10**, 794
11. Gibson, F. and Pittard, J. (1968). *Bact. Rev.*, **32**, 468
12. Davis, B. D. (1951). *J. Biol. Chem.*, **191**, 315
13. Shigeura, H. and Sprinson, D. B. (1952). *Fed. Proc.*, **11**, 286
14. Davis, B. D. and Mingioli, E. S. (1953). *J. Bact.*, **66**, 129
15. Davis, B. D. (1952). *J. Bact.*, **64**, 729, 749
16. Davis, B. D. (1948). *J. Amer. Chem. Soc.*, **70**, 4267
17. Chain, E. and Duthie, E. S. (1945). *Lancet*, **1**, 652
18. Lederberg, J. and Zinder, N. (1948). *J. Amer. Chem. Soc.*, **70**, 4267
19. Weiss, U., Davis, B. D. and Mingioli, E. S. (1953). *J. Amer. Chem. Soc.*, **75**, 5572
20. Salamon, I. I. and Davis, B. D. (1953). *J. Amer. Chem. Soc.*, **75**, 5567
21. Davis, B. D. and Mingioli, E. S. (1953). *J. Bact.*, **66**, 129
22. Weiss, U. and Mingioli, E. S. (1956). *J. Amer. Chem. Soc.*, **78**, 2894
23. Levin, J. G. and Sprinson, D. B. (1960). *Biochem. Biophys. Res. Comm.*, **3**, 157
24. Levin, J. G. and Sprinson, D. B. (1964). *J. Biol. Chem.*, **239**, 1142
25. Gibson, M. I. and Gibson, F. (1964). *Biochem. J.*, **90**, 248
26. Gibson, F. (1964). *Biochem. J.*, **90**, 256
27. Gollub, E., Zalkin, H. and Sprinson, D. B. (1967). *J. Biol. Chem.*, **242**, 5323
28. Mitsuhashi, S. and Davis, B. D. (1954). *Biochim. Biophys. Acta*, **15**, 54
29. Yaniv, H. and Gilvarg, C. (1955). *J. Biol. Chem.*, **213**, 787
30. Fewster, J. A. (1962). *Biochem. J.*, **85**, 388
31. Millican, R. C. (1963). *Anal. Biochem.*, **6**, 181
32. Morell, H. and Sprinson, D. B. (1968). *J. Biol. Chem.*, **243**, 676
33. Nester, E. W., Lorence, J. H. and Nasser, D. S. (1967). *Biochemistry*, **6**, 1553
34. Morell, H., Clark, M. J., Knowles, P. F. and Sprinson, D. B. (1967). *J. Biol. Chem.*, **242**, 82
35. Srinivasan, P. R., Shigeura, H. T., Sprecher, M., Sprinson, D. B. and Davis, B. D. (1956), *J. Biol. Chem.*, **220**, 477
36. Kalan, E. B., Davis, B. D., Srinivasan, P. R. and Sprinson, D. B. (1956). *J. Biol. Chem.*, **223**, 907
37. Horecker, B. L., Smyrniotis, P. Z., Hiatt, H. H. and Marks, P. A. (1955). *J. Biol. Chem.*, **212**, 827
38. Srinivasan, P. R., Sprinson, D. B., Kalan, E. B. and Davis, B. D. (1956), *J. Biol. Chem.*, **223**, 913

44 Biosynthesis of the Aromatic Amino Acids

39. Sprinson, D. B., Rothschild, J. and Sprecher, M. (1963). *J. Biol. Chem.*, **238**, 3170
40. Srinivasan, P. R. and Sprinson, D. B. (1959). *J. Biol. Chem.*, **234**, 716
41. Srinivasan, P. R., Sprinson, D. B. and Rothschild, J. (1963). *J. Biol. Chem.*, **238**, 2176
42. Bohm, B. A. (1965). *Chem. Rev.*, **65**, 435
43. Davis, B. D. and Weiss, U. (1953). *Arch. exptl Pathol. U. Pharmakol.*, **220**, 1
44. Mitsuhashi, S. and Davis, B. D. (1954). *Biochim. Biophys. Acta*, **15**, 268
45. Giles, N. H., Partridge, C. W. H., Ahmed, S. I. and Cass, M. E. (1967). *Proc. Acad. Nat. Sci.*, **58**, 1930
46. Ahmed, S. I. and Giles, N. H. (1969). *J. Bact.*, **99**, 231
47. Cain, R. B., Bilton, B. J. and Darrah, J. A. (1968). *Biochem. J.*, **108**, 797
48. Watson, J. D. (1965). *Molecular Biology of a Gene*. Amsterdam and New York; Benjamin
49. Srinivasan, P. R., Katagiri, M. and Sprinson, D. B. (1955). *J. Amer. Chem. Soc.*, **77**, 4943
50. Staub, M. and Denes, G. (1969). *Biochim. Biophys. Acta*, **178**, 588, 599
51. Moldovanyi, J. S. and Denes, G. (1968). *Acta Biochim. Biophys. Acad. Sci. Hung.*, **3**, 259
52. Nagano, H. and Zalkin, H⁻ (1970). *Arch. Biochem. Biophys.*, **138**, 58
53. Cleland, W. W. (1963). *Biochim. Biophys. Acta*, **67**, 104
54. DeLeo, A. B. and Sprinson, D. B. (1968). *Biochem. Biophys. Res. Comm.*, **32**, 373
55. Onderka, D. K. and Floss, H. G. (1969). *J. Amer. Chem. Soc.*, **91**, 5894
56. Floss, H. G., Onderka, D. K. and Carroll, M. (1972). *J. Biol. Chem.*, **247**, 736
57. Cohn, M., Pearson, J. E., O'Connell, E. L. and Rose I. A. (1970). *J. Amer. Chem. Soc.*, **92**, 4095
58. Straub, F. (1942). *Z. physiol. Chem.*, **275**, 63
59. Rose, I. A., O'Connell, E. L., Noce, P., Utter, M. F., Wood, H. G., Willard, J. M., Cooper, T. G. and Benziman, M. (1969). *J. Biol. Chem.*, **244**, 6130
60. Bondinell, W. E. and Sprinson, D. B. (1970). *Biochem. Biophys. Res. Comm.*, **40**, 1464
61. Rose, I. A. (1970). *J. Biol. Chem.*, **245**, 6052
62. Stubbe, J. A. and Kenyon, G. L. (1971). *Biochemistry*, **10**, 2669
63. Hanson, K. R. and Rose, I. A. (1963). *Proc. Acad. Nat. Sci.*, **50**, 981
64. Smith, B. W., Turner, M. J. and Haslam, E. (1970). *Chem. Commun.*, 842
65. Morrison, J. F. (1954). *Biochem. J.*, **56**, 99
66. Englard, S. (1960). *J. Biol. Chem.*, **235**, 1510
67. Anet, F. A. L. (1960). *J. Amer. Chem. Soc.*, **82**, 994
68. Sargent, D., Turner, M. J., Haslam, E. and Thompson, R. S. (1971). *J. Chem. Soc.* (C), 1489
69. Strominger, J. L. (1958). *Biochim. Biophys. Acta*, **30**, 645
70. Gunetileke, K. G. and Anwar, R. A. (1968). *J. Biol. Chem.*, **243**, 5770
71. Bondinell, W. E., Vnek, J., Knowles, P. F., Sprecher, M. and Sprinson, D. B. (1971). *J. Biol. Chem.*, **246**, 6191
72. Hill, R. K. and Newkome, G. R. (1969). *J. Amer. Chem. Soc.*, **34**, 740
73. Cornforth, J. W. (1968). *Agnew. Chem. Int. Edn Engl.*, **7**, 903
74. Adlersberg, M. and Sprinson, D. B. (1964). *Biochemistry*, **3**, 1855
75. Simmonds, S. (1950). *J. Biol. Chem.*, **185**, 525
76. Katagiri, M. and Sato, R. (1953). *Science*, **118**, 250
77. Davis, B. D. (1953). *Science*, **118**, 251
78. Weiss, U., Gilvarg, C., Mingioli, E. S. and Davis, B. D. (1954). *Science*, **119**, 774
79. Davis, B. D. (1955). *Harvey Lectures*, **50**, 230
80. Schwink, I. and Adams, E. (1959). *Biochim. Biophys. Acta*, **36**, 102

81. Cotton, R. G. H. and Gibson, F. (1965). *Biochim. Biophys. Acta*, **100**, 76
82. Cotton, R. G. H. and Gibson, F. (1967). *Biochim. Biophys. Acta*, **147**, 222
83. Cotton, R. G. H. and Gibson, F. (1968). *Biochim. Biophys. Acta*, **160**, 188
84. Koch, G. L. E., Shaw, D. C. and Gibson, F. (1970). *Biochim. Biophys. Acta*, **212**, 375, 387
85. Schmit, J. C. and Zalkin, H. (1969). *Biochemistry*, **8**, 174
86. Schmit, J. C., Artz, S. W. and Zalkin, H. (1970). *J. Biol. Chem.*, **245**, 4019
87. Davidson, B. E., Blackburn, E. H. and Dopheide, T. A. A. (1972). *J. Biol. Chem.*, **247**, 4441
88. Dayan, J. and Sprinson, D. B. (1968). *Fed. Proc.*, **27**, 290
89. Baker, T. I. (1966). *Biochemistry*, **5**, 2654
90. Baker, T. I. (1968). *Genetics*, **58**, 351
91. Lingens, F., Goebel, W. and Uessler, H. (1967). *Eur. J. Biochem.*, **1**, 363
92. Snell, E. E. (1943). *Arch. Biochem.*, **2**, 389
93. Fildes, P. (1941). *Brit. J. Exptl Path.*, **22**, 293
94. Tatum, E. L., Bonner, D. and Beadle, G. W. (1944). *Arch. Biochem.*, **3**, 477
95. Nye, J. F., Mitchell, H. K., Leifer, E. and Langham, W. H. (1949). *J. Biol. Chem.*, **179**, 783
96. Yanofsky, C. (1955). *J. Biol. Chem.*, **217**, 345
97. Yanofsky, C. (1956). *Biochim. Biophys. Acta*, **20**, 438
98. Yanofsky, C. (1956). *J. Biol. Chem.*, **223**, 171
99. Yanofsky, C. (1955). *Biochim. Biophys. Acta*, **16**, 594
100. Smith, O. H. and Yanofsky, C. (1960). *J. Biol. Chem.*, **235**, 2051
101. Doy, C. H., Rivera, A. and Srinivasan, P. R. (1961). *Biochem. Biophys. Res. Comm.*, **4**, 83
102. Wegman, J. and DeMoss, J. A. (1965). *J. Biol. Chem.*, **240**, 3781
103. Lingens, F. (1968). *Angew. Chem. Int. Edn Engl.*, **7**, 350
104. Tatum, E. L. and Bonner, D. (1944), *Proc. Nat. Acad. Sci.*, **30**, 30
105. Yanofsky, C. (1960). *Bact. Rev.*, **24**, 221
106. Goldberg, M. E., Creighton, T. E., Baldwin, R. L. and Yanofsky, C. (1966). *J. Mol. Biol.*, **21**, 71
107. Jackson, D. A. and Yanofsky, C. (1969). *J. Biol. Chem.*, **244**, 4526, 4539
108. Henning, U., Helinski, D. R., Chao, F. C. and Yanofsky, C. (1962). *J. Biol. Chem.*, **237**, 1523
109. Carlton, B. C. and Yanofsky, C. (1962), *J. Biol. Chem.*, **237**, 1531
110. Wilson, D. A. and Crawford, I. P. (1965). *J. Biol. Chem.*, **240**, 4801
111. Crawford, I. P. and Yanofsky, C. (1958). *Proc. Nat. Acad. Sci.*, **44**, 1161
112. Goldberg, M. E. and Baldwin, R. L. (1967). *Biochemistry*, **6**, 2113
113. Crawford, I. P. and Ito, J. (1964). *Proc. Nat. Acad. Sci.*, **51**, 390
114. Miles, E. W., Hatanaka, M. and Crawford, I. P. (1968). *Biochemistry*, **7**, 2742
115. York, S. S. (1972). *Biochemistry*, **11**, 2733
116. Mohler, W. C. and Suskind, S. R. (1960). *Biochim. Biophys. Acta*, **43**, 288
117. Ensign, S., Kaplan, S. and Bonner, D. M. (1964). *Biochim. Biophys. Acta*, **81**, 357
118. Hutter, R. and DeMoss, J. A. (1967). *J. Bact.*, **94**, 1896
119. Manney, T. R., Duntze, W., Janosko, N. and Salazar, J. (1969). *J. Bact.*, **99**, 590
120. Tsai, H. and Suskind, S. R. (1972). *Biochim. Biophys. Acta*, **284**, 324
121. Lingens, F., Sprössler, B. and Goebel, W. (1966). *Biochim. Biophys. Acta*, **121**, 164
122. Ratledge, C. A. (1964). *Nature*, **203**, 428
123. Srinivasan, P. R. and Rivera, A. (1963). *Biochemistry*, **2**, 1059
124. Srinivasan, P. R. (1965). *Biochemistry*, **4**, 2860
125. Tamir, H. and Srinivasan, P. R. (1970). *Proc. Nat. Acad. Sci.*, **66**, 547
126. Kane, J. F. and Jensen, R. A. (1970). *Biochem. Biophys. Res. Comm.* **41**, 328

46 Biosynthesis of the Aromatic Amino Acids

127. Queener, S. F. and Gunsalus, I. C. (1970). *Proc. Nat. Acad. Sci.*, **67**, 1225
128. Robb, F., Hutchinson, M. A. and Belser, W. L. (1971). *J. Biol. Chem.*, **246**, 6908
129. Hwang, L. H. and Zalkin, H. (1971). *J. Biol. Chem.*, **246**, 6899
130. Nagano, H., Zalkin, H. and Henderson, E. J. (1970). *J. Biol. Chem.*, **245**, 3810
131. Zalkin, H. and Henderson, E. J. (1971). *J. Biol. Chem.*, **246**, 6891
132. Bauerle, R. H. and Margolin, P. (1966). *Cold Spring Harbour Symp. Quant. Biol.*, **31**, 203
133. Baker, T. I. and Crawford, P. I. (1966). *J. Biol. Chem.*, **241**, 5577
134. Ito, J. and Yanofsky, C. (1966). *J. Biol. Chem.*, **241**, 4112
135. Ito, J., Cox, E. C. and Yanofsky, C. (1969). *J. Bact.*, **97**, 725
136. Gaertner, F. H. and DeMoss, J. A. (1969). *J. Biol. Chem.*, **244**, 2716
137. Umbarger, H. E. (1969). *Ann. Rev. Biochem.*, **38**, 358
138. Tristram, H. (1968). *Science Progress*, **56**, 449
139. Doy, C. H. (1968). *Rev. Pure Appl. Chem.*, **18**, 41
140. Brown, K. D. and Doy, C. H. (1963). *Biochim. Biophys. Acta.*, **77**, 170
141. Brown, K. D. and Doy. C. H. (1965). *Biochim. Biophys. Acta*, **104**, 377
142. Smith, L. C., Ravel, J. M., Lax, S. R. and Shive, W. (1962). *J. Biol. Chem.*, **237**, 3566
143. Wallace, B. J. and Pittard, J. (1967). *J. Bact.*, **93**, 237; **94**, 1279
144. Wallace, B. J. and Pittard, J. (1969). *J. Bact.*, **97**, 1234
145. Im, S. K., Davidson, H. and Pittard, J. (1971). *J. Bact.*, **108**, 400
146. Jensen, R. A. and Nasser, D. S. (1968). *J. Bact.*, **95**, 188
147. Jensen, R. A., Nasser, D. S. and Nester, E. W. (1967). *J. Bact.*, **94**, 1582
148. Pittard, J., Camakaris, J. and Wallace, B. J. (1969). *J. Bact.*, **97**, 1242
149. Doy, C. H. (1968). *Biochim. Biophys. Acta*, **159**, 352
150. Doy, C. H. (1970). *Biochim. Biophys. Acta*, **198**, 364
151. Doy. C. H. and Halsall, D. M. (1969). *Biochim. Biophys. Acta*, **185**, 432
152. Lingens, F., Goebel, W. and Uessler, H. (1967). *Eur. J. Biochem.*, **2**, 442
153. Doy, C. H. (1968). *Biochim. Biophys. Acta*, **151**, 293
154. Jensen, R. A. and Nester, E. W. (1965). *J. Mol. Biol.*, **12**, 468
155. Jensen, R. A. and Nester, E. W. (1966). *J. Biol. Chem.*, **241**, 3365, 3373
156. Nasser, D. S., Henderson, G. and Nester, E. W. (1969). *J. Bact.*, **98**, 44
157. Nakatsukasa, W. M. and Nester, E. W. (1972). *J. Biol. Chem.*, **247**, 5972
158. Giles. N. H., Case, M. E., Partridge, C. W. H. and Ahmed, S. I. (1967). *Proc. Nat. Acad. Sci.*, **58**, 1453
159. Lorence, J. H. and Nester, E. W. (1967). *Biochemistry*, **6**, 1541
160. Nasser, D. S., Henderson, G. and Nester, E. W. (1969). *Bacteriology*, **98**, 44
161. Ito, J. and Crawford, I. P. (1965). *Genetics*, **52**, 1303
162. Crawford, I. P. and Gunsalus, I. C. (1966). *Proc. Nat. Acad. Sci.*, **56**, 717
163. Nester, E. W. and Jensen, R. A. (1966). *J. Bact.*, **91**, 1594
164. Doy, D. H. and Cooper, J. M. (1966). *Biochim. Biophys. Acta*, **127**, 302
165. Zalkin, H. and Chen, S. H. (1972). *J. Biol. Chem.*, **247**, 5996
166. Imamoto, F., Ito, J. and Yanofsky, C. (1966). *Cold Spring Harb. Symp. Quant. Biol.*, **31**, 235
167. Creighton, T. E. and Yanofsky, C. (1966). *J. Biol. Chem.*, **241**, 980
168. Neish, A. C. (1960). *Ann. Rev. Plant. Physiol.*, **11**, 55
169. Brown, S. A. (1966). *Ann. Rev. Plant. Physiol.*, **17**, 223
170. Yoshida, S. (1969). *Ann. Rev. Plant. Physiol.*, **20**, 41
171. Weinstein, L. H., Porter, C. A. and Laurencot, H. J. (1959). *Contrib. Boyce Thompson Inst.*, **20**, 121
172. Weinstein, L. H., Porter, C. A. and Laurencot, H. J. (1961). *Contrib. Boyce Thompson Inst.*, **21**, 201
173. Larsen, P. O. (1966). *Biochim. Biophys. Acta*, **115**, 529

174. Dewick, P. M. and Haslam, E. (1969). *Biochem. J.*, **113**, 537
175. Calvin, M. and Bassham, J. A. (1962). *The Photosynthesis of Carbon Compounds*. New York; Benjamin
176. Towers, G. H. N. and Yoshida, S. (1963). *Can. J. Biochem. Physiol.*, **41**, 579
177. Gamborg, O. L. and Neish, A. C. (1959). *Can. J. Biochem. Physiol.*, **37**, 1277
178. McCalla, D. R. and Neish, A. C. (1959). *Can. J. Biochem. Physiol.*, **37**, 537
179. Delmer, D. P. and Mills, S. E. (1968). *Plant Physiol.*, **43**, 81
180. Wightman, F., Chisholm, M. D. and Neish, A. C. (1961). *Phytochemistry*, **1**, 30
181. Nair, P. M. and Vining, L. C. (1965). *Phytochemistry*, **4**, 401
182. Rosenberg, H., McLaughlin, J. L. and Paul, A. G. (1967). *Lloydia*, **30**, 100
183. Massicot, J. and Marion, L. (1957). *Can. J. Chem.*, **35**, 1
184. Leete, E., Bowman, R. M. and Manuel, M. F. (1971). *Phytochemistry*. **10**, 3029
185. McCalla, D. R. and Neish, A. C. (1959). *Can. J. Biochem. Physiol.*, **37**, 531
186. Minamikawa, T. (1967). *Plant Cell Physiol.* (Tokyo), **8**, 695
187. Minamikawa, T. and Uritani, I. (1967). *J. Biochem.* (Tokyo), **61**, 367
188. Nandy, M. and Ganguli, N. G. (1961). *Biochim. Biophys. Acta*, **48**, 608
189. Gamborg, O. L. (1966). *Can. J. Biochem.*, **44**, 791
190. Balinsky, D. and Davies, D. D. (1961). *Biochem. J.*, **80**, 292, 296, 300
191. Balinsky, D. and Dennis, A. W. (1970). In *Methods in Enzymology*, vol. XVIIA, 354. Edited by H. Tabor and C. W. Tabor. New York: Academic Press
192. Nandy, M. and Ganguli, N. N. (1961). *Arch. Biochem. Biophys.*, **92**, 399
193. Sanderson, G. W⁻ (1966). *Biochem. J.*, **98**, 248
194. Higuchi, T. and Shimada, M. (1967)., *J. Agr. Biol. Chem.* (Japan), **31**, 1179
195. Minamikawa, T., Oyama, I. and Yoshida, S. (1968). *Plant Cell Physiol.* (Japan), **9**, 451
196. Cotton, R. G. H. and Gibson, F. (1968). *Biochim. Biophys. Acta*, **156**, 187
197. Gamborg, O. L. (1965). *Can. J. Biochem.*, **43**, 723
198. Gamborg, O. L. and Wetter, L. R. (1963). *Can. J. Biochem. Physiol.*, **41**, 1733
199. Gamborg, O. L. and Simpson, F. J. (1964). *Can. J. Biochem.*, **42**, 583
200. Gamborg, O. L. and Keeley, F. W. (1966). *Biochim. Biophys. Acta*, **115**, 65
201. Delmer, D. P. and Mills, S. E. (1968). *Biochim. Biophys. Acta*, **167**, 431
202. Belser, W. L., Murphy, J. B., Delmer, D. P. and Mills, S. E. (1971). *Biochim. Biophys. Acta*, **237**, 1
203. Widholm, J. M. (1972). *Biochim. Biophys. Acta*, **261**, 44
204. Hulme, A. C. and Wooltorton, L. S. C. (1957). *J. Sci. Food Agric.*, **8**, 117
205. Goldschmid, O. and Quimby, G. R. (1964). *Tappi*, **47**, 528
206. Rohringer, R., Fuchs, A., Lunderstadt, J. and Samborski, D. J. (1967). *Can. J. Bot.*, **45**, 863
207. Boudet, A. (1969). *Compt. Rend.*, **269D**, 1966
208. Tresguerres, M. E. F., de Torrontegui, G., Ingledew, W. M. and Canovas, J. I. (1970). *Eur. J. Biochem.*, **14**, 445
209. Ornston, L. N. and Stanier, R. Y. (1966). *J. Biol. Chem.*, **241**, 3776
210. Whiting, G. C. and Coggins, R. A. (1967). *Biochem. J.*, **102**, 283
211. Virtanen, A. I. (1965). *Phytochemistry*, **4**, 207
212. Fowden, L. (1967). *Ann. Rev. Plant Physiol.*, **18**, 85
213. Morris, C. J., Thompson, J. F., Asen, S. and Irreverre, F. (1959). *J. Amer. Chem. Soc.*, **81**, 6069
214. Irreverre, F., Kny, H., Asen, S., Thompson, J. F. and Morris, C. J. (1961). *J. Biol. Chem.*, **236**, 1093
215. Kjaer, A. and Larsen, P. O. (1962). *Acta Chem. Scand.*, **16**, 242
216. Kjaer, A. and Larsen, P. O. (1963). *Acta Chem. Scand.*, **17**, 2397
217. Morris, C. J. and Thompson, J. F. (1965). *Arch. Biochem. Biophys.*, **110**, 506
218. Morris, C. J. and Thompson, J. F. (1967). *Arch. Biochem. Biophys.*, **119**, 269
219. Larsen, P. O. (1964). *Biochim. Biophys. Acta*, **93**, 200

220. Larsen, P. O. (1967). *Biochim. Biophys. Acta*, **141**, 27
221. Larsen, P. O., Onderka, D. K. and Floss, H. G. (1972). *Chem. Commun.*, 842
222. Young, I. G., Batterham, T. J. and Gibson, F. (1969). *Biochim. Biophys. Acta*, **177**, 389
223. Pridham, J. B. (1965). *Ann. Rev. Plant Physiol.*, **16**, 13
224. Guggenheim, M. Z. (1913). *Z. Physiol. Chem.*, **88**, 276
225. Evans, W. C. and Raper, H. S. (1937). *Biochem. J.*, **31**, 2155
226. Liss, I. (1961). *Flora*, **151**, 35

2

The Chemistry of Intermediates in the Shikimate Pathway

2.1 INTRODUCTION

Prior to the discovery of the shikimate pathway natural product chemists had assiduously pursued the chemistry of (−)-shikimic acid and (−)-quinic acid for over half a century. These substances were then regarded as of considerable intrinsic interest from the chemical point of view but as with many other natural products it was not apparent until much later that they served exceedingly important functions in living systems. Nevertheless the knowledge gained in this period provided the secure and invaluable chemical basis upon which the subsequent developments at the biochemical level took place during the 1950s and 1960s. The chemist will clearly also play an increasingly important role in the future as the details of catalysis and metabolic regulation in this pathway are formulated at the molecular level. Knowledge of the chemical properties and the detailed stereochemistry of intermediates and enzymes in the pathway will be an essential prerequisite for the fruitful collaboration of chemists and biologists necessary to solve these problems. In this chapter some of these chemical details are summarised for principal intermediates and related compounds in the pathway.

2.2 (−)-SHIKIMIC ACID

2.2.1 STRUCTURE, SYNTHESIS AND STEREOCHEMISTRY

(−)-Shikimic acid (1)* was first obtained by Eykmann[1−5] from the

---

* (−)-Shikimic acid is named as 3,4,5-trihydroxy-cyclohex-1-ene-1-carboxylic acid. Numeration of the ring carbon atoms is carried out from the locus of the carboxyl group and *through* the double bond.

fruit of *Illicium anisatum, Illicium religiosum* and other related plants. Later work, based primarily on chromatographic methods of analysis has shown the acid to be present in the leaves and fruits of many plants and several studies of its distribution in the plant kingdom have been reported[1, 6-12]. A number of methods for the isolation and purification of substantial quantities of (−)-shikimic acid have been described and generally this is most satisfactorily carried out using the fruit of species of *Illicium*[13-15]. The acid is crystallised from methanol, ethanol, or glacial acetic acid, melts at 184–186° and its solutions are strongly laevorotatory ($[\alpha]_D$−176°, EtOH). Some conjugates of (−)-shikimic acid with various phenolic acids have been observed in plant tissues[16, 17] but few have been isolated. Probably the best characterised of these substances is 5-O-caffeoylshikimic acid (3), dactylifric acid, obtained from dates, *Phoenix dactylifera*[18]. Shikimic acid-3-phosphate (4), the intermediate in the shikimate pathway immediately following shikimic acid itself, is most conveniently isolated[19] from cultures of *Aerobacter aerogenes* A 170–40 as its barium salt. Weiss and Mingioli obtained the free acid (4) by decomposition of the brucine salt[20].

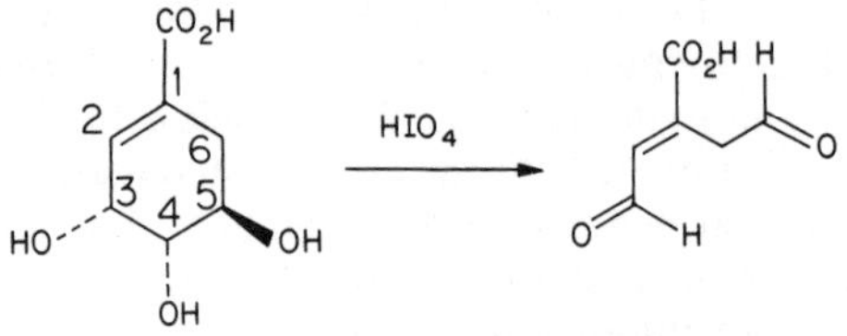

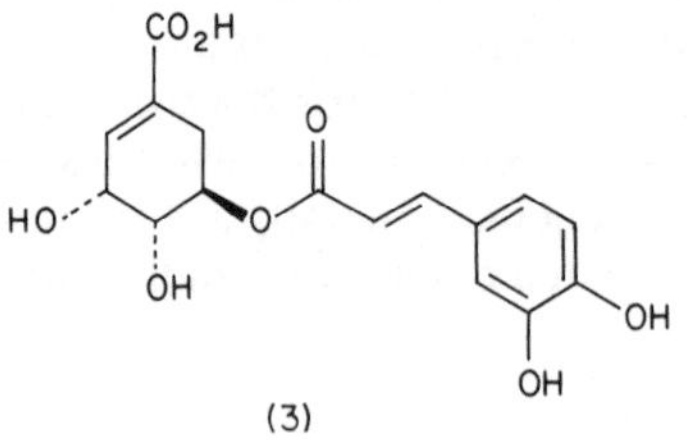

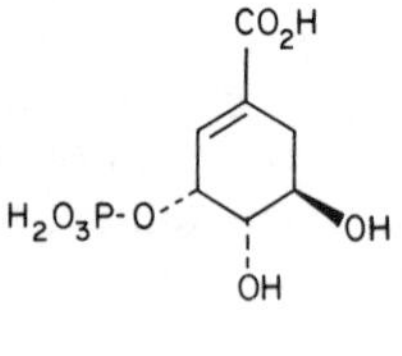

Colorimetric methods for the quantitative determination of (−)-shikimic acid have been described[21-25] and these normally involve periodate oxidation to give the dialdehyde (2) followed by treatment with aniline[21], alkali[22] or thiobarbituric acid[23-25] to produce coloured species. The thiobarbituric acid chromogen formed in the Saslaw and Waravdekar method[23] has a very high molar extinction coefficient at 660 nm but it is unstable and this has limited its usefulness in analytical work. A more stable chromogen is developed by treating the oxidation product (2) with thiobarbituric acid in strong acid solution and this forms the basis of the recommended analytical procedure[24, 25]. The method is however not specific for (−)-shikimic acid and for example 2-deoxy sugars also react under these conditions to give coloured products. Paper chromatography provides the simplest technique for purely qualitative work and extensive compilations of solvent systems, R_F values and distinctive spray reagents have been recorded for (−)-shikimic acid and (−)-quinic acid and their derivatives[1, 26, 27].

(−)-Shikimic acid (1) was recognised, following the early work of Eykmann[3–5] and Chen[28], as a trihydroxycyclohexene carboxylic acid. More detailed studies of its chemistry were carried out by Fischer and Dangschat[29, 30, 31] who demonstrated the relative positions of the functional groups in the molecule by degradation of the methyl ester (5) to aconitic acid (6) and the dihydro derivative (7) to tricarballylic acid (8), *Figure 2.1*. The absolute configuration at each of the chiral centres (positions 3, 4 and 5) in (−)-shikimic acid was determined by degradation[31] to 2-deoxy-D-arabinohexono-γ-lactone (10). This sequence of degradative reactions also provided confirmatory evidence for the location of the double bond and the three hydroxyl groups. More recently further elegant work on the chemistry of (−)-shikimic acid has been carried out by Grewe and his collaborators who prepared (*Figure 2.1*) both shikimyl alcohol (11)[32] and the corresponding aldehyde (12)[33]. Grewe and his colleagues[34] also investigated the reaction between (−)-shikimic acid and diazomethane. Brief treatment with the reagent gave the methyl ester but more prolonged exposure to an excess of diazomethane led to addition to the olefinic linkage and the formation of isomeric pyrazolines[34, 35]. Distillation of the pyrazolines (14,15) which resulted from the action of diazomethane on (13) gave (16), a derivative of 6-methylshikimic acid[34].

Weiss and Mingioli[20] established the structure of shikimic acid-3-phosphate, an intermediate on the shikimate pathway, as (4) by a comparison with a synthetic sample of shikimic acid-5-phosphate (17) prepared from (9), *Figure 2.1*. Both acids reacted with periodic acid to show the presence of an α-glycol grouping but the dialdehydes formed in this reaction had markedly different spectral characteristics. The presence of a *cis* α-glycol grouping in (17) was confirmed by the relatively rapid reaction with periodate and the formation of a copper complex and an acetonide derivative. In contrast the natural phosphate ester (4) reacted slowly with periodate and did not form a complex with copper acetate nor an

52

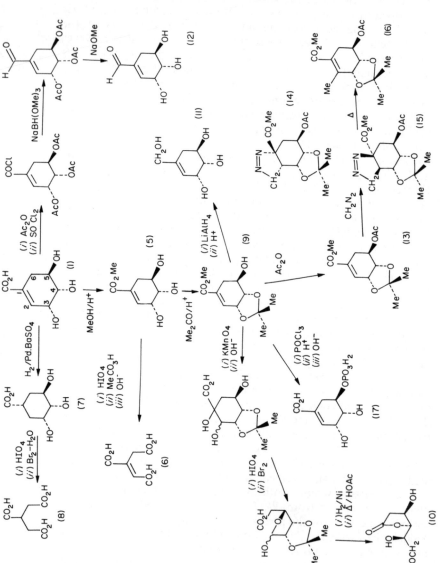

Figure 2.1. Some reactions of (−)-shikimic acid[20, 28–35]

53

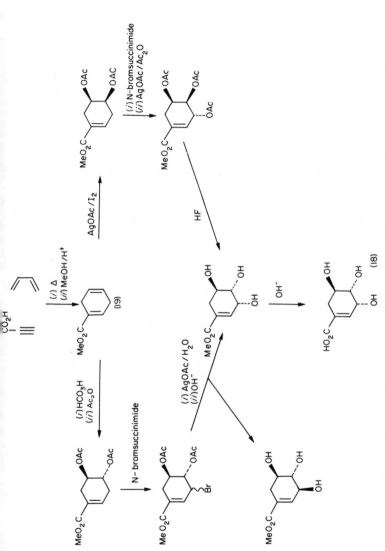

Figure 2.2. Synthesis of (±)-shikimic acid (18)[40, 41]; all formulae represent racemic modifications. Resolution of (±)-shikimic acid was achieved by fractional crystallisation of the diastereoisomeric salts of (—)-2-phenyl-ethylamine with 3,4-cyclohexylidene shikimic acid followed by acid hydrolysis

acetonide derivative and these observations were interpreted as showing a *trans* α-glycol grouping to be present in (4). Shikimic acid-3-phosphate (4) was, however, partially isomerised on acid treatment to give the 4-phosphate by migration of the ester group whilst the synthetic phosphate (17) was stable to these conditions. This observation confirmed the presence of a *cis*-hydroxyl group adjacent to the phosphate ester in (4) and further substantiated this formulation of the metabolite. A rational synthesis of (4) has not been completed.

Syntheses of (±) and (−)-shikimic acid have been completed by several groups: McCrindle, Overton and Raphael[36, 37], Smissman, Suh, Oxman and Daniels[38, 39], Grewe and his collaborators[40, 41] and by Bestmann and Heid[42]. The two syntheses due to Grewe[40, 41] both of which are based on the intermediate dihydrobenzoic acid (19) are shown in *Figure 2.2*.

Hall[43] has utilised the proton magnetic resonance spectrum of (−)-shikimic acid in deuterium oxide to determine its precise conformation in aqueous solution. The assignment of the spectrum was straightforward, the signal at lowest field was assigned with certainty to the olefinic proton and the multiplets at highest field to the C-6 methylene protons. Allocation of the remaining multiplets then followed. The first order coupling constants and chemical shifts of the various protons are shown in *Table 2.1* and from these the dihedral angles for the various interacting protons were inferred using the Karplus equation. These values were compared

Table 2.1   THE PROTON MAGNETIC RESONANCE SPECTRUM OF (−)-SHIKIMIC ACID IN D_2O AT 60 MHz (Hall [43])

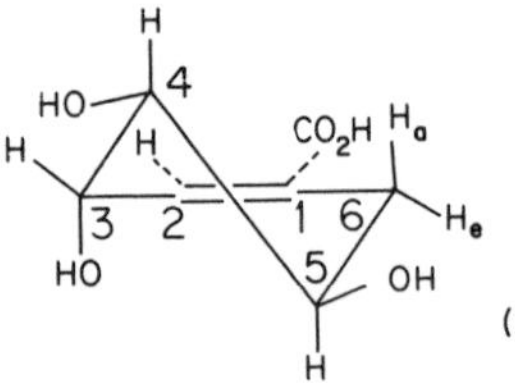

(20)

τ *values*

H-2	H-3	H-4	H-5	H-6e	H-6a
3·12	5·59	6·31	6·00	7·23	7·86

J *values* (Hz)

$J_{2,3}$	$J_{2,6}$	$J_{3,4}$	$J_{3,6}$	$J_{4,5}$	$J_{5,6a}$	$J_{5,6e}$	$J_{6a,6e}$
4·0	1·8	3·9	1·5	8·4	6·2	5·0	18·5

with those measured from models of the energetically most favourable half-chair and boat conformations for (−)-shikimic acid. From these observations it was readily deduced that (−)-shikimic acid adopts a half-chair conformation in aqueous solution which approximates most closely to the conformation (20). Similar measurements on a range of other shikimic acid derivatives has shown[44] that these adopt in almost all cases analogous half-chair conformations.

2.2.2 ISOTOPICALLY LABELLED DERIVATIVES

The availability of isotopically labelled derivatives of (−)-shikimic acid is of considerable importance to the experimental study of metabolism by the shikimate pathway and methods have been devised for the preparation of samples of (−)-shikimic acid specifically labelled with deuterium, tritium or ^{14}C at several points in the molecule (*Table 2.2*). Sprinson and his collaborators[45] were the first

Table 2.2 ISOTOPICALLY LABELLED DERIVATIVES OF (−)-SHIKIMIC ACID

Compound	Reference
G-[^{14}C]-(−)-shikimic acid	26, 46, 47, 59, 60
6-[^{14}C]-(−)-shikimic acid	49
7-[^{14}C]-(−)-shikimic acid	49
2,3,4,5-[^{14}C$_4$]-(−)-shikimic acid	50
1,6-[^{14}C$_2$]-(±)-shikimic acid	51
2-[^{2}H]-(−)-shikimic acid	56, 57
2-[^{3}H]-(−)-shikimic acid	49
3-[^{3}H]-(−)-shikimic acid	55
6R and 6s-[^{3}H]-(−)-shikimic acid	48
6R and 6s-[^{2}H]-(−)-shikimic acid*	52

* These compounds are produced as racemates with 6S and 6R-[^{2}H]-(−)-shikimic acid respectively

to observe that selectively labelled ^{14}C derivatives of (−)-shikimic acid could be prepared using the *Escherichia coli* 83-24 mutant and specifically labelled (^{14}C)-D-glucose derivatives as substrates. Thus using either 1- or 6-[^{14}C]-D-glucose, which are the most readily available isotopically labelled D-glucose derivatives, (−)-shikimic acid is produced in which the isotopic carbon is almost exclusively partitioned (∼ 90 per cent) between positions 2 and 6 in the ring. The same method has also been utilised with generally labelled [^{14}C]-D-glucose as substrate to prepare generally labelled [^{14}C]-(−)-shikimic acid[46], and this form of labelling of the metabolite may be achieved with plants using the same substrate or by

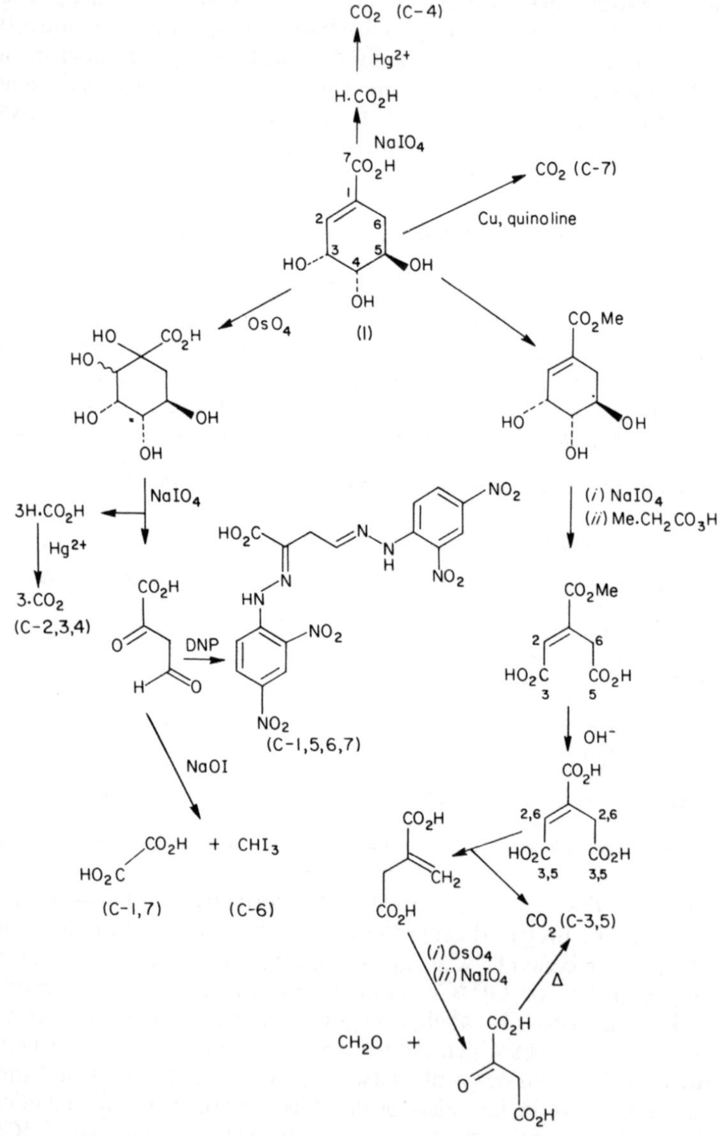

Figure 2.3. Chemical degradation of (−)-shikimic acid to determine its isotopic content[45]

photosynthesis in a $[^{14}C]$ carbon dioxide enriched atmosphere[26, 47]. Determination of the extent of isotopic labelling at any one position of the (−)-shikimic acid molecule may be derived by chemical degradation. The method evolved by Sprinson and his group[45] gives the isotopic content at C-4, 6 and 7 directly and the remainder by difference (*Figure 2.3*); that of Dewick and Haslam[26] gives C-6 and 7 directly and C-2 by difference (*Figure 2.4*).

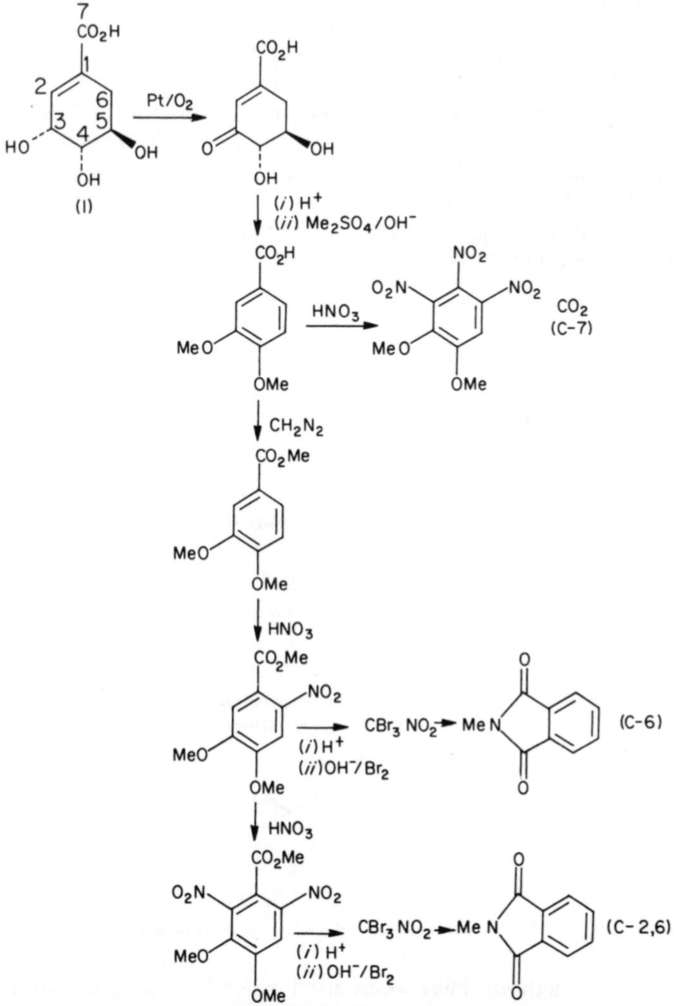

Figure 2.4. Chemical degradation of (−)-shikimic acid to determine its isotopic content[26]

Biochemical methods have also been employed for the synthesis of several specifically labelled (^{3}H, ^{14}C) samples of (−)-shikimic acid (*Figure 2.5*). Aged cell free extracts of *Escherichia coli* 83-24 when coupled to an enzyme system designed for the cyclic generation of NADPH (NADP$^+$, L-*iso*citrate and *iso*citrate dehydrogenase) are able to convert mixtures of D-erythrose-4-phosphate (22) and phosphoenolpyruvate (21) to (−)-shikimic acid (1)[48]. In this way samples of 4-[^{3}H]-D-erythrose phosphate, (E) and (Z)-3-[^{3}H]-phosphoenolpyruvate, 1-[^{14}C] and 3-[^{14}C]-phosphoenolpyruvate have been employed to furnish (−)-shikimic acid (1) labelled respectively 2-[^{3}H], 6(s)-6-[^{3}H], 6(R)-6-[^{3}H], 7-[^{14}C] and 6-[^{14}C][48, 49]. A minor variation of this method was used by Sprinson and his co-workers[50] to obtain 2,3,4,5-[^{14}C$_4$]-(−)-shikimic acid from 4,5,6,7-[^{14}C$_4$]-D-altroheptulose-1,7-diphosphate. In this particular reaction the substrate is first cleaved at the C-3 to C-4 linkage to give a three and a four carbon fragment which are then used in the enzyme synthesis of (−)-shikimic acid.

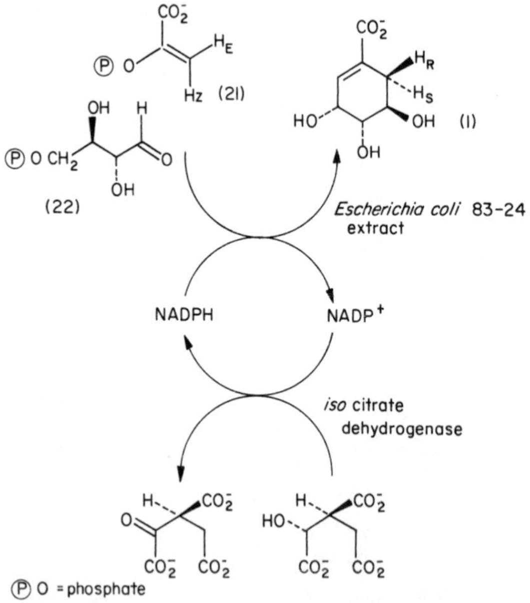

Figure 2.5. Biochemical synthesis of (−)-shikimic acid[48−50]

Chemical methods have been used wholly or in part for the synthesis of other specifically labelled derivatives of (±)-shikimic acid. Thus the synthesis of (±)-shikimic acid due to Raphael,

Smissman and their collaborators[36-39] has been employed[51, 52] to prepare ($\pm$)1,6-[$^{14}C_2$], and the stereoisomeric ($\pm$)-6-[^2H]-shikimic acids. The first step in this synthesis involves the condensation of *trans, trans*-1,4-diacetoxybutadiene (23) and acrylic acid or its methyl ester (24, R = H or Me) to give (25, R = H or Me). The intermediate (25) is then transformed to ($\pm$)-shikimic acid by hydroxylation of the double bond and finally elimination of one acetoxy group to give the $\alpha\beta$ unsaturated carboxyl system. Using 2,3-[$^{14}C_2$]-methylacrylate, Chabannes, Pichat, Herbert and Pacheco[51] prepared ($\pm$)1,6-[$^{14}C_2$]-shikimic acid by this route. Raphael and his group chose[36, 37] the all *cis* configuration for the intermediate (25) on the basis of the Alder rules and this assignment of the relative stereochemistry was shown to be the correct one by subsequent proton magnetic resonance analysis[37, 53, 54]. Hill and Newkome[52] also confirmed this point by the preparation and analysis of the deuterio analogues (28 and 29) formed from methyl-α, *trans*-β-dideuterioacrylate (26) and methyl-*cis*-β-deuterio-acrylate (27) respectively. Subsequent completion of the synthesis from (28) and (29) gave ($\pm$)-shikimic acid in which the natural isomer is labelled respectively with deuterium (85 per cent) in the 6s (30) and 6R (31) positions (*Figure 2.6*).

Leduc, Dansette and Azerad[55] prepared 3-[^3H]-($-$)-shikimic

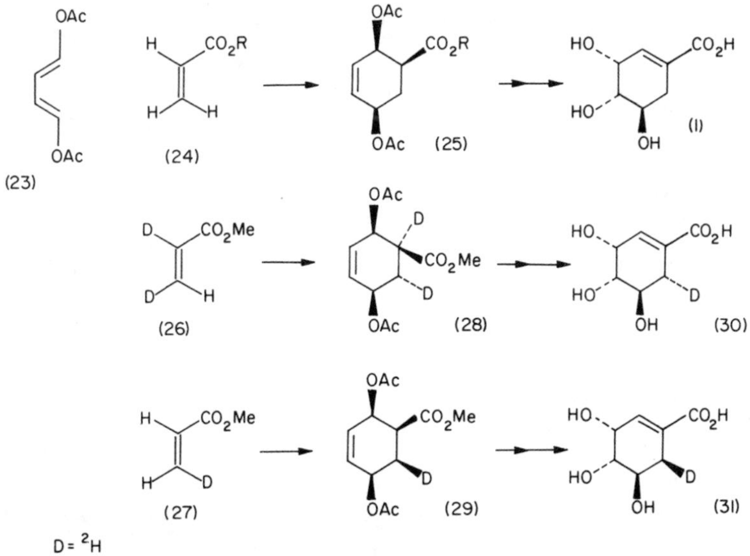

Figure 2.6. Chemical synthesis of the stereoisomeric 6-[^2H]-($\pm$)-shikimic acids[52]. All formulae show racemic modifications

acid from methyl-3-dehydroshikimic acid by sodium borotritiide reduction. The correct diastereoisomer was separated by crystallisation and hydrolysis gave the appropriately labelled metabolite. Both chemical and biochemical procedures were used to prepare 2-[^{2}H]-(−)-shikimic acid (40) from (−)-shikimic acid (1)[56–58], *Figure 2.7*. Solid phase bromination of (−)-shikimic acid (1) gave the dibromo acid (32) and this when treated with silver carbonate in aqueous solution gave 6-bromo-6-deoxyquinic acid, isolated as the lactone (33), by an $S_N i$ reaction. Catalytic reduction of the triacetyl-lactone (34) with deuterium gas gave 2(RS)-triacetyl-2-deuterioquinide (35) which was converted into the acid (36) by base and thence by catalytic oxidation to 2(RS)-2-deuterio-3-dehydroquinic acid (37). The proton magnetic spectrum of the acid (37) showed 0.3 atom of deuterium in the pro-2s position of (37). Stereoselective labelling of 3-dehydroquinic acid was also obtained by equilibration of the parent acid (38) in deuterium oxide at pH 7.0 and 2(s)-2-deuterio-3-dehydroquinic acid (39, 0.6–0.7 atom deuterium) was obtained by this method. Treatment of the deuterio acids (37 or 39) with an enzyme extract from *Escherichia coli* 83-24 and

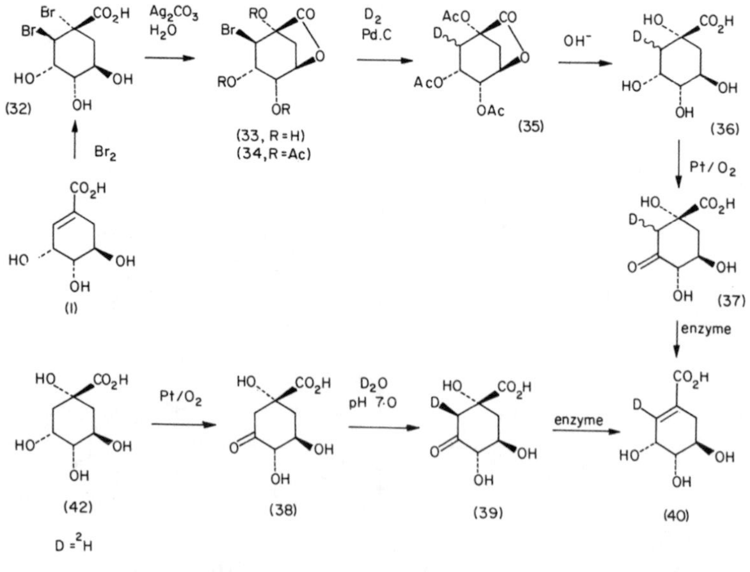

Figure 2.7. Preparation of 2-[^{2}H]-(−)-shikimic acid[56–58]

an enzyme system for the cyclic generation of NADPH gave samples of (−)-shikimic acid labelled at C-2 with deuterium (40, 0.3 and 0.6 atom deuterium respectively).

2.2.3   3-DEHYDROSHIKIMIC ACID

3-Dehydroshikimic acid (41) was first isolated by Salamon and Davis[61] from cultures of *Escherichia coli* 83-2 by chromatography on charcoal. Improved preparative procedures have since been described by several groups[62-66] and in all of these the product is separated, prior to crystallisation, by ion-exchange chromatography. Heyns and Gottschalk[62] and Knowles, Haworth and Haslam[63,64] observed that the platinum catalysed oxidation of (−)-shikimic acid (1) occurred with the selective oxidation of the quasi-axial 3-hydroxyl group (20) to give 3-dehydroshikimic acid (4) directly. A similar selective oxidation forms the basis of the method of Grewe and Jeschke[65]. Nitric acid oxidation of quinic acid (42) gives 3-dehydroquinic acid (38) and acid catalysed dehydration then yields (41). Whiting and Coggins[66] have described an enzymic

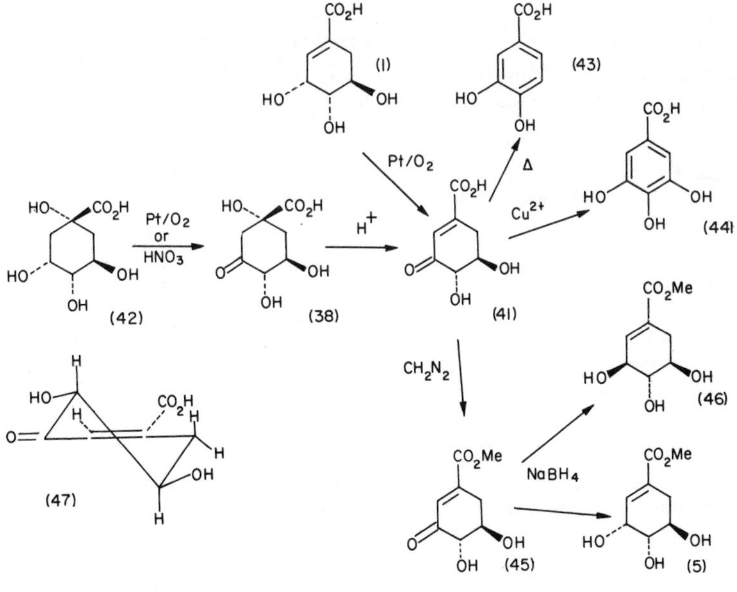

Figure 2.8. Preparation and properties of 3-dehydroshikimic acid[61-66]

62 The Chemistry of Intermediates

method for the preparation of 3-dehydroshikimic acid (41) from shikimic acid (1). These workers surveyed 66 strains of *Acetomonas oxydans* (formerly *Acetobacter suboxydans*) and isolated five strains capable of oxidising both (−)-quinic and (−)-shikimic acid to their respective 3-dehydro derivatives. The preparative procedure outlined by Whiting and Coggins[66] utilised strain CR 49 of this organism and 3-dehydroshikimic acid (41) was isolated after a ten days' incubation of the organism with (−)-shikimic acid as substrate.

3-Dehydroshikimic acid (41) readily undergoes conversion to protocatechuic acid (43) by treatment with acid or upon heating[61, 65]. This latter property gives rise to a characteristic double melting point 151–152° followed by 198–200°. As an α-hydroxy ketone the acid also readily reduces Fehlings and Tollens solution and under controlled conditions gallic acid (44) may be isolated as one of the products of oxidation[63]. Because of its ease of aromatisation few well-characterised derivatives of 3-dehydroshikimic acid have been prepared. Grewe and Jeschke[65] have described the methyl ester (45) which results from brief treatment with diazomethane at −70°. Reduction of the methyl ester with sodium borohydride yields methyl shikimate (5) and its epimer at the 3- position (46).

Proton magnetic resonance studies of the acid (41) in aqueous solution show that the molecule adopts a half-chair conformation (47) analogous to that of (−)-shikimic acid[67]. First order coupling constants and chemical shifts for 3-dehydroshikimic acid (41) and its methyl ester are shown in *Table 2.3*.

Table 2.3 THE PROTON MAGNETIC RESONANCE SPECTRA OF 3-DEHYDROSHIKIMIC ACID (41) AND ITS METHYL ESTER IN D_2O AT 100 MHz

Compound \ Proton	H-2	H-4	H-5	H-6e	H-6a
3-Dehydro-shikimic acid	τ 2.81 $J_{2,6a}$ 2.5	τ 5.30 $J_{4,5}$ 10.5	τ 5.51 $J_{5,6a}$ 4.5 $J_{5,6e}$ 9.5	τ 6.45 $J_{6a, 6e}$ 18.0	τ 6.99
Methyl 3-dehydroshikimate	τ 2.94 $J_{2,6a}$ 2.5	τ 5.44 $J_{4,5}$ 10.5	τ 5.66 $J_{5,6e}$ 4.5 $J_{5,6a}$ 9.5	τ 6.56 $J_{6a,6e}$ 18.0	τ 7.11

τ values relative to TMS. Coupling constants (J) in Hz.

2.2.4   5-ENOLPYRUVYLSHIKIMIC ACID-3-PHOSPHATE

The early studies concerning this compound centred on its dephosphorylated form 5-enolpyruvylshikimic acid (48) which was shown to accumulate, in addition to (−)-shikimic acid and shikimic acid-3-phosphate, in the media of several multiple aromatic bacterial auxotrophs blocked beyond shikimate[68]. The compound (48) was readily hydrolysed with acid to give equimolar quantities of (−)-shikimic acid and pyruvic acid and gave a positive reaction to periodate. Somewhat later it was observed that a product with similar properties was formed from phosphoenolpyruvate and shikimate-3-phosphate by extracts of *Escherichia coli*. Detailed study of this reaction showed, however, that the first formed product was in fact 5-enolpyruvylshikimic acid-3-phosphate (49) and that 5-enolpyruvylshikimic acid was a by-product produced by dephosphorylation[69–71]. Only (49) was found to be active in the subsequent biosynthetic transformations to the various aromatic amino acids and was the true biosynthetic intermediate[71]. The correct chemical formulation of 5-enolpyruvylshikimic acid-3-phosphate (49) undoubtedly owes a great deal to the suggestion of R. B. Woodward[69] who reasoned that formation of L-phenyl-alanine and L-tyrosine and their immediate precursors the corresponding phenylpyruvic acids and prephenic acid (51) from (49) could best be rationalised if (49) was first dehydrated to give (50), chorismic acid, which then underwent a Claisen rearrangement.

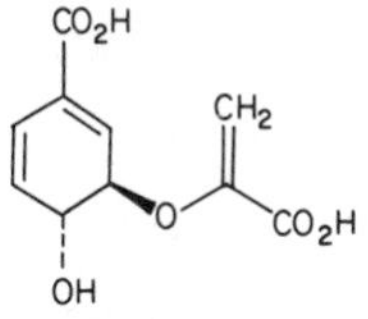

(50)

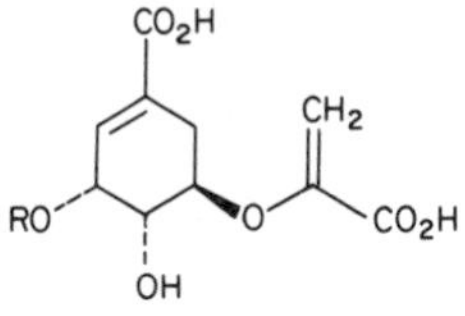

(48, R = H)
(49, R = PO₃H₂)

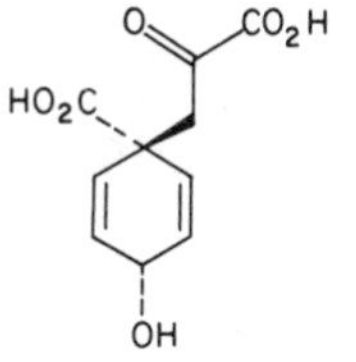

(51)

5-Enolpyruvylshikimic acid-3-phosphate (49) is best prepared[72] enzymically from shikimic acid-3-phosphate and phosphoenol-pyruvate using an ammonium sulphate fraction from *Escherichia coli* K-12 mutant 58-278 and is normally isolated as its barium salt. Its proton magnetic resonance spectrum (potassium salt in deuterium oxide) has been reported by Sprinson and his collaborators[73] and analysis of the various coupling constants indicated that the molecule adopts a half-chair conformation quite analogous to that shown by (−)-shikimic acid itself (20). Mass spectrometric analysis of (49) was carried out by Sprinson's group[73] with a pentamethyl derivative prepared by treatment of the silver salt with methyl iodide.

2.3 (−)-QUINIC ACID

Although it does not occur as an intermediate on the shikimate pathway (−)-quinic acid (42) is related very closely both chemically and biochemically to (−)-shikimic acid (1). The historical development of the chemistry of both substances is intimately linked.

2.3.1 NOMENCLATURE

Some uncertainty surrounds the nomenclature of (−)-quinic acid and the proposal[74] of the IUPAC–IUB commissions on the nomenclature of Organic Chemistry and Biochemistry that it should be named according to the rules and procedures recommended for cyclitols has not met with universal acceptance[75]. Nevertheless they have been adopted here and are briefly outlined below.

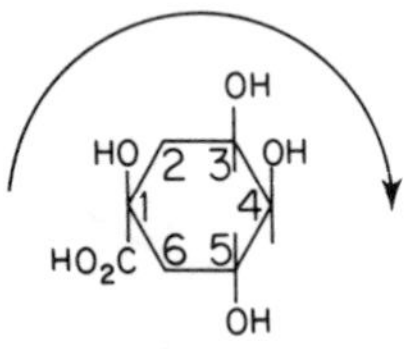

(52) 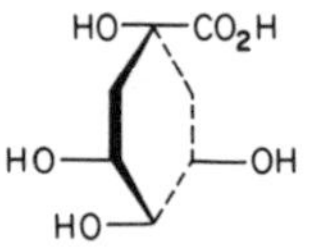   (53) 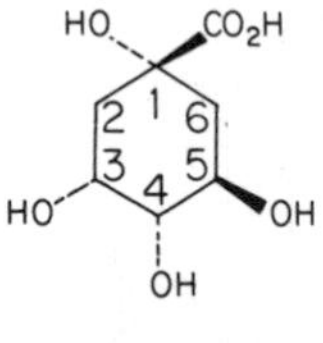   (54)

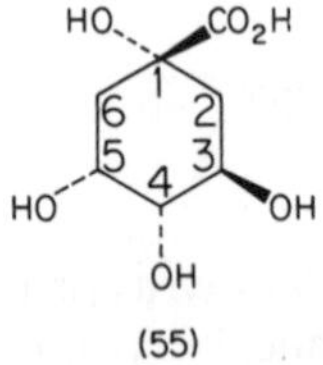

(55)

For (−)-quinic acid the trivial name is preferred but application of the recommended procedures leads to a different numeration of the ring carbon atoms (54) to that established in the chemical and biochemical literature (55). This numeration derives from the systematic name for (−)-quinic acid, 1L-1(OH),3,4/5-tetrahydroxy-cyclohexane carboxylic acid. Numbering is clockwise when the number of substituents above the plane of the ring is the same as or greater than the number below that plane. Thus the numeration of the ring of (−)-quinic acid derives from the projection (52) of the molecule. The absolute configuration is specified from a vertical Fischer–Tollens type projection (53) in which the hydroxyl group at the lowest numbered chiral centre projects to the left (1L). The relative configurations at the various ring positions are described by the fraction which precedes the systematic name. The numerator consists of the locants of those substituents (or 'set') that lie above the plane of the ring (52). The denominator contains the locants of the other set.

2.3.2   STRUCTURE, SYNTHESIS AND STEREOCHEMISTRY

Several micro-organisms are able to convert (−)-quinic acid (42) to 3-dehydroquinic acid (38) and thus use the acid as an alternative carbon source for growth. In the plant kingdom where paper chromatographic methods have shown it to be widely distributed[1, 76–81], (−)-quinic acid (42) may perform a similar function. It is found in vegetative tissues of plants both in the free state and in combined forms such as the various hydroxycinnamoyl and galloyl esters. Typical of these are chlorogenic acid (56)[79], theogallin (57) in *Camellia sinensis*[80], and the pentagalloyl ester which is the principal constituent of the vegetable tannin (Tara) from *Caesalpinia spinosa*[81]. Free (−)-quinic acid (42) itself was first isolated[82] in 1790 by Hofmann from Cinchona species and is today most conveniently obtained from this source and occasionally the coffee bean.

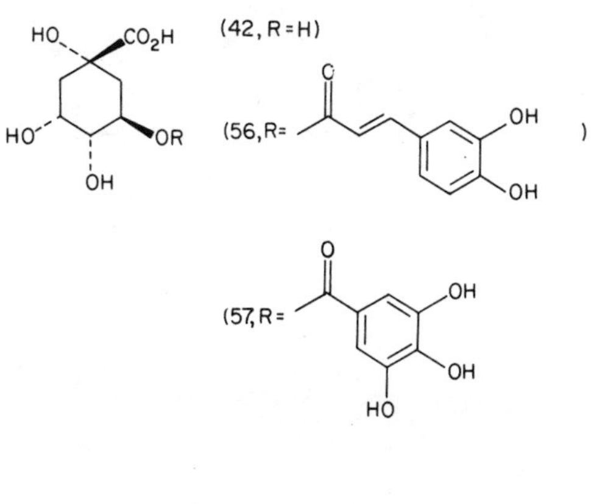

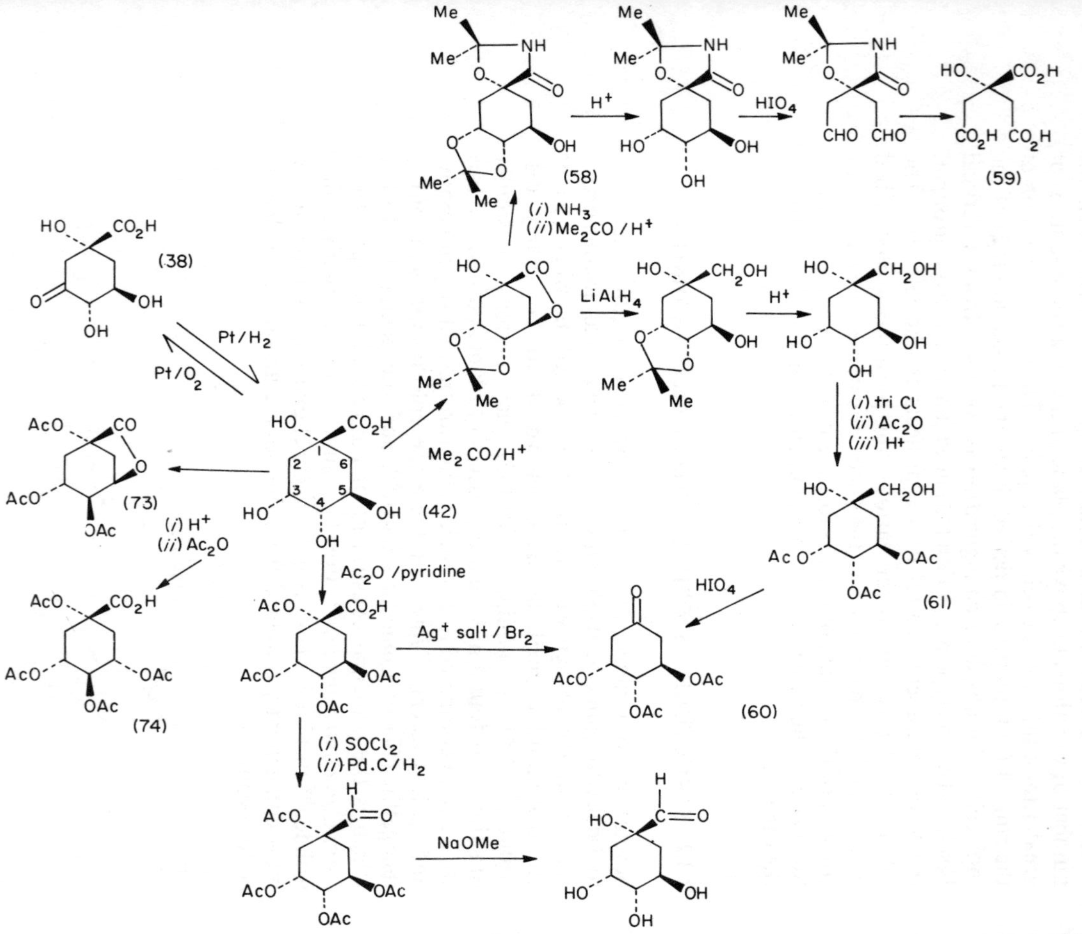

(58)
H+
HIO4
(59)
(i) NH3
(ii) Me2CO / H+
LiAlH4
H+
(38)
Pt/H2
Pt/O2
(i) tri Cl
(ii) Ac2O
(iii) H+
Me2CO/H+
(73)
(42)
(i) H+
(ii) Ac2O
Ac2O /pyridine
(61)
HIO4
Ag+ salt / Br2
(74)
(60)
(i) SOCl2
(ii) Pd.C / H2
NaOMe

The general chemistry of (−)-quinic acid (42) has been extensively studied by various groups including Fischer[29, 82–85], Karrer[86, 87] and Grewe[15, 58, 88–93] and their collaborators (*Figure 2.9*). Early uncertainties surrounding its structural formulation were finally resolved by Fischer and Dangschat[83] who carried out the sequence of reactions (42) → (59) shown in *Figure 2.9* which converted (−)-quinic acid to citric acid and thus established the position of the fourth hydroxyl group, α to the carboxyl function. Total syntheses of racemic quinic acid have been reported by four groups[42, 89, 95, 96].

The close chemical and biochemical relationship which exists between (−)-quinic acid and (−)-shikimic acid has prompted several successful attempts to chemically interconvert these two substances[15, 84, 85, 90, 92]. These have not only provided further proof of structure but also of the absolute stereochemistry of the four chiral centres in the molecule of (−)-quinic acid (42). Grewe and Lorenzen[15] converted (−)-shikimic acid (1) to (−)-quinic acid (42) via the dibromo derivative (32) and some details of the sequence of steps have been referred to above in the discussion of the chemistry of (−)-shikimic acid (1). The triacetoxycyclohexanone (60) is a key intermediate not only in the conversion of (−)-quinic acid to (−)-shikimic acid but also in the elegant synthesis of these two natural products in optically pure forms from D-arabinose (63) described by Bestmann and Heid[42] in 1971. In this synthesis the intermediate (60) is derived from the sugar in several stages which are summarised in *Figure 2.10*. The cyclic ylide (65) was derived from the 1,5-ditosyl-2,3,4-tri-o-benzyl-D-arabitol (64) and methylene triphenylphosphorane by a previously described procedure[97] and then transformed to the crystalline (1R, 2S, 3R)-5-methylene-1,2,3-cyclohexanetriol (66) by standard methods. Acetylation and treatment with osmium tetroxide and sodium metaperiodate afforded the ketone (60). Grewe and Vangermain[92] alternatively prepared the ketone (60) from (−)-quinic acid (42) by two methods (*Figure 2.9*). In the first of these the triacetylquinyl alcohol (61) was treated with sodium metaperiodate and in the second a Hunsdiecker reaction on the silver salt of tetra-acetylquinic acid (62) gave (60) directly.

Further steps in the synthetic and the conversion sequences from the ketone (60) are identical. A stereoselective cyanohydrin reaction gives the triacetylquinic acid nitrile (67, 85 per cent). Acetylation and reaction with hydrogen bromide in acetic acid gives the amide (68) and treatment with liquid N_2O_3 provides tetra-acetylquinic acid (62). To prepare (−)-shikimic acid (1) from (67) the cyanohydrin is dehydrated to give triacetylshikimonitrile (69, 30 per cent) and hydrolysis then gives the natural product. The

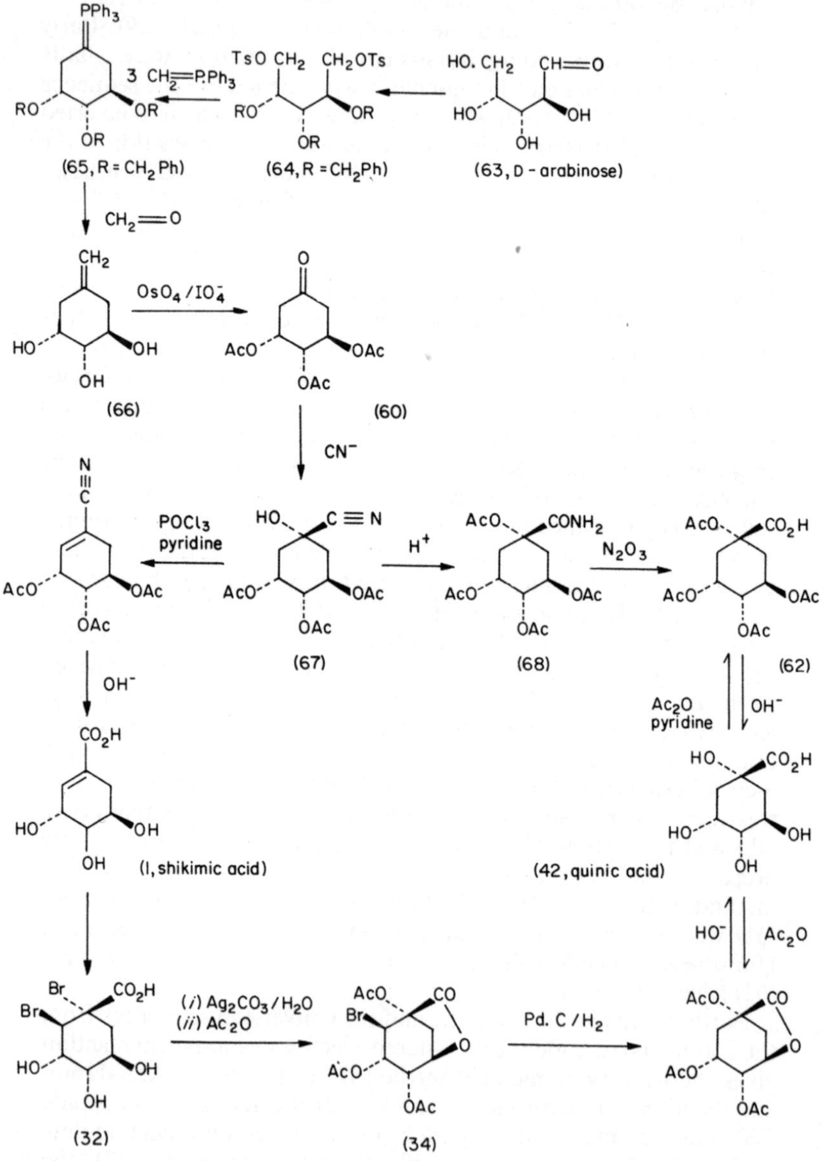

Figure 2.10. Synthesis of (−)-quinic and (−)-shikimic acid from D-arabinose[42]. Interconversions of (−)-quinic and (−)-shikimic acid[92]

synthetic route of Bestmann and Heid[42] from D-arabinose and the conversion sequence from the ketone (60) should provide excellent methods for the preparation of optically pure (−)-quinic and (−)-shikimic acid labelled with isotopes of carbon and hydrogen at various positions in the molecules.

Orloff[98] favoured the assignment of a chair conformation (71) to (−)-quinic acid in which the carboxyl group occupied an axial position on the grounds that the acid readily formed a γ-lactone, quinide (72). However, optical rotary measurements[99] and the evidence of proton magnetic resonance spectroscopy (*Table 2.4*)[100, 101] later showed that this conclusion was incorrect and demonstrated quite clearly that (−)-quinic acid in solution preferentially adopts a chair conformation in which the carboxyl group occupies an equatorial position (70). Formation of the γ-lactone, quinide (72), therefore involves an inversion of the preferred conformation to (71) and this change in shape is reflected in the proton magnetic resonance characteristics of quinide (*Table 2.4*) and its derivatives. A notable feature of the proton magnetic resonance spectra of (−)-quinic acid and several of its derivatives is that although no molecular symmetry axis of rotation relates H-2 eq. and H-2 ax. and they are therefore chemically non-equivalent, they are isochronous at 60, 100 and 220 MHz with no apparent geminal coupling and identical couplings to H-3. Indeed quinide (72) itself provides the simplest derivative of (−)-quinic acid for which the proton magnetic resonance signals due to each of the C-2 methylene protons may be analysed at 100 MHz. Paradoxically the C-6 methylene protons of quinide (72) are isochronous at 100 MHz. Corse and Lundin[75] calculated the energy difference between the two chair forms of (−)-quinic acid (70) and (71), and showed that the chair form (70) in which the carboxyl group is equatorially disposed is favoured to the extent of 5.67 KJ mole^{-1}

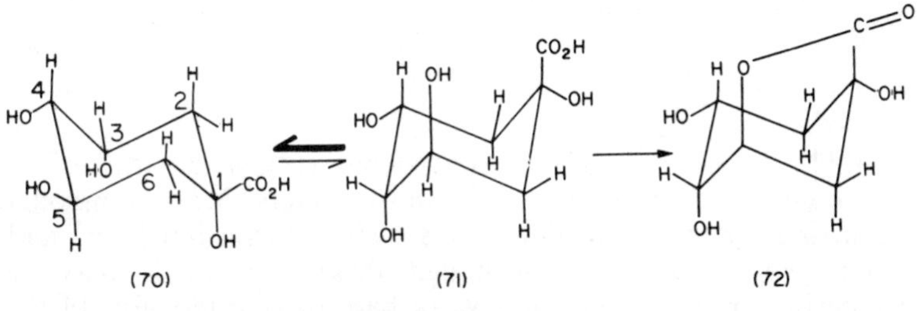

(70)   (71)   (72)

(−)-Quinic acid may be isomerised by the action of sulphuric acid in acetic acid and derivatives of (−)-epiquinic acid and scylloquinic acid, which are two further diastereoisomers of the

Table 2.4 PROTON MAGNETIC RESONANCE SPECTRA OF $(-)$-QUINIC ACID AND QUINIDE IN DEUTERIUM OXIDE (100 MHz). FIRST ORDER COUPLING CONSTANTS AND CHEMICAL SHIFTS[100]

proton compound	H-2e	H-2 ax	H-3	H-4	H-5	H-6e	H-6ax
(-) quinic acid (70)	τ 7·30 $J_{2,3}$ 4·0	τ 7·30	τ 5·21 $J_{3,4}$ 4·0	τ 5·85 $J_{4,5}$ 10·0	τ 5·35 $J_{5,6e}$ 4·0 $J_{5,6ax}$ 10·0	τ 7·05 $J_{6e\,6ax}$ 13·0	τ 7·50
(-) quinide (72)	τ 7·88 $J_{2e,2ax}$ 1·5 $J_{2e,3}$ 6·5	τ 8·10 $J_{2ax,3}$ 11·5	τ 6·15 $J_{3,4}$ 5·0	τ 5·82 $J_{4,5}$ 5·0	τ 5·10	τ 7·55 $J_{5,6e}$ 2·0	τ 7·55 $J_{5,6ax}$ 2·0

τ values relative to TMS   J values in Hz

1,3,4,5-tetrahydroxycyclohexane-1-carboxylic acid structure, have been obtained. The three major products isolated by Gorin[94] and Corse and Lundin[75] were $(-)$-epiquinide triacetate (73), $(\pm)$ epiquinide triacetate and tetra-acetoxy scylloquinic acid (74), *Figure 2.9*. The calculated energy difference between the two chair forms of $(-)$-epiquinic acid (75 and 76) of 1.64 KJ mole^{-1} implies that neither chair form is strongly preferred[75]. However, for scylloquinic acid the chair form in which the carboxyl group adopts an axial position (77) is clearly favoured $(-13.1$ KJ mole$^{-1})$.

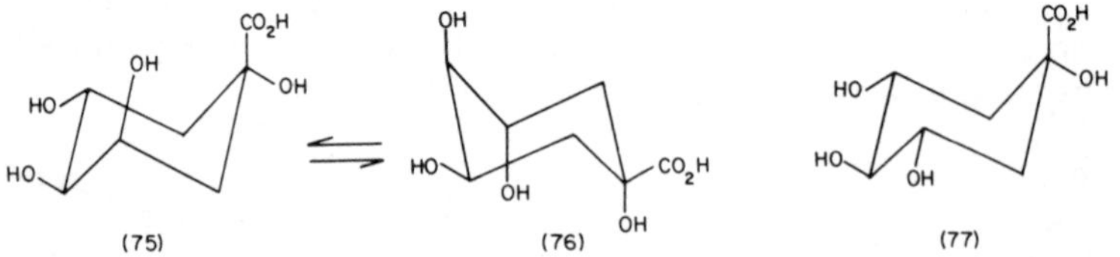

Acyl (hydroxybenzoyl and hydroxycinnamoyl) derivatives of $(-)$-quinic acid are frequently found in plants. Most commonly occurring are the 5-acyl derivatives such as theogallin (57)[80] and chlorogenic acid (56)[79]. Although these compounds may be isomerised by the action of a weak base to give mixtures of the corresponding 3- and 4-acyl derivatives location of the acyl group on the hydroxyl at position 5 appears to be thermodynamically the most favourable[78].

2.3.3 3-DEHYDROQUINIC ACID

The first cyclic intermediate on the shikimate pathway is 3-dehydro-quinic acid (38) and it was first obtained by Weiss, Davis and Mingioli[102] from cultures of *Escherichia coli* mutant 170-27 by charcoal chromatography and precipitation of the brucine salt. Probably the first chemical preparation of this biochemical inter-mediate was reported by Hesse in 1859 who oxidised (−)-quinic acid (42) with bromine water[103, 104]. Several procedures have since been reported for the preparation of this metabolite[62-66]. Yields of almost 80 per cent of 3-hydroquinic acid (38) were reported by Whiting and Coggins[66] after a seven-day incubation of (−)-quinic acid (42) with *Acetomonas oxydans* CR-49 and similar selective oxidation of the axial 3-hydroxyl group in (−)-quinic acid with nitric acid[65] or platinum and oxygen[62-64] forms the basis of chemical methods of preparation.

The acid reduces Fehlings solution and forms protocatechuic acid by heating with acid or by melting. Grewe and Jeschke[65]

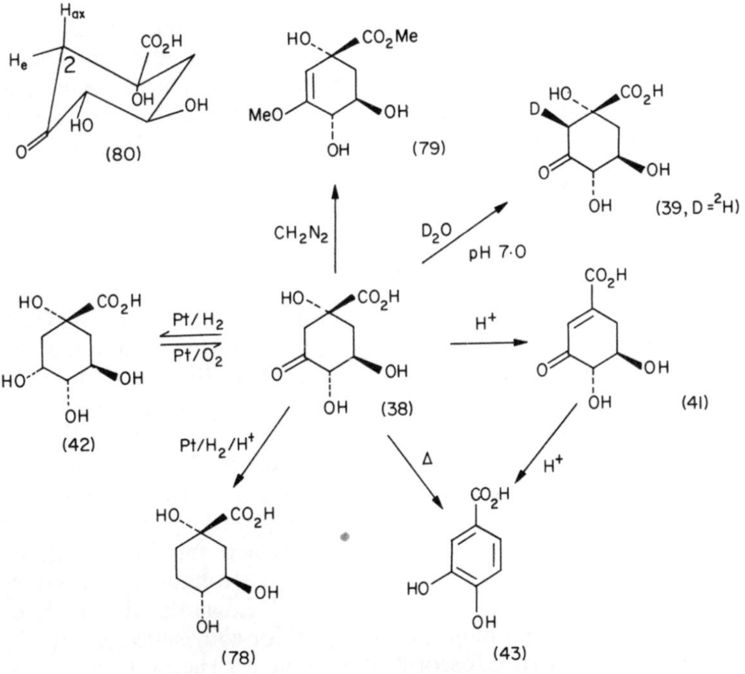

Figure 2.11. Preparation and properties of 3-dehydroquinic acid[56, 57, 65, 102]

characterised the acid by reduction in neutral solution to $(-)$-quinic acid (42) and by reduction in acid solution to 3-deoxyquinic acid (78). The only crystalline derivative of this acid which has been reported is the enol ether (79) formed by the action of diazomethane[57]. The proton magnetic resonance spectral characteristics of 3-dehydroquinic acid (38) demonstrate that both the free acid and the anion adopt chair conformations (e.g. 80) in solution, analogous to that of $(-)$-quinic acid (70), in which the carboxyl group has an equatorial position. Calculations also show that this conformation is the energetically preferred one[75]. An unusual feature of the proton magnetic resonance spectra is the higher field resonance of the equatorial (pro R) proton at the 2 position compared to the axial (pro s) proton[56, 57] although similar observations have been made with respect to the protons adjacent to the carbonyl group in other cyclohexanones[105]. At pH 7.0 in deuterium oxide, the exchange of deuterium for hydrogen at C-2 and C-4 in 3-dehydroquinic acid (38) is strongly site and stereoselective. A monodeuterio 3-dehydroquinic acid (39) may be isolated[56, 57] in which the axial (pro s) proton at the 2 position has been largely (60–70 per cent) exchanged for deuterium.

Partial conversion of 3-dehydroquinic acid to 3-dehydroshikimic acid (38) can be achieved by mild acid or base treatment and in contrast to the parallel enzymic conversion, which forms an integral part of the shikimate pathway, the mechanism for this reaction has been shown[106] to involve a *trans* elimination of H_{2ax} and the hydroxyl at C-1.

2.4   CHORISMIC ACID

Chorismic acid (81) is the major branch point on the shikimate pathway and the trivial name adopted for this compound was derived[107], on ecclesiastical advice, from the Greek word meaning separating. Several procedures for the isolation of chorismic acid either as its barium salt or the free acid have been described[107–111] and all utilise the *Aerobacter aerogenes* mutant 62-1. Chorismic acid (81) is readily accumulated by cell suspensions of this multiply-blocked strain grown in media containing L-tryptophan to repress anthranilate synthetase. The free acid, which may be crystallised, melts at 148–149° and its solutions are strongly laevorotatory. Samples retain solvent tenaciously and Edwards and Jackman obtained the correct molecular formula for chorismic acid by high resolution mass spectroscopy. A complete structural and stereochemical analysis of the molecule was made by Gibson[108, 112, 113]

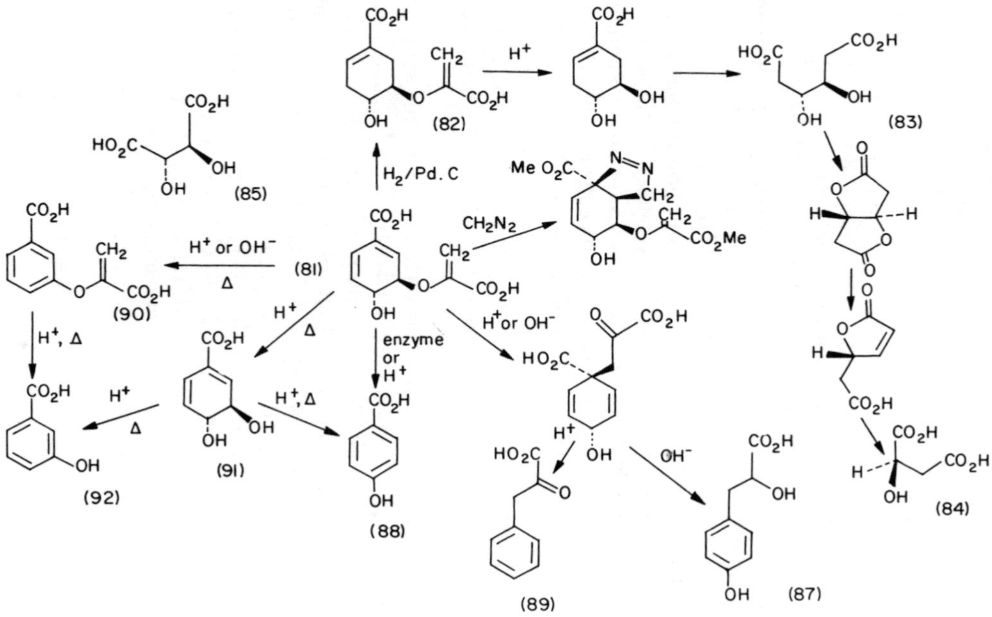

73

Figure 2.12. Some reactions of chorismic acid[108, 109, 112, 113, 115]

and by Edwards and Jackman[109] on the basis of chemical and spectral data.

Infra-red and ultra-violet measurements showed chorismic acid to be a derivative of 3,4-dihydrobenzoic acid and analysis of the infra-red spectrum was also consistent with the location of an enolpyruvyl grouping in the molecule. Partial hydrogenation of chorismic acid (81)[109] gave the dihydro derivative (82), acid treatment and ozonolysis of which gave the known (+)-$\beta\beta'$-dihydroxyadipic acid (83). Posternak and Susz[114] have correlated the stereochemistry of this acid with (+)-D-malic acid (84) and the R,R configuration at C-3 and C-4 in chorismic acid which this observation suggests was confirmed by Gibson and Young[112] who obtained (−)-(S,S)-tartaric acid (85) by ozonolysis of chorismic acid and oxidative decomposition of the ozonide. A range of products were isolated or detected[108, 113] in solutions of chorismic acid (81) which had been heated. In basic solution prephenic acid (86), 4-hydroxybenzoic acid (88), 4-hydroxyphenyl-lactic acid (87) and 3-(1'-carboxyvinyloxy)-benzoic acid (90) were observed and in acid media the formation of 3- and 4-hydroxybenzoic acids (92 and 88), phenylpyruvic acid (89) and 3,4-dihydro-3,4-dihydroxybenzoic acid (91) was noted. The production of each of these products was rationalised in terms of the structure (81) for chorismic acid and the known reactions of breakdown products such as prephenic acid (86), *Figure 2.12.*

The proton magnetic resonance spectrum of chorismic acid (81) was fully analysed by Edwards and Jackman[109] and this analysis confirmed the structure and relative stereochemistry assigned to the molecule. Furthermore the magnitude of the coupling constant between the protons H_3 and H_4 (11.7 Hz) as well as both the allylic coupling constants (1.9 Hz) suggested that the most extensively populated conformation of the molecule in solution is that in which the two substituents at C-3 and C-4 are quasi-equatorial (93). With this information in mind Edwards and Jackman[109] postulated that the role of the enzyme chorismate mutase, which promotes the transformation to prephenic acid (86), is to invert the

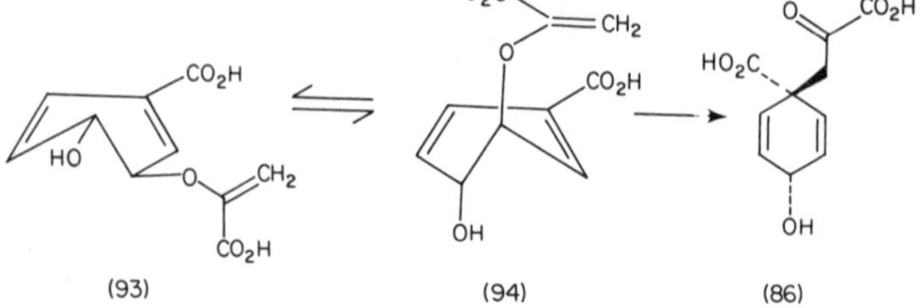

(93)   (94)   (86)

conformation to (94) as well as to orient the enolpyruvyl side chain correctly for the rearrangement.

2.5   PREPHENIC ACID

Prephenic acid (86) was first isolated[116] as its barium salt from the cultures of *Escherichia coli* mutants which were unable to convert chorismic acid (81) to the aromatic amino acids. Mutants of *Neurospora crassa* also produce[117, 118] prephenic acid when cultured under neutral conditions but the method of Dayan and Sprinson[119] is the one which has been recommended for the isolation of this metabolite. Cells of a L-tyrosine auxotroph of *Salmonella typhimurium* which has chorismate mutase but no prephenate dehydratase activity are used to accumulate prephenic acid under resting conditions. The medium used is devoid of L-tyrosine but contains L-phenylalanine and L-tryptophan.

Davis and his collaborators[116] deduced the structure of prephenic acid (86) from analytical data, the absorption of 3–4 moles of hydrogen during catalytic hydrogenation, its conversion by enzymic or acid catalysis to phenylpyruvic acid (89) and the identification of phenyl lactic acid (97) as a product of the reaction with sodium borohydride followed by acid catalysed decarboxylation. Prephenic acid (86) has a plane of symmetry and is therefore optically inactive. The relative steric arrangement of the carboxyl group at C-1 and the hydroxyl at C-4 on the ring was deduced from a consideration of its mode of formation from chorismic acid (81) and from a chromatographic comparison of the DNP derivatives of natural tetrahydroprephenic acid (95) and the DNP derivatives of the two synthetic tetrahydroprephenic acids corresponding in structure to (95) and (96)[120–123].

Prephenic acid (86) is readily transformed with acid to give phenyl pyruvic acid (89) but is somewhat more stable towards alkali. However, on heating or prolonged standing in dilute alkaline solution it is converted to *p*-hydroxyphenyl lactic acid (87). Several proposals have been put forward to account for the formation of this product under these conditions and one of these[124] is shown in *Figure 2.13*.

Synthetic approaches to prephenic acid have succeeded in producing the acid in admixture with *epi*-prephenic acid and dihydroprephenic acid[124, 125]. This mixture of acids was obtained when a crude mixture of the isomers (98 and 99)[126] was subject to ketal hydrolysis at pH 1.8 in water at 20°. Phenylpyruvic acid was removed by ion-exchange chromatography and the mixture of barium salts

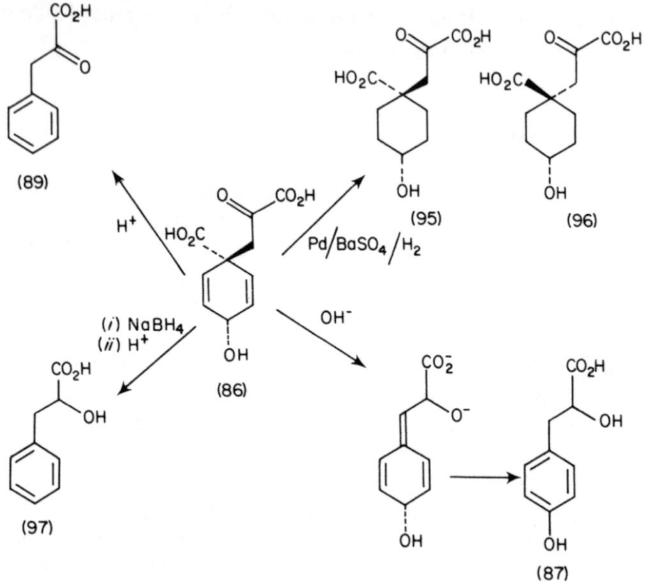

Figure 2.13. Reactions of prephenic acid[116, 120-126]

of (86) and its isomer (100) was isolated. Spectroscopic examination showed the mixture to contain approximately 20 per cent of the isomers (86 and 100) and the remainder of the product was tentatively identified as dihydroprephenic acid.

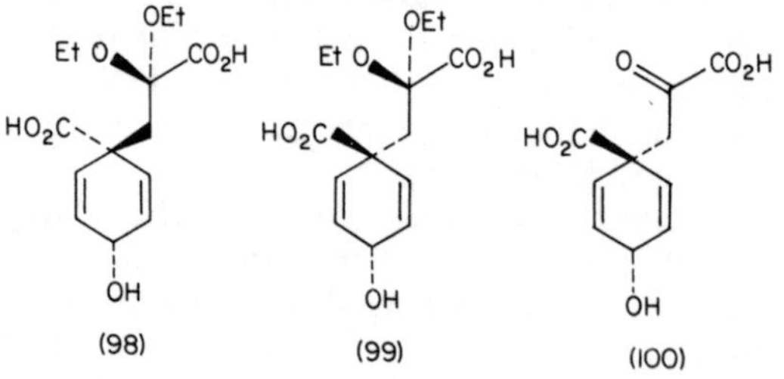

REFERENCES

1. Bohm, B. A. (1965). *Chem. Rev.*, **65**, 435
2. Dangschat, G. (1955), *Moderne Methoden der Pflanzenanalyse.* Vol. II. p. 91. Edited by K. Paech and M. V. Tracey, Berlin, Heidelberg; Springer
3. Eykmann, J. F. (1885). *Rec. Trav. Chim.*, **4**, 32
4. Eykmann, J. F. (1886). *Rec. Trav. Chim.*, **5**, 299
5. Eykmann, J. F. (1891). *Chem. Ber.*, **24**, 1278

6. Hasegawa, M., Nakagawa, T. and Yoshida, S. (1957). *J. Japan Forestry Soc.*, **39**, 159
7. Hattori, S., Yoshida, S. and Hasegawa, M. (1954). *Physiol. Plantarum*, **7**, 283
8. Plouvier, V. (1959). *Compt. Rend.*, **249**, 1563
9. Plouvier, V. (1960). *Compt. Rend.*, **250**, 1721
10. Plouvier, V. (1961). *Compt. Rend.*, **252**, 599
11. Plouvier, V. (1964). *Compt. Rend.*, **258**, 2921
12. Whiting, G. C. (1957). *Nature*, **179**, 531
13. Freudenberg, K. and Geiger, J. (1952). *Annalen Chem.*, **575**, 145
14. Lane, J. F., Koch, W. T., Leeds, N. S. and Gorin, G. (1952). *J. Amer. Chem. Soc.*, **74**, 3211
15. Crewe, R. and Lorenzen, W. (1953). *Chem. Ber.*, **86**, 928
16. Goldschmid, O. and Hegert, H. L. (1961). *Tappi*, **44**, 858
17. Hanson, K. R. and Zucker, M. (1963). *J. Biol. Chem.*, **238**, 1105
18. Maier, V. P., Metzler, D. M. and Huber, A. F. (1964). *Biochem. Biophys. Res. Comm.*, **14**, 124
19. Sprinson, D. B. and Knowles, P. F. (1970). *Methods in Enzymology*, Volume XVIIA, p. 351. Edited by H. Tabor and C. W. Tabor. London and New York; Academic Press
20. Weiss, U. and Mingioli, E. S. (1956). *J. Amer. Chem. Soc.*, **78**, 2894
21. Yoshida, S. and Hasegawa, M. (1957). *Arch. Biochem. Biophys.*, **70**, 377
22. Gationde, M. K. and Gordon, M. W. (1958). *J. Biol. Chem.*, **230**, 1043
23. Saslaw, L. D. and Waravdekar, V. S. (1960). *Biochim. Biophys. Acta*, **37**, 367
24. Millican, R. C. (1963). *Anal. Biochem.*, **6**, 181
25. Millican, R. C. (1970). *Methods in Enzymology*. Volume XVIIA, p. 352. Edited by H. Tabor and C. W. Tabor. London and New York; Academic Press
26. Dewick, P. M. and Haslam, E. (1969). *Biochem. J.*, **113**, 537
27. Cartwright, R. A. and Roberts, E. A. H. (1955). *Chem. and Ind.*, 230
28. Scee Yee Chen. (1929). *Amer. J. Pharm.*, **101**, 550
29. Fischer, H. O. L. and Dangschat, G. (1934). *Helv. Chim. Acta*, **17**, 1196, 1200
30. Fischer, H. O. L. and Dangschat, G. (1935). *Helv. Chim. Acta*, **18**, 1204, 1206
31. Fischer, H. O. L. and Dangschat, G. (1937). *Helv. Chim. Acta*, **20**, 705
32. Grewe, R., Jensen, H. and Schnoor, M. (1956). *Chem. Ber.*, **89**, 898
33. Grewe, R. and Buttner, H. (1958). *Chem. Ber.*, **91**, 2452
34. Grewe, R. and Bokranz, A. (1955). *Chem. Ber.*, **88**, 49
35. Keana, J. F. W. and Kim, C. U. (1970). *J. Org. Chem.*, **35**, 1093
36. McCrindle, R., Overton, K. H. and Raphael, R. A. (1960). *J. Chem. Soc.*, 1560
37. McCrindle, R., Overton, K. H. and Raphael, R. A. (1968). *Tetrahedron Lett.*, 1847
38. Smissman, E. E., Suh. J. T., Oxman, M. and Daniels, R. (1959). *J. Amer. Chem. Soc.*, **81**, 2910
39. Smissman, E. E., Suh, J. T., Oxman, M. and Daniels, R. (1962). *J. Amer. Chem. Soc.*, **84**, 1040
40. Grewe, R. and Hinrichs, I. (1964). *Chem. Ber.*, **97**, 443
41. Grewe, R. and Kersten, S. (1967). *Chem. Ber.*, **100**, 2546
42. Bestmann, H. J. and Heid, H. A. (1971). *Angew. Chem. Internatl Edn*, **10**, 336
43. Hall, L. D. (1964). *J. Org. Chem.*, **29**, 297
44. Thompson, R. S. and Haslam E., unpublished observations
45. Srinivasan, P. R., Shigeura, H. T., Sprecher, M., Sprinson, D. B. and Davis, B. D. (1956). *J. Biol. Chem.*, **220**, 477
46. Millican, R. C. (1962). *Biochim. Biophys. Acta*, **57**, 407
47. Weinstein, L. H., Porter, C. A. and Laurencot, H. A. (1959). *Contrib. Boyce Thompson Inst.*, **20**, 121
48. Floss, H. G., Onderka, D. K. and Carroll, M. (1972). *J. Biol. Chem.*, **247**, 736

49. Zenk, M. H. and Scharf, K. H. (1971). *J. Labelled Compounds*, **7**, 525
50. Srinivasan, P. R., Sprinson, D. B., Kalan, F. B. and Davis, B. D. (1956). *J. Biol. Chem.*, **223**, 907
51. Chabannes, B., Pichat, L., Herbert, M. and Pacheco, H. (1965). *J. Labelled Compounds*, **1**, 102
52. Hill, R. K. and Newkome, G. R. (1969). *J. Amer. Chem. Soc.*, **91**, 5893
53. Hill, R. K. and Newkome, G. R. (1968). *Tetrahedron Lett.*, 1851
54. Smissman, E. E. and Li, J. P. (1968). *Tetrahedron Lett.*, 4601
55. Leduc, M. M., Dansette, P. M. and Azerad, R. G. (1971). *Eur. J. Biochem.*, **15**, 428
56. Smith, B. W., Turner, M. J. and Haslam, E. (1970). *Chem. Commun.*, 842
57. Turner, M. J., Haslam, E., Thompson, R. S. and Sargent, D. (1971). *J. Chem. Soc.*, (C), 1489
58. Grewe, R. and Lorenzen, W. (1953). *Chem. Ber.*, **86**, 928
59. Podojil, M. and Gerber, N. N. (1967). *Biochemistry*, **6**, 2701
60. Larsen, P. O. (1966). *Biochim. Biophys. Acta*, **115**, 259
61. Salamon, I. I. and Davis, B. D. (1953). *J. Amer. Chem. Soc.*, **75**, 5567
62. Heyns, K. and Gottschalk, H. (1961). *Chem. Ber.*, **94**, 343
63. Knowles, P. F., Haworth, R. D. and Haslam, E. (1961). *J. Chem. Soc.*, 1854
64. Knowles, P. F., Haworth, R. D. and Haslam, E. (1963). *Methods in Enzymology*. Volume VI, p. 498. Edited by S. Colowick and N. Kaplan. London and New York; Academic Press
65. Grewe, R. and Jeschke, J. P. (1956). *Chem. Ber.*, **89**, 2080
66. Whiting, G. C. and Coggins, R. A. (1967). *Biochem. J.*, **102**, 283
67. Thompson, R. S. and Haslam, E., unpublished observations
68. Davis, B. D. and Mingioli, E. S. (1953). *J. Bact.*, **66**, 129
69. Levin, J. G. and Sprinson, D. B. (1960). *Biochem. Biophys. Res. Commun.*, **3**, 157
70. Levin, J. G. and Sprinson, D. B. (1962). *Federation Proc.*, **21**, 88
71. Levin, J. G. and Sprinson, D. B. (1964). *J. Biol. Chem.*, **239**, 1142
72. Knowles, P. F., Levin, J. G. and Sprinson, D. B. (1970). *Methods in Enzymology*. Volume XVIIA, p. 360. Edited by H. C. Tabor and S. Tabor. London and New York; Academic Press
73. Bondinell, W. E., Vnek, J., Knowles, P. F., Sprecher, M. and Sprinson, D. B. (1971). *J. Biol. Chem.*, **246**, 6191
74. 'Tentative rules for cyclitols'. (1968). *Eur. J. Biochem.*, **5**, 1
75. Corse, J. E. and Lundin, R. E. (1970). *J. Org. Chem.*, **35**, 104
76. Bate-Smith, E. C. (1954). *Chem. and Ind.*, 1457
77. Panizzi, L. and Scarpati, M. (1954). *Gazzetta*, **84**, 792, 806
78. Haslam, E., Makinson, G. K., Naumann, M. O. and Cunningham, J. (1964). *J. Chem. Soc.*, 2137
79. Herrmann, K. (1956). *Naturwiss.*, **43**, 109
80. Roberts, E. A. H. and Meyers, M. (1958). *J. Sci. Food Agric.*, **11**, 701
81. Haslam, E., Haworth, R. D. and Keen, P. C. (1962). *J. Chem. Soc.*, 3814
82. Fischer, H. O. L. (1921). *Chem. Ber.*, **54**, 775
83. Fischer, H. O. L. and Dangschat, G. (1932). *Chem. Ber.*, **65**, 1009
84. Fischer, H. O. L. and Dangschat, G. (1938). *Naturwiss.*, **26**, 562
85. Fischer, H. O. L. and Dangschat, G. (1950). *Biochim. Biophys. Acta*, **4**, 149
86. Karrer, P., Widmer, R. and Riso, P. (1925). *Helv. Chim. Acta*, **8**, 195
87. Karrer, P. and Link, K. P. (1927). *Helv. Chim. Acta*, **10**, 794
88. Grewe, R. and Nolte, E. (1952). *Annalen Chem.*, **575**, 1
89. Grewe, R., Lorenzen, W. and Vining, L. (1954). *Chem. Ber.*, **87**, 793
90. Grewe, R., Buttner, H. and Burmeister, G. (1957). *Angew. Chem.*, **69**, 61
91. Grewe, R. and Haendler, H. (1962). *Annalen Chem.*, **658**, 113

92. Grewe, R. and Vangermain, E. (1965). *Chem. Ber.*, **98**, 104
93. Fischer, M., Freidrichsen, W., Grewe, R. and Haendler, H. (1968). *Chem. Ber.*. **101**, 3313
94. Gorin, P. A. J. (1963). *Canad. J. Chem.*, **41**, 2417
95. Smissman, E. and Oxman, M. A. (1963). *J. Amer. Chem. Soc*, **85**, 2184
96. Wolinsky, R., Novak, R. and Vasileff, R. (1964). *J. Org. Chem.*, **29**, 3596
97. Bestman, H. J. and Kranz, E. (1969). *Chem. Ber.*, **102**, 1802
98. Orloff, H. (1954). *Chem. Rev.*, **54**, 347
99. Gaffield, W., Waiss, A. C. and Corse, J. (1966). *J. Chem. Soc.*, (C), 1885
100. Turner, M. J. and Haslam, E. (1971). *J. Chem. Soc.*, (C), 1496
101. Corse, J., Lundin, R. E., Sondheimer, E. and Waiss, A. C. (1966). *Phytochemistry*, **5**, 767
102. Weiss, U., Davis, B. D. and Mingioli, E. S. (1953). *J. Amer. Chem. Soc.*, **75**, 5572
103. Hesse, O. (1859). *Annalen Chem.*, **112**, 52
104. Grewe, R. and Winter, G. (1959). *Angew. Chem.*, **71**, 163
105. Butterworth, R. F., Collins, P. M. and Overend, W. G. (1969). *Chem. Commun.*, 378
106. Turner, M. J. and Haslam, E., unpublished observations
107. Gibson, M. I. and Gibson, F. (1964). *Biochem. J.*, **90**, 248
108. Gibson, F. (1964). *Biochem. J.*, **90**, 256
109. Edwards, J. M. and Jackman, L. M. (1965). *Austral. J. Chem.*, **18**, 1227
110. Gibson, F. (1968). *Biochemical Preparations*, **12**, 94
111. Gibson, F. (1970). *Methods in Enzymology*. Volume XVIIA, p. 362. Edited by H. C. Tabor and S. Tabor. London and New York; Academic Press
112. Young, I. G. and Gibson, F. (1969). *Biochim. Biophys. Acta*, **177**, 348, 401
113. Young, I. G., MacDonald, C. G. and Gibson, F. (1969). *Biochim. Biophys. Acta*, **192**, 62
114. Posternak, Th. and Susz, J-Ph. (1956). *Helv. Chim. Acta*, **39**, 2032
115. Haslam, E. and Ife, R., unpublished observations
116. Weiss, U., Gilvarg, C., Mingioli, E. S. and Davis, B. D. (1954). *Science*, **119**, 774
117. Metzenberg, R. L. and Mitchell, H. K. (1956). *Arch. Biochem. Biophys.*, **64**, 51
118. Gamborg, O. L. and Simpson, F. J. (1964). *Canad. J. Biochem.*, **42**, 583
119. Dayan, J. and Sprinson, D. B. (1970). *Methods in Enzymology*. Volume XVIIA, p. 559. Edited by H. C. Tabor and S. Tabor. London and New York; Academic Press
120. Plieninger, H. and Keilich, G. (1961). *Z. Naturforsch.*, **16B**, 81
121. Plieninger, H. and Grasshoff, H. J. (1957). *Chem. Ber.*, **90**, 1973
122. Plieninger, H. and Keilich, G. (1959). *Chem. Ber.*, **92**, 2897
123. Levin, J. G. and Sprinson, D. B. (1960). *Biochem. Biophys. Res. Commun.*, **3**, 157
124. Plieninger, H. (1962). *Angew. Chem. Internatl Edn*, **1**, 367
125. Plieninger, H. (1962). *Angew. Chem. Internatl Edn*, **1**, 61
126. Pleininger, H. (1961). *Angew. Chem.*, **73**, 543

3

Metabolites of the Shikimate Pathway

3.1 INTRODUCTION

The evidence that (−)-shikimic acid plays a central role in aromatic biosynthesis was obtained by Davis[1-4] with a variety of nutritionally deficient mutants of *Escherichia coli*. In one group of mutants with a multiple requirement for L-tyrosine, L-phenylalanine, L-tryptophan and p-aminobenzoic acid and a partial requirement for p-hydroxybenzoic acid, (−)-shikimic acid substituted for all the aromatic compounds. The quintuple requirement for aromatic compounds which these mutants displayed arises from the fact that, besides furnishing a metabolic route to the three aromatic α-amino acids, the shikimate pathway also provides in micro-organisms a means of synthesis of other essential metabolites, and in particular, the various isoprenoid quinones involved in electron transport and the folic acid group of co-enzymes. The biosynthesis of both of these groups of compounds is discussed below. In addition the biosynthesis of a range of structurally diverse metabolites, which are derived from intermediates and occasionally end-products of the pathway, is outlined. These metabolites are restricted to certain types of organism and their function, if any, is in the majority of cases obscure.

3.2 ISOPRENOID QUINONES

Investigations of the process of electron transport in the terminal oxidation systems of diverse organisms has led to the discovery of a series of lipid soluble isoprenoid quinones which participate in the multi-enzyme respiratory complex[5, 6]. The quinones appear to have a definite spatial orientation in the complex and probably function by 'shuttling' electrons between other respiratory co-enzymes. Suggestions have been made that they may, in addition, play a more direct role in the process of oxidative phosphorylation but further evidence seems necessary to establish this point[7]. The quinones fall into four biogenetically distinct classes—the ubiquinones, plastoquinones and related chromanols which are all based on benzoquinone and the phylloquinones and menaquinones based on naphthoquinone. Because of the very close similarities of structure and of physiological activity consideration is given here to the biosynthesis of isoprenoid quinones from all forms of living matter ranging from the microbial, through plants to higher organisms.

Ubiquinones (co-enzyme Q) are widely distributed in nature[8] and with the exception of gram-positive bacteria and blue–green algae they have been detected in all groups of living organisms which have been examined. The sub-cellular distribution of the ubiquinones shows them to be localised in the mitochondria in animals and plants and the cell membranes of non-photosynthetic bacteria. The ubiquinones are all derivatives of 5,6-dimethoxy-3-methyl-2-*trans*-polyprenyl-1,4-benzoquinone (1, where $n = 1$ to 12) and generally most organisms synthesise a series of ubiquinones in

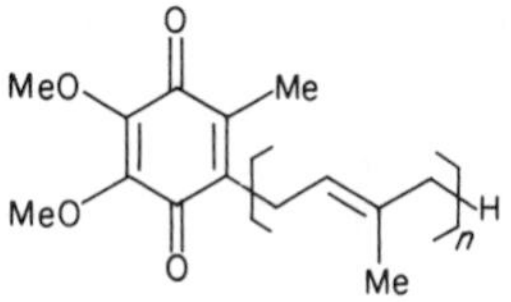

(1)

which particular homologues, usually those where $n = 8$, 9 or 10, predominate[9]. Variations on this structural pattern such as amino substitution in the benzoquinone nucleus[10] and reduction[11] or epoxidation[12] of isolated double bonds in the polyprenyl chain have also been recognised.

Plastoquinones, Plastochromanols and Tocopherols (vitamin E) are found in the chlorophyll containing tissues of higher plants and algae. Experimental evidence has been obtained to indicate that plastoquinones are intimately involved in photosynthetic electron transport[13] and the tocopherols are similarly frequently found

located in the photosynthetic regions of the organism. The plastoquinones are a family of polyprenylquinones of the general structure
(2)[14, 15] which differ only in the nature of their polyprenyl side
chains. The most widely distributed plastoquinones are plasto-
quinone-9 (2, $n = 9$), the plastoquinone C in which side chain
hydroxylation has occurred[16] and a cyclic derivative of plasto-
quinone-9, plastochromanol-8 (3)[17].

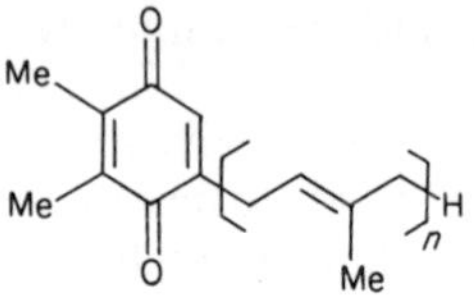

(2) 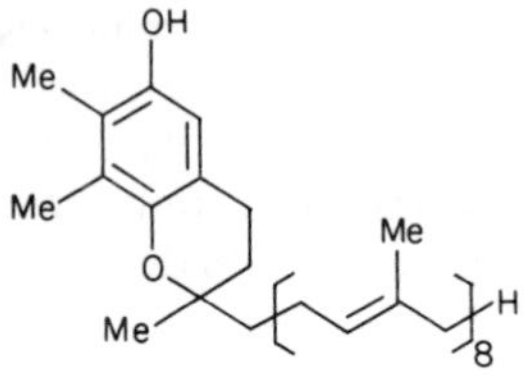   (3)

Related to the plastoquinones and plastochromanols are the
members of the vitamin E group the tocopherols (4–7) and tocotrienols and four members of each family have been found in nature.
The most frequently found member of the tocopherols is α-toco-
pherol (4). The tocotrienols (α, β, γ and δ) correspond in structure
to the various tocopherols but the side chain is unsaturated (8).

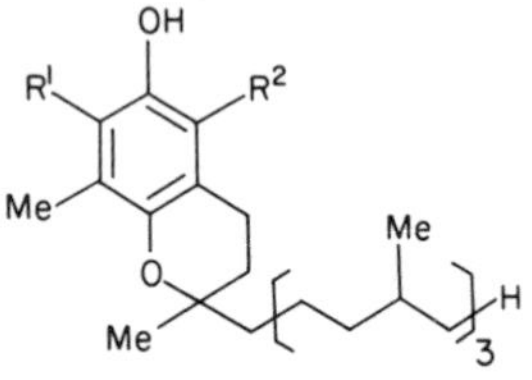

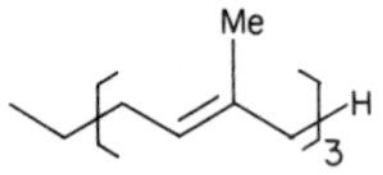

(8)

$(4, R^1 = R^2 = Me, \alpha)$
$(5, R^1 = H; R^2 = Me, \beta)$
$(6, R^1 = Me; R^2 = H, \gamma)$
$(7, R^1 = R^2 = H, \delta)$

The Phylloquinones and Menaquinones (vitamins K_1 and K_2) are two
chemically related classes of isoprenoid quinones based on the
naphthoquinone nucleus. Biologically they are differentiated by
the fact that the biosynthesis of the menaquinones (vitamin K_2) is
is confined to bacteria and some fungi[11, 19–21] and that of the
phylloquinones (vitamin K_1) is restricted to higher plants and
algae[11]. The principal phylloquinone present in higher plants is

the phylloquinone (9) which has been located in both photosynthetic and non-photosynthetic tissue.

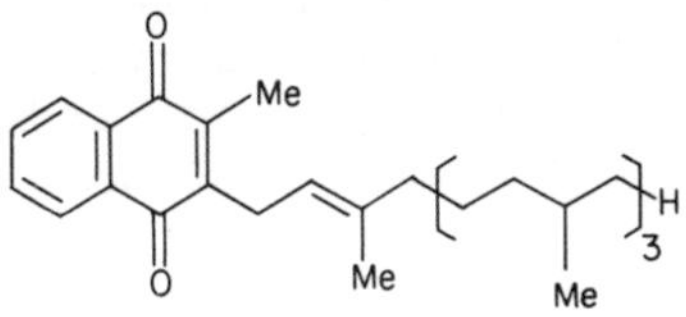

(9) 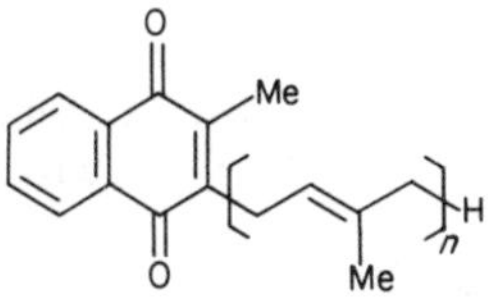   (10)

The menaquinones bear many similarities to the ubiquinones and are of widespread occurrence in bacteria. They possess the general structure (10, where n varies from 1 to 13). Some organisms contain a series of menaquinones and for example the series of menaquinones -1 to 8 (10, $n = 1$ to $n = 8$) have been identified in *Staphylococcus aureus*. Quinones of this category have also been described which lack the nuclear methyl group or have a partially reduced polyprenyl side chain.

3.2.1 GENERAL FEATURES OF ISOPRENOID QUINONE BIOSYNTHESIS

Mevalonic acid (11) is firmly established as the specific biosynthetic precursor of isoprenoid units in natural products and the pathway by which it is formed and utilised in such processes has been amply reviewed and discussed[23-25]. Experimental evidence has been obtained to show the involvement of mevalonic acid (11) in the building of the polyprenyl side chain of ubiquinone in higher plants, mammals and microbial systems, of the plastoquinones and phylloquinones in higher plants and of the menaquinones in bacteria[22, 26-35]. In several cases little is known about the manner in which the isoprenoid side chains are assembled and at what stage isoprenylation of the quinone nucleus, or its precursor, occurs. It seems probable nevertheless that the pattern may be similar to that established for the ubiquinones. Here it is widely assumed that polyprenylpyrophosphates are synthesised independently of the aromatic nucleus and at a subsequent stage the polyprenylpyrophosphate of appropriate length couples with *p*-hydroxybenzoic acid (12) to form an intermediate of structure (13). Evidence in support of this proposal is the frequent co-occurrence of polyprenyl alcohols themselves (14) and the isolation of polyprenylpyrophosphate synthetases from living systems[15, 30, 31].

The polyprenylpyrophosphate synthetase from *Micrococcus*

lysodeikticus, which is able to synthesise polyprenols from two to ten isoprene units in length[31], when added to cell free extracts of rat kidney, brain or liver[32] or of the bacterium *Rhodospirillum rubrum*[33], gave an enzyme preparation which was able to carry out the iso-prenylation of *p*-hydroxybenzoic acid to give compounds of the type (13). In later work with rat liver, Rudney and Momose obtained

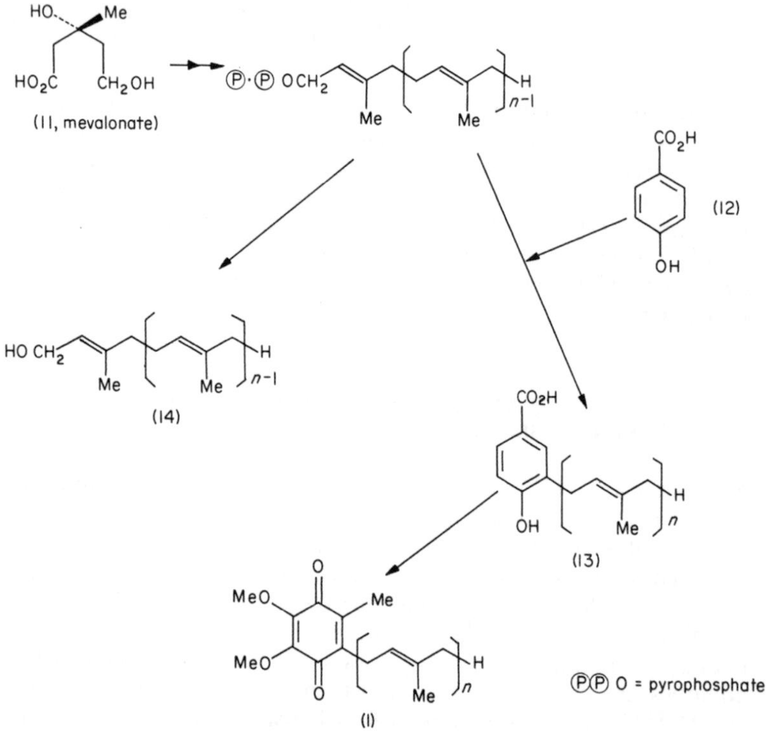

polyprenyl-*p*-hydroxybenzoic acid synthesis without supplementation and demonstrated that a system of enzymes for the synthesis of polyprenylpyrophosphates exists in the inner mitochondrial membrane of the rat liver[34]. Hamilton and Cox[35] have made similar observations with cell extracts of a multiple aromatic auxotroph of *Escherichia coli* K-12. Such extracts converted U-[^{14}C]-*p*-hydroxybenzoic acid into a mixture of [^{14}C]-3-octaprenyl-4-hydroxybenzoic acid (13, $n = 8$) and 2-octaprenylphenol. An octaprenol, farnesyl-farnesylgeraniol accumulated when the organism was grown in the absence of precursors for the quinone nuclei of ubiquinone and menaquinone. It was concluded from these observations that in the biosynthesis of ubiquinone-8, the polyprenyl side chain is added to p-hydroxybenzoic acid as a C-40 unit, the active form of which is converted by cell extracts into farnesylfarnesylgeraniol.

3.2.2   UBIQUINONE (CO-ENZYME Q) BIOSYNTHESIS

In 1950, Davis[1, 2] showed that p-hydroxybenzoic acid had vitamin-like activity for multiple aromatic auxotrophs of *Escherichia coli* but it was not until 1964 that Gibson and Cox[36, 37] showed that this growth factor requirement was associated with the formation of ubiquinones in these mutant organisms. Subsequent work has revealed that p-hydroxybenzoic acid (12), a derivative of the shikimate pathway, is a focal metabolite for ubiquinone biosynthesis in a very wide range of organisms[9, 38, 39]. Thus [^{14}C]-p-hydroxybenzoic acid, and in several cases the corresponding aldehyde also, has proved to be a most effective precursor of the benzoquinone

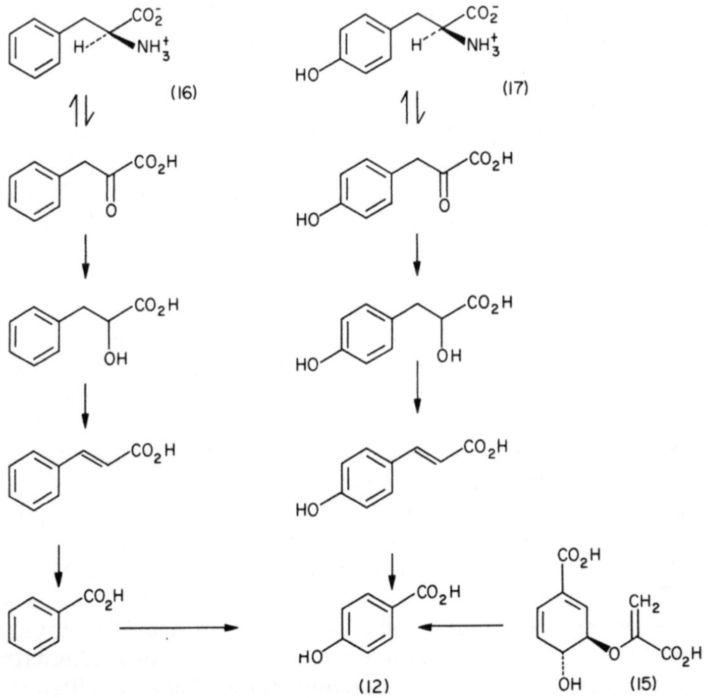

Figure 3.1. Pathways for the formation of p-hydroxybenzoic acid[48-50]

ring of ubiquinones in mammals[12, 40], bacteria[36, 39, 41, 42], yeasts[43], moulds[43, 44], algae[45, 46] and higher plants[45, 47]. In bacteria, p-hydroxybenzoic acid is formed directly from chorismic acid (15), *Figure 3.1*, and the reaction has been demonstrated enzymically in cell free systems[48]. In mammals p-hydroxybenzoic acid is produced via the aromatic amino acids (L-phenylalanine, 16 and L-tyrosine, 17) by a pathway that involves the corresponding phenyl-lactic and cinnamic acids and finally β-oxidation of the side chain[49]. A similar pathway has been formulated for higher plants after work with maize (*Zea mays*) and French bean (*Phaseolus vulgaris*)[50] and preliminary observations suggest that a similar route is utilised by fungi and algae[43, 45].

Since 1965, the bulk of the work on the biosynthesis of ubiquinones has been devoted to the determination of the sequence of biochemical events in the metabolic pathway between p-hydroxy-benzoic acid and the quinones. In 1966, Folkers and his collaborators[55] proposed a pathway for the photosynthetic bacterium *Rhodospirillum rubrum* based upon the isolation of a number of decaprenylated phenols and quinones (18, 20, 21, 22, $n = 10$) from the lipid extracts of this organism[51–56]. Several of the proposed intermediates were not isolated but their presence was postulated on the basis of reasonable chemical analogy and the knowledge that the proposed steps of nuclear C and O methylation occur via the intermediacy of S-methyladenosine[57, 58]. Although only two of the steps have been validated at the enzymic level[33] (12 → 13, 13 → 18, *Figure 3.2*) this scheme has provided an extremely useful basis for the investigation of the pathway of ubiquinone biosynthesis in other organisms and argument now prevails regarding the universality of this pathway.

Whistance and Threlfall[59–61] have surveyed a large number of organisms of varying types for the presence of the intermediates postulated by Folkers and his group[55] (*Figure 3.2*). In a number of instances evidence was obtained for the compounds (18, 20, 21 and 22) but in *Pseudomonas ovalis* an additional compound, tentatively assigned the structure of 2-polyprenyl-1,4-benzoquinone, was identified. This compound does not fit readily into the proposed biogenetic pathway of Folkers[55] and it may represent an alternative mode of metabolism of 2-polyprenyl phenol (18). Olson, Nowicki and Dialameh[62] have suggested a variation of the Folkers scheme for the biosynthesis of ubiquinone-9 in the rat in which 2-nonaprenyl-phenol (18, $n = 9$) itself is on a side pathway. In their alternative scheme, which was based on isolation studies, they visualised that p-hydroxybenzoic acid was metabolised in the form of its thiol co-enzyme A ester (24) and they suggested that this was necessary to

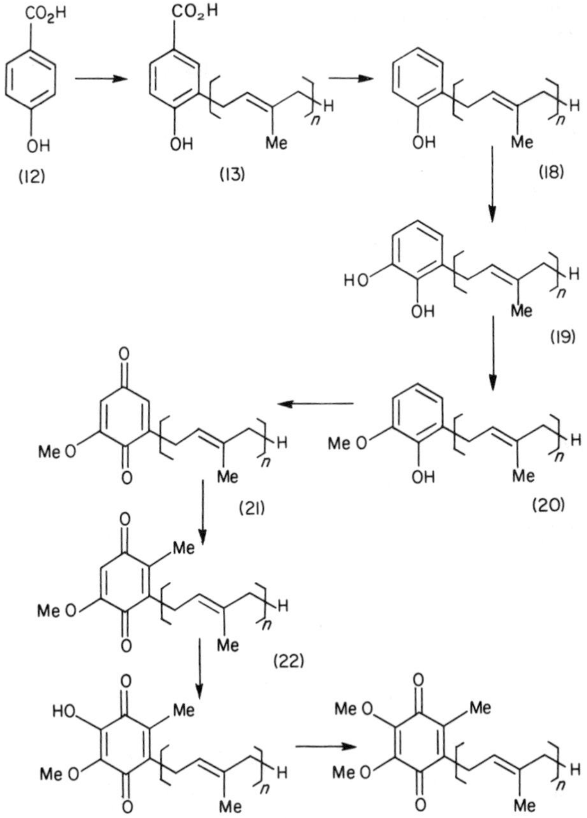

Figure 3.2. Biosynthesis of ubiquinone (co-enzyme Q) in *Rhodospirillum rubrum* — Folkers[55]

direct the alkylation (nonaprenylation) and the first hydroxylation steps. Then in concert with O-methylation, or in series with it, the removal of the carboxyl group occurs, probably by reduction (*Figure 3.3*).

Approaches such as those described above which rely on the isolation and detection of polyprenyl substituted phenols and quinones in lipid extracts are restricted by the minute quantities of compounds present and the difficulties, frequently encountered, in establishing a precursor relationship between the suspected intermediate and ubiquinone. Gibson and his colleagues[63–66] approached the problem of ubiquinone biosynthesis in *Escherichia coli* K-12 by the isolation and examination of mutants unable to form ubiquinone. Five classes of ubiquinone deficient mutants were described which during growth accumulate respectively 3-octa-prenyl-4-hydroxybenzoic acid (13, $n = 8$), 2-octaprenylphenol (18, $n = 8$), 2-octaprenyl-6-methoxy-1,4-benzoquinone (21, $n = 8$), 2-octaprenyl-3-methyl-6-methoxy-1,4-benzoquinone (22, $n = 8$) and 2-octaprenyl-3-methyl-5-hydroxy-6-methoxy-1,4-benzoquinone (23, $n = 8$). Cell free extracts of a wild type *Escherichia coli* were in addition shown to be able to convert chorismate to 2-octaprenyl phenol (15 → 18, $n = 8$)[67]. Using this mutant technique and a genetic analysis, Gibson put forward a scheme for the biosynthesis of

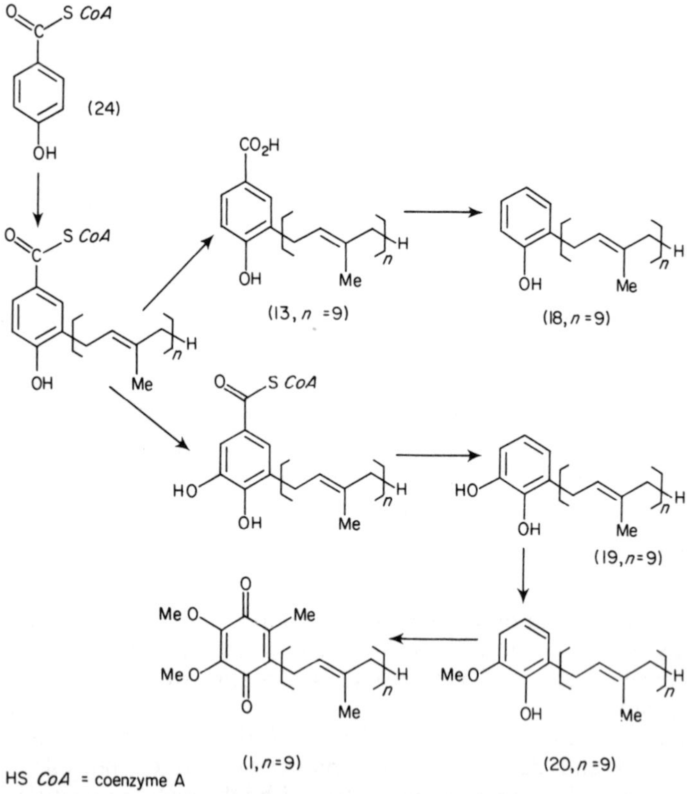

Figure 3.3. Suggested pathway of biosynthesis of ubiquinone-9 in the rat[62]

ubiquinone-8 in *Escherichia coli* (*Figure 3.4*) which bears many similarities to that put forward earlier by Folkers[55]. If identical pathways operate for the biosynthesis of ubiquinones in both *Escherichia coli* and *Rhodospirillum rubrum*, then the work of the

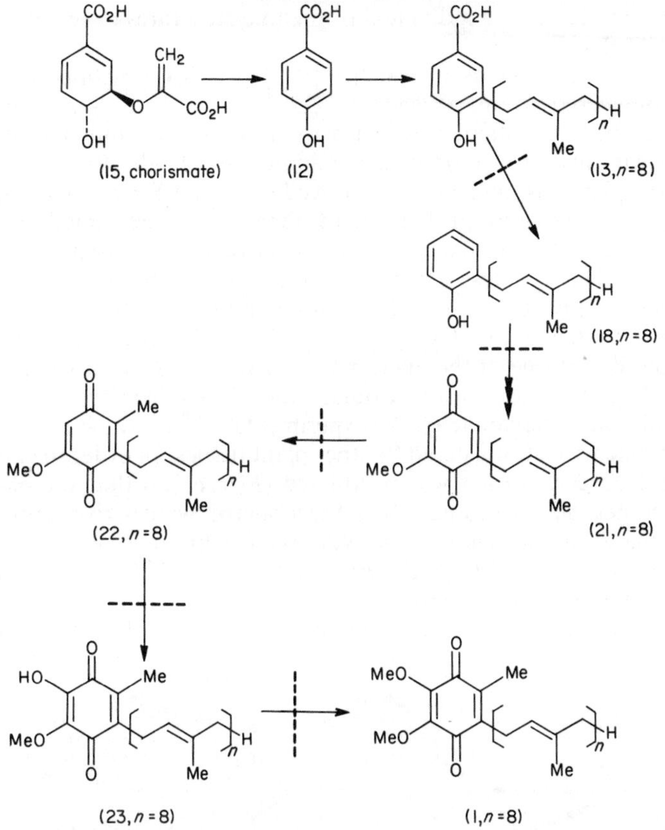

CO_2H

CH_2

CO_2H

OH

(15, chorismate)

CO_2H

OH

(12)

CO_2H

OH   Me

(13,n=8)

OH   Me

(18,n=8)

O

MeO

O   Me

(22,n=8)

O

MeO

O   Me

(21,n=8)

O

HO   Me

MeO

O   Me

(23,n=8)

O

MeO   Me

MeO

O   Me

(1,n=8)

Figure 3.4. Biosynthesis of ubiquinone-8 in *Escherichia coli*[63-67]. Mutants which were isolated and identified were blocked in one of the reactions shown

two groups is quite complimentary since the intermediates (13 and 23) which were postulated but not isolated by Folkers and his collaborators were isolated by Gibson's group using the mutant technique.

3.2.3 BIOSYNTHESIS OF PLASTOQUINONES, PLASTOCHROMANOLS AND TOCOPHEROLS

The biosynthesis of this group of compounds is confined almost exclusively to higher plants and algae and whilst the steps in the formation of the quinone or aromatic nucleus are in most cases fairly well understood the pathways utilised for the subsequent development of the individual isoprenoid structures have still to be finally elucidated.

From experiments in which L-[^{14}C], [^{3}H$_3$]-Me-methionine was administered to maize shoots (*Zea mays*), Threlfall and Whistance[58, 68, 69] obtained evidence to show that all but one of the nuclear methyl groups in plastoquinone-9 (2, $n = 9$) and α and γ-tocopherol (4 and 6) were derived from the S-methyl group of L-methionine (most probably in the form of S-adenosylmethionine). The origin of the remaining methyl group and the aromatic or quinone nucleus of these compounds was revealed by a study of the incorporation of D L-β-[^{14}C]-tyrosine and D L-β-[^{14}C]-phenylalanine[68, 69]. Chemical degradation of the ^{14}C labelled metabolites showed that one of the nuclear C-methyl groups was formed from the β-carbon atom of the aromatic amino acid side chain. Similar results were derived from the experiments[68, 69] in which L-U-[^{14}C]-tyrosine was administered to other plant tissues. In order to explain these observations it was postulated (*Figure 3.5*) that the plastoquinones, plastochromanols and tocopherols were derived from the aromatic amino acids via *p*-hydroxyphenylpyruvic acid (25) and homogentisic acid (26). A series of experiments[61, 68, 69] in which the two proposed intermediates, specifically labelled with ^{14}C, were incorporated into plastoquinone-9 in maize shoots, provided

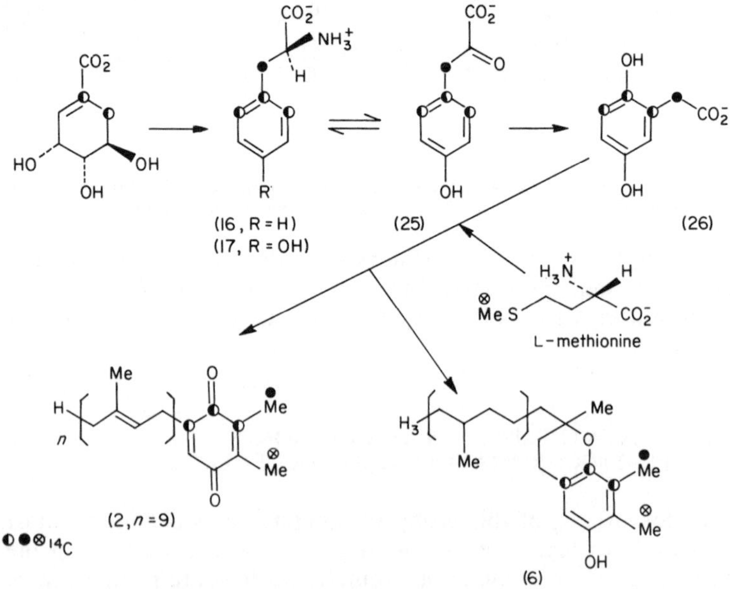

Figure 3.5. Biosynthesis of plastoquinone-9 and γ-tocopherol in maize[9, 68, 69, 71]

more direct proof of this hypothesis. Moreover, the observation that p-hydroxyphenylacetic acid is not a specific precursor for the plastoquinones was interpreted as showing that the formation of homogentisic acid from p-hydroxyphenylpyruvic acid occurs in a manner quite analogous to that observed in mammals and some micro-organisms[69, 70]—namely a synchronous side chain rearrangement accompanying the introduction of the second hydroxyl group into the aromatic nucleus. Further evidence to support this scheme of biosynthesis was obtained using D L-1,6-[$^{14}C_2$]-shikimic acid (27) as a precursor of plastoquinone-9 (2, $n = 9$) and γ-tocopherol (6) in maize[9, 71]. Degradation of the quinone (2, $n = 9$) and γ-tocopherol (6) by Kuhn–Roth oxidation gave acetic acid which, for both isoprenoid metabolites, contained 25 per cent of the total radioactivity incorporated. This result not only confirmed the idea that homogentisic acid (26) is an intermediate in the pathway but also that its α-carbon atom gives rise to the nuclear methyl group which is *meta* to the polyprenyl side chain in this range of compounds. Although there has been some speculation[9] on the sequence of events between homogentisic acid (26) and the plastoquinones and tocopherols there is as yet little experimental evidence to support any of these proposals.

3.2.4   BIOSYNTHESIS OF PHYLLOQUINONES AND MENAQUINONES (VITAMIN K_1 AND K_2)

Examination of some of the features of the biosynthesis of phylloquinones such as (9) in higher plants and of menaquinones (10) in bacteria has shown that the polyprenyl chains and nuclear C methyl groups arise unexceptionally from (R)-mevalonic acid (11) and L-methionine respectively[28, 29, 57]. The aspect of biosynthesis which has commanded most attention is the building of the naphthoquinone nucleus from (−)-shikimic acid and almost all of the critical and pertinent observations upon this intriguing problem have been made in relation to menaquinones from bacterial sources.

Using standard tracer techniques (−)-shikimic acid has been established as an efficient precursor of the naphthoquinone nuclei of both menaquinones and phylloquinones[37, 72–79]. Several groups have deduced that in this process (−)-shikimic acid is utilised as an intact seven carbon unit in which carbon atoms 1–6 of the metabolite form the benzene ring of the naphthoquinone, and the carboxyl group (C-7) contributes *equally* to each of the quinone carbonyl groups. Thus degradation of menaquinone-8 (10, $n = 8$,

Figure 3.6) from *Escherichia coli* after the incorporation of U-[^{14}C]-(−)-shikimic acid gave phthalic acid (28) which retained 90 per cent of the total radioactivity of the original quinone[77]. Decarboxylation of (28) using the Schmidt procedure gave carbon dioxide which contained 7 per cent of the total activity. Similar results were obtained by Guerin, Leduc and Azerad[76] with the menaquinones of *Mycobacterium phlei*; and in work with a *Bacillus megaterium* mutant, Zenk and Floss and their collaborators[79]

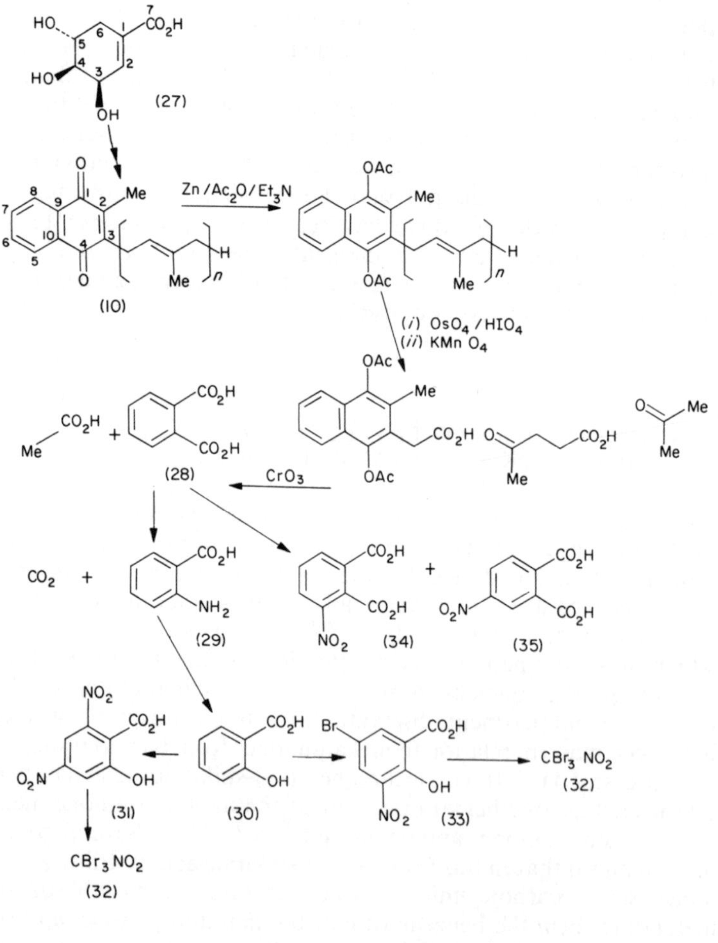

Figure 3.6. Degradation of bacterial menaquinones derived from feeding experiments with (−)-shikimic acid[37, 76−79]

showed that 6-[^{14}C] and 7-[^{14}C]-(−)-shikimic acid were readily incorporated into vitamin K_2 (23 and 36 per cent efficiency respectively). Degradation of the menaquinone to phthalic acid (28) and finally anthranilic acid (29) and carbon dioxide, showed that the isotopic label was retained almost entirely (92 per cent) in the anthranilic acid from 6-[^{14}C]-(−)-shikimic acid as precursor, but that with 7-[^{14}C]-(−)-shikimic acid the amino acid (29) retained only half of the original isotopic carbon and the remainder was present in the carbon dioxide.

Further evidence on the mode of incorporation of the molecule of (−)-shikimic acid into the naphthoquinone nucleus was obtained from experiments with specifically labelled precursors[76, 79, 80]. Thus vitamin K_2 obtained from *Bacillus megaterium* after the administration of 6-[^{14}C]-(−)-shikimic acid was degraded to 3-nitro-5-bromosalicylic acid (33) and thence to bromopicrin (32). The activity in the bromopicrin should be derived equally from that at C-5 and C-8 of the original menaquinone nucleus and analysis showed it to contain 46 per cent of the total isotopic carbon incorporated into the quinone. Similarly, degradation of mena-quinone-9-[H_2] obtained in *Mycobacterium phlei* from 3-[^{3}H]-(−)-shikimic acid gave a mixture of 3- and 4-nitrophthalic acids (34 and 35) in which the ratio of tritium retained was 1:2. These and other analogous experimental results are in agreement with the proposal that C-9 and C-10 of the naphthoquinone are derived exclusively from C-1 and C-2 of (−)-shikimic acid (27).

The origin of the remaining three carbon atoms of the quinone nucleus were traced[77, 78] to C-2, 3 and 4 of L-glutamic acid (36) and on the basis of this evidence a scheme was proposed[9, 77, 79, 80] in which a carbanion (38) is generated at C-2 of L-glutamic acid via α-ketoglutarate (37) and its adduct with thiamine pyrophosphate. This then condenses with either (−)-shikimic acid, chorismic acid or *iso*chorismic acid[81, 82] (40, *Figure 3.7*) to yield eventually *o*-succinylbenzoic acid (39) as the first intermediate of an aromatic type on the biosynthetic pathway. The proposal may be formulated in terms of a condensation with chorismic acid (15), and an analog-ous sequence utilising (−)-shikimic acid (27) is also possible. However, a condensation with *iso*chorismic acid (40) is the most attractive from the chemical point of view and is shown in *Figure 3.7*.

Leduc, Dansette and Azerad[80] and Bentley and his collaborators[77] have found *o*-succinylbenzoic acid (39) specifically labelled with ^{14}C to be efficiently incorporated into bacterial menaquinones. However, the subsequent stages in the biosynthetic pathway are not clear and they have been the subject of conflicting views. Thus Leistner,

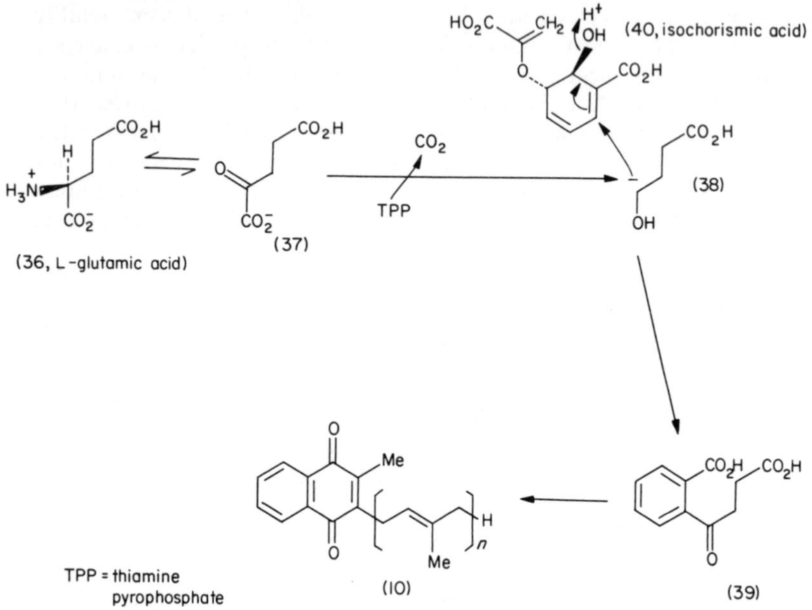

Figure 3.7. Biosynthesis of the naphthoquinone nucleus of bacterial menaquin-ones[76-80]

Schmitt and Zenk[74] showed α-naphthol to be a precursor of vitamin K_2 in *Bacillus megaterium* and Guerin, Leduc and Azerad[76] corroborated this finding using a mutant strain of *Aerobacter aerogenes*. The French workers[76] also demonstrated that this organism was able to utilise both [14C]-2-methyl-1,4-naphtho-quinone and [14C]-1,4-naphthoquinone as efficient precursors of bacterial menaquinones. Conversely, Bentley and his group[77] were unable to demonstrate the utilisation of either of these latter two compounds for menaquinone biosynthesis in *Mycobacterium phlei* or *Escherichia coli* and a similar conclusion was also reached regard-ing the status of α-naphthol. Clearly further work is desirable to elucidate the nature of these areas in the pathway.

3.3 PHENOLIC METABOLITES

Micro-organisms and particularly fungi produce a large number of phenolic secondary metabolites which range in complexity from the simple 6-methylsalicylic acid (41) from *Penicillium griseofulvum* to the complex polycyclic products such as the perinaphthenone pigment atrovenetin (42) from *Penicillium atrovenetum*. Biosynthetic studies have shown that secondary metabolites of this type are largely if not exclusively derived from polyketide intermediates obtained from acetate and malonate[83, 84]. In contrast, the number of phenolic microbial products derived via the shikimate pathway is small. Several of these are formed by transformation of the aromatic amino acids and are discussed later. Others are produced from intermediates on the pathway and such metabolites are described below.

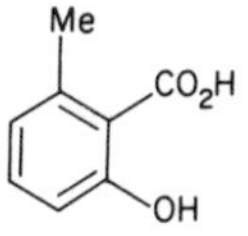

(41)

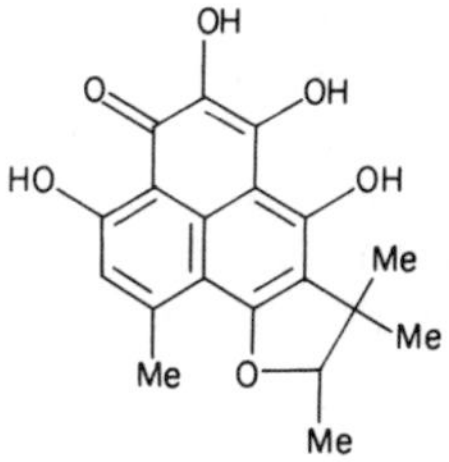

(42, atrovenetin)

3.3.1   2,3-DIHYDROXYBENZOIC ACID AND ITS DERIVATIVES

A number of species of bacteria when grown on a medium deficient in iron excrete relatively high levels of compounds containing the 2,3-dihydroxybenzoyl grouping. Thus under these conditions 2,3-dihydroxybenzoic acid (43) and 2,6-N,N′-bis (2,3-dihydroxybenzoyl)-L-lysine have been isolated from *Azobacter vinelandii*[85], N-2,3-dihydroxy benzoylglycine from *Bacillus subtilis*[86] and 2,3-dihydroxybenzoic acid (43), N-2,3-dihydroxybenzoyl serine (48) and the cyclic trimer of this substance, enterochelin (45), from *Escherichia coli*[87, 88, 89, 90]. 2,3-Dihydroxybenzoic acid is also formed in *Streptomyces griseus*[91], *Streptomyces rimosus*[92] and *Aerobacter aerogenes*[93] in iron deficient media and enterobactin, which is identical to enterochelin, has been obtained from cultures of *Salmonella typhimurium*[94].

Cultures of *Escherichia coli* and *Aerobacter aerogenes* grown under low iron conditions have been shown to contain, during the early exponential phase of growth, 2,3-dihydroxybenzoic acid (43), N-2,3-dihydroxybenzoyl serine (48) and a cyclic trimer enterochelin (45). Subsequently a linear dimer (47) and linear trimer (46) of N-2,3-dihydroxybenzoyl serine are formed apparently from the cyclic trimer enterochelin (*Figure 3.8*). Gibson has suggested that enterochelin probably acts as an iron sequestering compound and that the hydrolysis to the linear trimer (46) and linear dimer (47) by enterochelin esterase is irreversible and is dependent on the uptake of an iron (III)–enterochelin complex by the organism. Enterochelin esterase activity—which probably plays an important role in the utilisation of iron by the cell—is repressed by iron in the growth medium in parallel to the repression of the synthesis of enterochelin (45) from serine, ATP and 2,3-dihydroxybenzoate (*Figure 3.8*)[90, 95, 96]. Incorporation experiments with radioactively labelled serine and 2,3-dihydroxybenzoyl serine (48) precursors indicated that the latter is not an intermediate in enterochelin biosynthesis[97]. Recent studies by Gibson and his collaborators[81, 82, 98–102] have shown that 2,3-dihydroxybenzoic acid (43) is formed from chorismic acid (15) and it is generally assumed at this stage that enterochelin is a biosynthetic end-product on the pathway leading from chorismic acid (15).

Enzyme systems have been isolated from *Escherichia coli* and *Aerobacter aerogenes*[98] which convert chorismic acid to 2,3-dihydroxybenzoic acid. The enzyme from *Aerobacter aerogenes* requires magnesium and NAD^+ for activity and is strongly repressed by low concentrations of iron or cobalt[98]. Omission of the pyridine nucleotide from the enzymic reaction mixture allowed the conversion of chorismic acid to an intermediate (44), which itself was converted to 2,3-dihydroxybenzoic acid on the subsequent addition of NAD^+[102, 104]. The intermediate compound was isolated and identified as 5,6-dihydrocyclohexa-1,3-diene-1-carboxylic acid (44) for which the trivial name 2,3-dihydro-2,3-dihydroxybenzoic acid was suggested[104]. The structure of the intermediate was established by physical means and by its conversion when heated in acid or base to 3-hydroxybenzoic acid (major product) and salicylic acid (minor product). Proton magnetic resonance spectra showed $J_{2,3}$ to be 8.9 Hz which is consistent with a *trans* quasidiaxial relationship of H-2 and H-3 in (44). The absolute configuration at C-2 and C-3 was established[101] as s, s by ozonolysis of the diene and decomposition of the ozonide to give (R, R)-(+)-tartaric acid (49) whose absolute configuration was determined by comparison of its optical rotary dispersion curve with an authentic specimen.

Subsequently a further intermediate in the conversion of chorismic acid to 2,3-dihydroxybenzoic acid was isolated and identified using enzyme extracts of *Aerobacter aerogenes*. The intermediate was identified as 2-hydroxy-3-(1′-carboxyvinloxy)-2,3-dihydrobenzoic acid (40) and was given the trivial name *isochorismic acid*. The compound was converted enzymically to equimolar amounts of

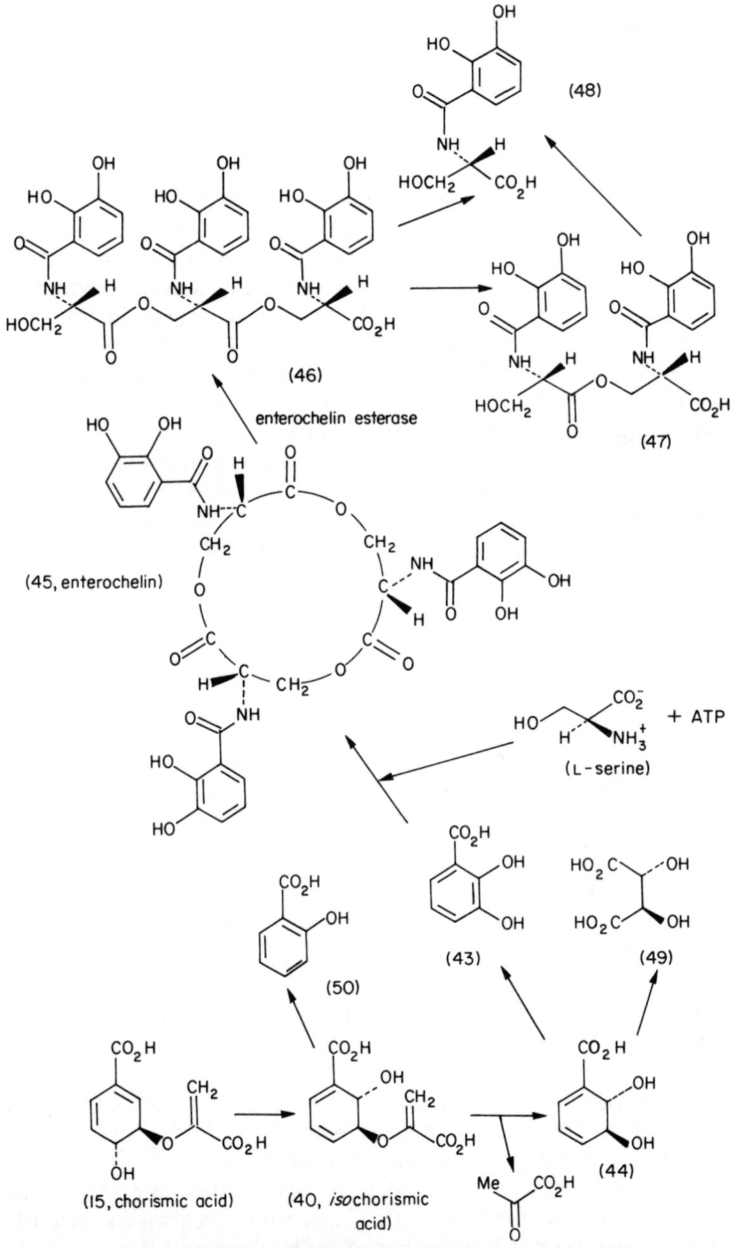

Figure 3.8. Biosynthesis of salicylic acid, 2,3-dihydroxybenzoic acid and entero-chelin[81, 82, 95-102]. Enzymic degradation of enterochelin

2,3-dihydro-2,3-dihydroxybenzoic acid (44) and pyruvic acid and combined with a consideration of its proton magnetic resonance spectrum this led to the structural formulation (40). Ozonolysis of the *iso*chorismic acid (40) and oxidative decomposition of the ozonide gave (R, R)-(+)-tartaric acid (49) which proved, as in (44), the *trans* relationship of substituents at C-2 and C-3 and the s, s configuration. The proton magnetic resonance spectrum of *iso*-chorismic acid in DMSO is, however, noteworthy for the very small coupling constant between H-2 and H-3, and this shows that the preferred conformation of the molecule in this solvent medium is, in contrast to (44), that with the two oxygen substituents in a quasi-*trans* diaxial conformation. Details of the mechanism of the isomerisation of chorismic acid (15) to *iso*chorismic acid (40) have not been elucidated but it is interesting to note that it has all the characteristics of a 1,5-sigmatropic shift.

*Iso*chorismic acid (40) readily decomposes in aqueous solution at pH 7 and ambient temperatures to give a mixture of salicylic acid (50) and 3-carboxyphenylpyruvic acid (51). Salicylic acid (50) is formed by elimination of pyruvic acid and Gibson and his collaborators interpreted[82] the production of 3-carboxyphenylpyruvic acid (51) in terms of a Claisen rearrangement followed by dehydration—steps which are directly analogous to the conversion of chorismic acid to phenylpyruvic acid in the shikimate pathway.

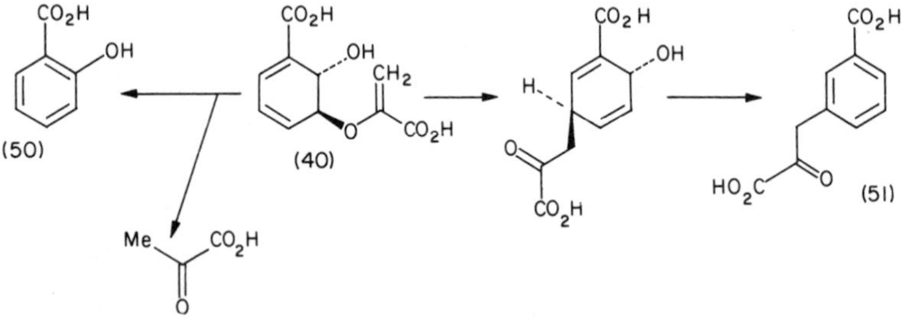

These smooth *in vitro* transformations of *iso*chorismic acid (40) led Gibson[82] to suggest that *iso*chorismic acid was probably a precursor *in vivo* of salicylic acid (50) in mycobacteria and of the *m*-carboxyaromatic amino acids which have been isolated from the non-protein fraction of higher plants[103, 104, 105]. Evidence to sustain this latter proposal has been obtained[104, 105, 106] and has been discussed earlier. Support for the further suggestion that salicylic acid is formed biosynthetically from *iso*chorismic acid (40) in mycobacteria has been obtained by Hudson and Bentley[107–110] and by Ratledge[111–112].

3.3.2 SALICYLIC ACID AND THE MYCOBACTINS

Salicylic acid is excreted by cells of *Mycobacterium smegmatis* grown on an iron deficient medium[113] and many mycobacteria, in response to conditions of iron deprivation, form a series of hydroxamates— the mycobactins[114–116]. All of the mycobactins so far isolated may be divided into two structural classes depending on whether they contain salicylic acid (40) or 6-methylsalicylic acid (41) as the aromatic component of the structure. Hudson and Bentley[107, 108, 110] examined the biosynthesis of salicylic acid in *Mycobacterium smegmatis*—both the extracellular acid and that which forms part of the structure of the growth promoting metabolite—mycobactin-S (52). Generally labelled [^{14}C]-(−)-shikimic acid administered to this organism gave salicylic acid from both sources which contained one seventh of its activity in the carboxyl group, thus suggesting that the carbon skeleton of (−)-shikimic acid is incorporated directly into the phenolic acid. Similar observations were made by Ratledge[111] on the extracellular acid produced by the same organism. In an extension of these experiments Hudson[109] ex- amined the biosynthesis of the aromatic units in the two myco- bactins F and H which are produced concurrently by *Mycobacterium fortuitum*. These mycobactins are identical in every respect except that of their aromatic components; F contains a salicylic acid unit and H a 6-methylsalicylic acid unit. These two acids are not normally co-metabolites and hence with this organism a unique opportunity was presented to study the biogenesis of these structurally similar aromatic compounds in parallel. This work confirmed previous observations that the biosynthesis of 6-methyl- salicylic acid and salicylic acid in mycobacteria proceeds by totally differing routes. Thus 6-methylsalicylic acid was shown to be of polyketide origin and salicylic acid to be derived from (−)-shikimic acid or a closely related intermediate.

Marshall and Ratledge[112], from an enzymic study of salicylic

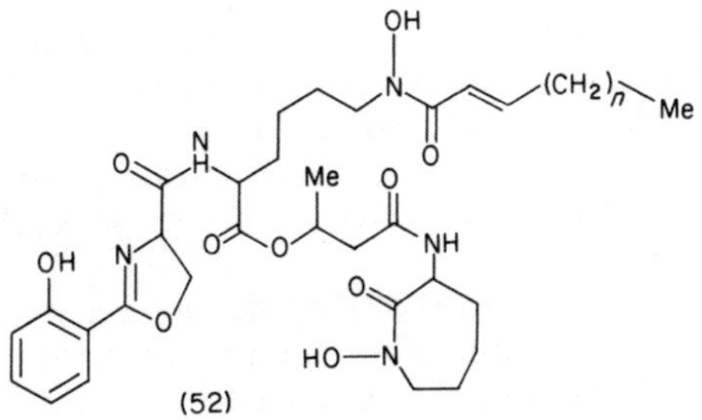

(52)

acid metabolism in *Mycobacterium smegmatis*, have shown that *iso*chorismic acid is the immediate precursor of salicylic acid in mycobacteria and that there are no further intermediates formed in the process (*Figure 3.8*). Cell free extracts of the organism grown under iron deficiency contained the enzyme—salicylate synthetase— which catalysed the transformation of *iso*chorismic acid to salicylic acid. The enzyme required no additional cofactors and it was shown that 2,3-dihydro-2,3-dihydroxybenzoic acid was not an inter- mediate in the process. The enzymes of salicylic acid biosynthesis were repressed by the presence of iron but salicylic acid showed no repression or inhibitory effects on these enzymes. The decline of salicylic acid in the culture medium of *Mycobacterium smegmatis* closely parallels the rate of mycobactin synthesis in the cell. Salicylic acid is in fact excreted by the cells as an early metabolic product only to be re-adsorbed and subsequently incorporated solely into the mycobactin molecule[108, 110].

3.3.3 PROTOCATECHUIC ACID AND THE β-KETOADIPATE PATHWAY

Benzoic acid and *p*-hydroxybenzoic acid can be used as sole sources of carbon and energy by many species of aerobic bacteria belonging to several different genera. The most frequent mode of attack on these two compounds involves their conversion to β-ketoadipate (54) which is cleaved after activation to yield succinate and acetyl co-enzyme A[117–119]. The initial steps are mediated by specific aryl hydroxylases which convert benzoic acid to catechol, and *p*-hydroxy- benzoic acid to protocatechuic acid (*Figure 3.9, a, b*). A detailed analysis of the conversion of catechol and protocatechuic acid in *Pseudomonas putida* has shown that the reaction pathways pass through two series of chemically analogous intermediates (*Figure 3.9*)[117, 118]. Metabolic convergence occurs with the formation of the lactone (53), the immediate precursor of β-ketoadipate (54). Two specific sets of enzymes control the conversion of both catechol and protocatechuic acid to this lactone and a common enzyme converts the lactone to β-ketoadipate (54). In bacteria these enzymes are typically inducible and aspects of the regulation of these pathways have been studied in *Pseudomonas putida*[118], several fungi[120–123] and *Moraxella calcoacetica* (syn. *Acinetobacter calco- aceticus*)[124–127].

The biological aromatisation of alicyclic compounds has long been known and the conversion of (−)-quinic acid (55) to phenolic compounds, and particularly protocatechuic acid, by moulds,

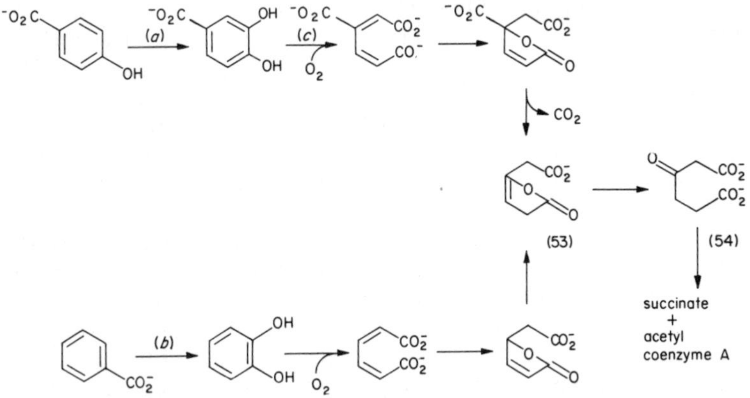

Figure 3.9. Bacterial degradation of benzoate and *p*-hydroxybenzoate by the β-ketoadipate pathway

yeasts and bacteria has been one of the reactions most widely studied[128-131]. Several of the organisms which have been examined in detail are in fact able to use (−)-quinic acid (55) or (−)-shikimic acid (27) as sole sources of carbon for growth. They do so by catabolism of the substrate to protocatechuic acid (58) and the β-ketoadipate pathway. Thus organisms such as *Neurospora crassa* possess both synthetic and degradative enzyme systems for the utilisation of 3-dehydroquinate (56), an intermediate on the shikimate pathway. The inducible degradative system may be part of a detoxification mechanism for (−)-quinic acid and related compounds as well as serving to make available to the organism an alternative source of carbon and energy. In *Neurospora crassa* an inducible 3-dehydroquinate dehydratase has been observed[132, 133] which catalyses one reaction in an inducible degradative aromatic pathway (*Figure 3.10*)—the conversion of 3-dehydroquinate (56) to 3-dehydroshikimate (57)—which is identical to the same step in the synthetic pathway prior to chorismate. In addition it has been possible to select *Neurospora crassa* strains which are able to grow on (−)-quinic acid (55) and enzymic analyses have provided evidence for an inducible quinate dehydrogenase in these strains. Catabolism of the substrates (55, 56 and 57) then occurs via protocatechuate and the β-ketoadipate pathway and an enzyme, 3-dehydroshikimate dehydratase, which catalyses the conversion of 3-dehydroshikimate to protocatechuate (58) has been identified in *Neurospora crassa*[134]

Similarly wild type *Moraxella calcoacetica* (syn. *Acinetobacter calco-aceticus*) can readily use (−)-shikimic acid as a sole source of carbon and energy and under these conditions a transient accumulation of protocatechuic acid (58) occurs in the growth medium. The (−)-shikimic acid is catabolised via protocatechuic acid and the β-ketoadipate pathway and mutants unable to synthesise protocatechuate oxygenase (*Figure 3.9, c*) are unable to grow with (−)-shikimic acid[125–127]. Analysis of cell free extracts of *Moraxella calcoacetica* reveal that the organism synthesises two enzymes which oxidise shikimate to 3-dehydroshikimate. One is $NADP^+$ linked and is part of the synthetic apparatus of the shikimate pathway, the other enzyme is not linked to a pyridine nucleotide coenzyme and its synthesis is dependent on the levels of (−)-shikimic acid and protocatechuic acid supplied. It appears to be part of the inducible enzyme system responsible for the conversion of shikimate to protocatechuate (*Figure 3.10*). *Moraxella calcoacetica*

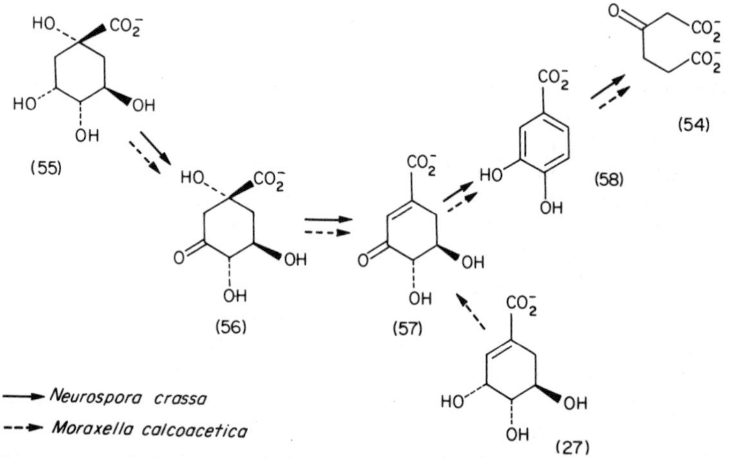

Figure 3.10. Catabolism of quinate and shikimate by the β-ketoadipate pathway in micro-organisms[123–124]

also degrades (−)-quinic acid by the β-ketoadipate pathway after conversion of the substrate to protocatechuic acid[126, 127]. A nutritionally induced dehydrogenase, which initially metabolises the quinate to 3-dehydroquinate in this process, has distinctly different properties from the quinate dehydrogenases of other bacteria[133, 135] and it appears to be the same enzyme that is concerned in the oxidation of shikimate by this organism.

3-Dehydroshikimate dehydratase, the enzyme which catalyses the conversion of 3-dehydroshikimate (57) to protocatechuate (58) in *Neurospora crassa* has been studied by several groups and some features of the mechanism deduced[134, 136-137]. Using a mutant strain of *Neurospora crassa* blocked in the conversion of 3-dehydroshikimate (57) to shikimate (27) in the synthetic pathway Gross, Tatum and Gafford[136] observed the incorporation of 1- and 6-[14C]-D-glucose into protocatechuic acid. It was observed that the isotope from these two substrates was incorporated almost exclusively into C-2 and C-6 of protocatechuic acid. Comparison of this labelling pattern with the corresponding labelling pattern observed in (−)-shikimic acid[138] in *Escherichia coli* with the same substrates led to the postulate that the 3-keto group of (57) is lost in the conversion to protocatechuate (58). This proposal was shown to be incorrect when in later work Gross[134] observed the conversion of (−)-shikimic acid (27), uniformly labelled with 14C at C-2, 3, 4 and 5, to protocatechuic acid (58). The product was degraded to β-ketoadipic acid (54) using an extract of *Neurospora crassa* and this was then transformed by chemical means to laevulinic acid (60), iodoform (62) and succinic acid (63). Gross found that the distribution of radioactivity in the β-ketoadipic acid (54) and the products (60, 61, 62 and 63) was as shown in *Figure 3.11*. These observations are only consistent with the loss of the hydroxyl group at C-5 of (57) in the conversion to protocatechuic acid and it was concluded that the enzyme catalyses the elimination of water from the enol form of 3-dehydroshikimate (59), *Figure 3.11*.

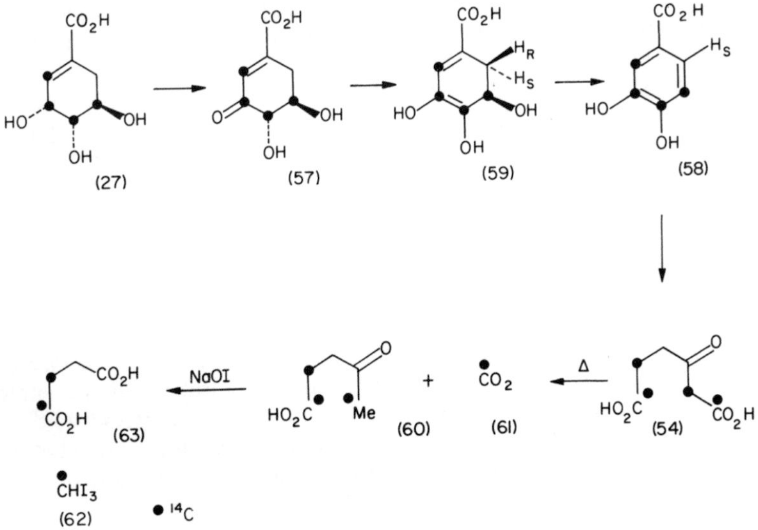

Figure 3.11. Biosynthesis of protocatechuate in *Neurospora crassa*[134, 137]

The steric course of this reaction has now been determined[137] using samples of 6R and 6s-6-[³H]-3-dehydroshikimic acid as substrates. The results of these experiments show that it is the 6-pro R hydrogen that is eliminated in this reaction which therefore involves the rather unusual *syn* elimination of the elements of water. In a parallel study of the chemical conversion of 3-dehydroshikimic acid to protocatechuic acid—a reaction which occurs upon pyrolysis or heating with concentrated acid—the results indicated a non-stereo-specific reaction mechanism which involved a large isotope effect

3.3.4   GALLIC ACID

Gallic acid (3,4,5-trihydroxybenzoic acid, 64) is metabolised along with protocatechuic acid by the mould *Phycomyces blakesleeanus* when grown on a glucose medium[139, 140] and this is the only authenticated case of the synthesis of this particular phenol by a micro-organism. Brucker[141] has suggested that both the gallic acid and the protocatechuic acid are formed in this case by deamination and oxidative degradation of the aromatic amino acids L-phenylalanine and L-tyrosine. Knowles, Haworth and Haslam[142] concluded, however, from growth experiments with the mould that gallic acid is formed directly from 3-dehydroshikimic acid (57) by dehydrogenation. In support of this proposal, they showed that 3-dehydroshikimic acid (57) was converted *in vitro* to gallic acid by mild oxidising agents such as Fehlings solution. In later work, Smith and Haslam[143] used tracer studies, with 1- and 6-[¹⁴C]-D-glucose as substrates, to establish that dehydrogenation of 3-dehydro-

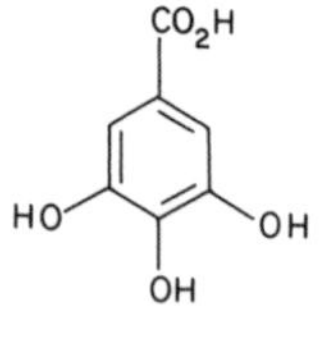

$$CO_2H$$

(64)

shikimic acid was the route of biosynthesis of gallic acid in *Phycomyces blakesleeanus*, and similarly that dehydration of the same substrates is the mode of formation of protocatechuic acid in this organism. No evidence has been obtained to show whether the phenolic acids produced by the mould are subsequently catabolised.

3.4 MISCELLANEOUS METABOLITES

3.4.1 p-AMINOBENZOIC ACID AND THE FOLIC ACID CO-ENZYMES

Co-enzymes derived from tetrahydrofolic acid (70) are frequently involved in the biosynthetic transfer of one carbon fragment (usually at the oxidation level of formate, formaldehyde or methanol) and they complement two other cofactors, S-adenosylmethionine and methylcobolamine, which are also used biosynthetically for this purpose. Thus the studies[144–147] which have revealed the metabolic origin of each of the atoms of the purine ring system (65) have demonstrated that carbon atoms 2 and 8 are derived from formic acid, respectively via N^{10}-formyltetrahydrofolic acid (71) and N^5, N^{10}-methenyltetrahydrofolic acid (partial formula 72).

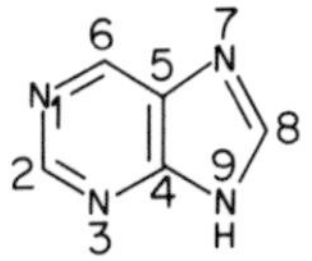

(65)

Tetrahydrofolic acid itself (70) is composed of three structural fragments, a reduced pteridine, p-aminobenzoic acid and L-glutamic acid and the principal steps in the biosynthetic pathway are shown in *Figure 3.12*. The starting material guanosine monophosphate (66) is converted to the dihydropteridine (67) which as its pyrophosphate condenses with p-aminobenzoic acid (73) to give the dihydropteroic acid (68). Finally the ATP dependent formation of the peptide bond with L-glutamic acid gives dihydrofolic acid (69), which forms the tetrahydro derivative (70) by a pyridine nucleotide linked reduction. Polyglutamic acid derivatives of tetrahydrofolic acid containing up to seven amino acid residues linked as γ-glutamyl peptides also occur in living systems and these are formed by sequential addition of further L-glutamic acid residues to (70).

The biosynthesis of p-aminobenzoic acid, a structural component of the folic acid co-enzymes and of the antibiotic amicetin[149], is at present still under investigation and full details are not yet available[150]. Weiss and Srinivasan[151, 152] showed that p-aminobenzoic acid (73) was formed by cell free extracts of baker's yeast from L-glutamine and shikimic acid-3-phosphate and that the amide group was the precursor of the aryl amino group in (73). In later work[153], cell free extracts of *Aerobacter aerogenes* 62-1 were found

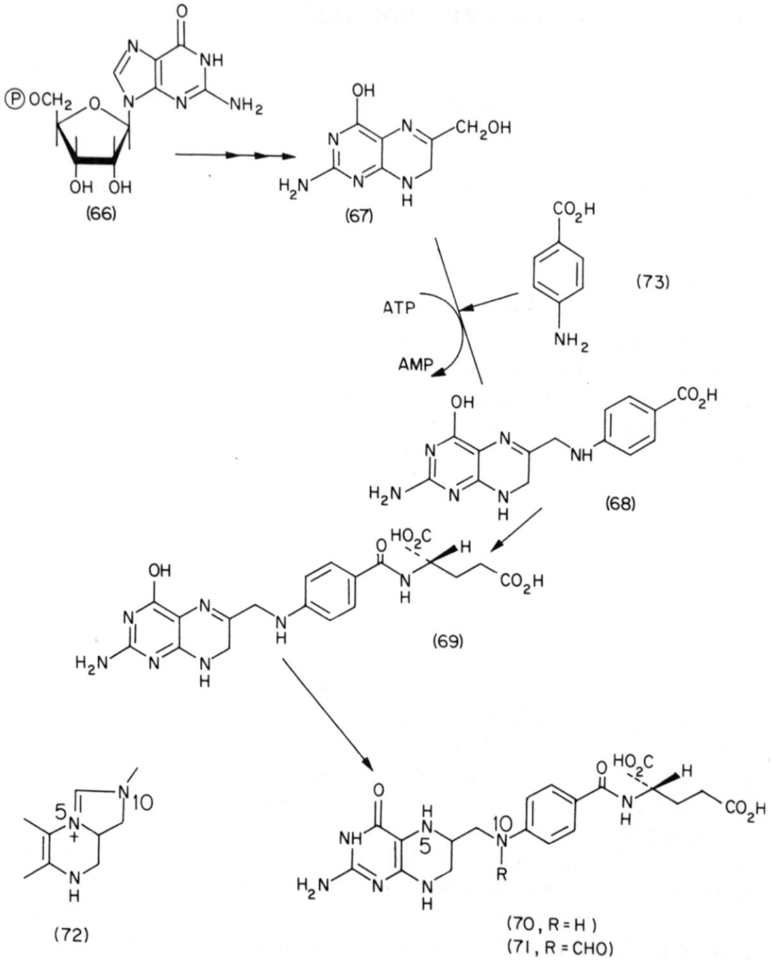

Figure 3.12. Biosynthesis of tetrahydrofolic acid[148]

to convert chorismic acid to p-aminobenzoic acid in the presence of L-glutamine and evidence for the formation of an intermediate in this process has been discussed by two groups[154, 155]. Lingens and his collaborators[155] produced a p-aminobenzoate deficient mutant of the polyauxotrophic mutant *Aerobacter aerogenes* 62-1 by treatment with 1-methyl-3-nitro-1-nitrosoguanidine. This mutant produced an intermediate in the biosynthesis of p-aminobenzoic acid which gave the amino acid (73) on heating in water. On the basis of chemical analogy and acceptability, *iso*chorismic acid (40) would appear to be the most probable substrate, related to chorismic acid, in which amination could occur at the 4- position. A *possible* pathway for the formation of p-aminobenzoic acid (73) from chorismic acid (15) is given below.

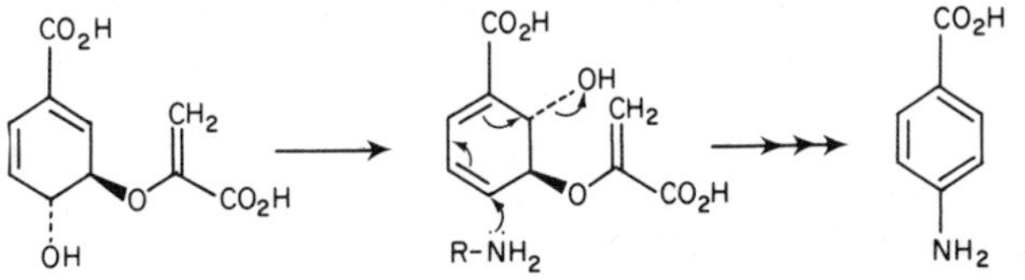

The synthesis of p-aminobenzoic acid (73) and anthranilic acid from chorismic acid (15) is probably very closely related in many systems. The manner in which the two synthetic pathways are co-ordinated in *Bacillus subtilis* has been revealed by the genetic and enzymic studies of Jensen and his collaborators[156]. In *Bacillus subtilis* anthranilate synthetase is a molecular complex composed of two non-identical sub-units. All L-tryptophan auxotrophs known to lack anthranilate synthetase activity show a modification in sub-unit E, the gene product of trp E. Kane and Jensen[157] have also described a trp X mutant which genetically defines another compo-nent of the anthranilate synthetase complex, sub-unit X—molecular weight $\sim 16\,000$. Crude extracts composed from cultures of trp X and trp E mutants were able to complement one another and recon-stitute the anthranilate synthetase activity. Sub-unit X is also a common sub-unit for the p-aminobenzoate synthetase of the same organism and Jensen suggested that it probably acts as a 'specifier protein'. A situation was envisaged in which some unidentified sub-unit of p-aminobenzoate synthetase competes with sub-unit E for association with the sub-unit X to form respectively p-amino-benzoate synthetase or anthranilate synthetase. The arrangement was described as one of metabolic interlock in which the co-ordination of the two metabolic pathways from chorismic acid (15) to anthranilic acid and p-aminobenzoic acid (73) is controlled by partitioning of the small protein sub-unit X to different enzyme complexes.

3.4.2 PHENAZINES

The phenazine ring system is common to some thirty microbial metabolites and antimicrobial assay shows many of them to be active against gram-positive bacteria, fungi and actinomycetes. The phenazine metabolites may be broadly divided into three groups: those with none, one or two carbon substituents on the carbon skeleton of the heterocyclic ring system. Typical of the first category are pyocyanin (74) and related 1-phenazinols from *Pseudomonas aeruginosa*[158] and iodinin (75) and its derivatives from *Pseudomonas iodinum, Streptomyces thioluteus, Brevibacterium crystalliodinum* and *Waksamania aerata* sp.[159–161]. Iodinin (75), which derives its name from its resemblance in bulk to iodine, was the first naturally occurring N-oxide to be described. Representative of the phenazines with one carbon substituent are the phenazine monocarboxylic acids in *Pseudomonas aureofaciens*[162–163], aeruginosin B (76) the first example of an aromatic sulphonic acid from natural sources[164], and chloraphin, a 3:1 molecular complex of phenazine-1-carboxylic acid amide and its dihydro derivative[165]. Phenazines bearing two carbon substituents are typified by griseolutein (77) and griseoluteic acid (78) from *Streptomyces griseoluteus*[166]

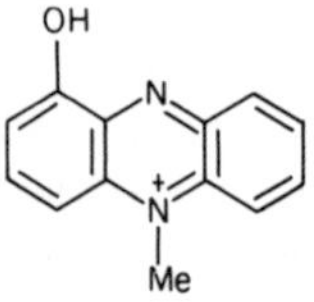

(74, pyocyanin)

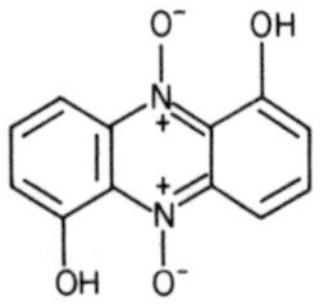

(75, iodinin)

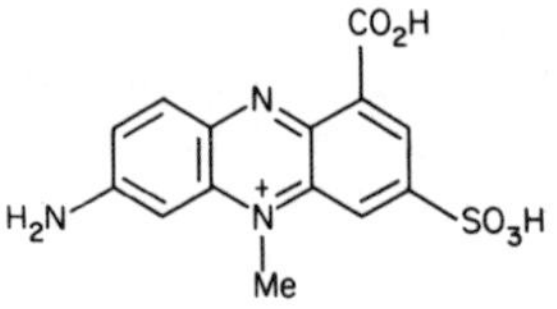

(76, aeruginosin-B)

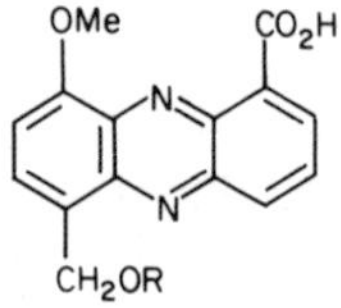

(77, R = CO · CH$_2$OH , griseolutein)
(78, R = H , griseoluteic acid)

Considerable effort has been expended on discovering how the phenazine carbon-nitrogen skeleton is constructed by microorganisms. Carter and Richards[167] suggested initially that formation of the phenazine skeleton takes place by the coupling of two anthranilic acid molecules. This observation was based on the isolation of chloraphin labelled with ^{14}C after feeding anthranilic acid labelled in the carboxyl group with ^{14}C to *Pseudomonas chloraphis*. The chloraphin was converted to phenazine-1-carboxamide which retained nearly 80 per cent of its radioactivity in the carboxyl group. Carter and Richards[167] concluded from this evidence that one C$_6$ ring of the phenazine nucleus originated from anthranilic acid. However, subsequent work has provided sparse evidence to support the suggestion that anthranilic acid has a key role to play in phenazine biosynthesis.

Various groups have for example examined the biogenesis of pyocyanin (74) with ^{14}C labelled precursors and, although no degradative studies were carried out, a variety of substrates such as glycerol, dihydroxyacetone[168], L-alanine, pyruvic acid, L-leucine[169] and (−)-shikimic acid[170] were shown to be efficient precursors of the phenazine ring system and L-methionine of the quaternary N-methyl group[171]. In other studies, generally labelled [^{14}C]-(−)-shikimic acid was also shown to be readily incorporated into the carbon skeleton of phenazine-1-carboxylic acid[172], 2-hydroxyphenazine[173] and iodinin[174]. The incorporation of substrate into iodinin was diminished by feeding inactive L-phenylalanine at the same time. On the other hand, unlabelled anthranilic acid administered with generally labelled [^{14}C]-(−)-shikimic acid enhanced the total activity of the iodinin isolated and Podojil and Gerber concluded[174] that it seemed unlikely that anthranilic acid was a direct precursor of the phenazine ring. Subsequently, Podojil and Gerber[175] isolated iodinin after administering 1,6-[[14]C$_2$]-(±)-shikimic acid and carried out a specific degradation of the isotopically labelled metabolite (*Figure 3.13*).

The radioactivities associated with the various fragments in the degradation of iodinin are shown in the annexed scheme (*Figure 3.13*), and from a consideration of these results, Podojil and Gerber[175] proposed that the phenazine nucleus was elaborated from two molecules of (−)-shikimic acid coming together as shown (*Figure 3.13*) to yield (79) as the hypothetical key biosynthetic intermediate to all phenazines. Further details of the condensation mechanism have yet to be elucidated.

Other work in this field has concentrated on the later stages of biosynthesis and the sequence of transformations which occur once the phenazine ring has been synthesised. Thus [^{14}C]-1,6-phenazinediol and [^{14}C]-1,6-phenazinediol-5-oxide were both incorporated into iodinin (75) in *Brevibacterium iodinum* in high yield (10 per cent and 15 per cent respectively) confirming the idea that they are immediate precursors of iodinin (75). Holliman and his collaborators have similarly examined the terminal steps of the biosynthesis of pyocyanin (74) and related pigments in *Pseudomonas aureofaciens*. On the basis of the known reactions of the 5-methylphenazinium ion (83), such as its light induced hydroxylation[176] to give pyocyanin (74) and its reaction with ammonia[177] to give the 2-amino-10-methylphenazinium ion (84), it was proposed by Holliman[177] that the betaine, 5-methylphena-

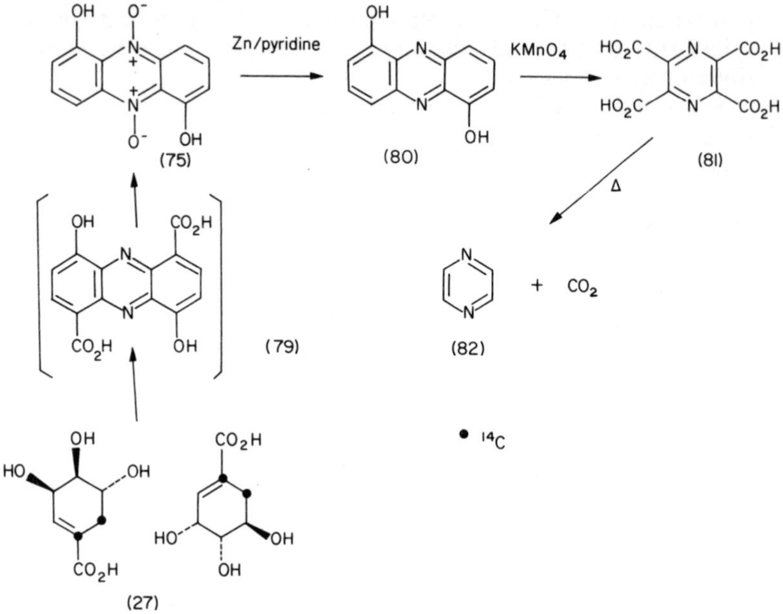

	Radioactivity (%)	
	Observed	*Calculated*
Iodinin (75)	100	–
1,6 – phenazinediol (80)	100	–
pyrazine tetracarboxylic acid (81)	72·9	75
pyrazine (82)	23·9	25
carbon dioxide	49·1	50

Figure 3.13. Biosynthesis and degradation of iodinin[174, 175] derived from 1,6-[^{14}C$_2$]-(±)-shikimic acid

zinium-1-carboxylate (85), was a common intermediate in the biosynthesis of pyocyanin (74) and aeruginosin A (86).

Support for these proposals was obtained by both *in vitro* and *in vivo* studies[178-180]. Thus reaction of the phenazinium ion (85) as its hydrochloride with ammonia gave aeruginosin A (86) in yields which were strongly dependent on the concentration of ammonia (64 per cent with 15N, and 0 per cent with 0.009N ammonia). When the phenazinium salt (85) was added to liquid cultures of *Pseudomonas aeruginosa* between the second and third day of growth, aeruginosin A (86) was readily formed. It was, moreover, observed that significant amination of the salt (85) occurred at ammonia concentrations much lower than the free ammonia concentrations required for amination of the salt *in vitro*. Thus a total (salts plus free) ammonia concentration of 0.007N in the medium gave yields of aeruginosin A (86) from the betaine (85) varying from 13–22 per cent and it was concluded that amination of (85) in the culture of *Pseudomonas aeruginosa* was enzyme mediated.

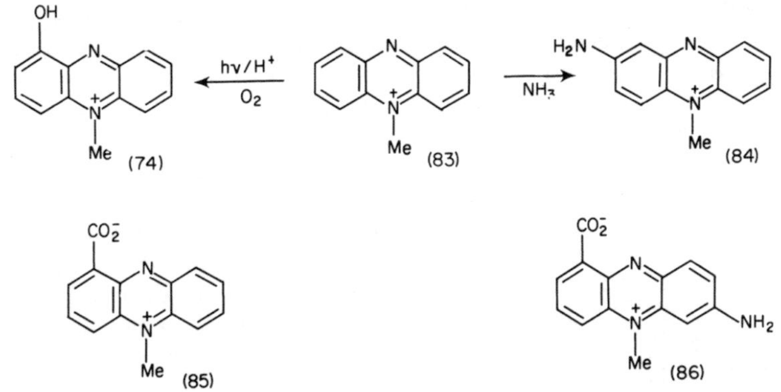

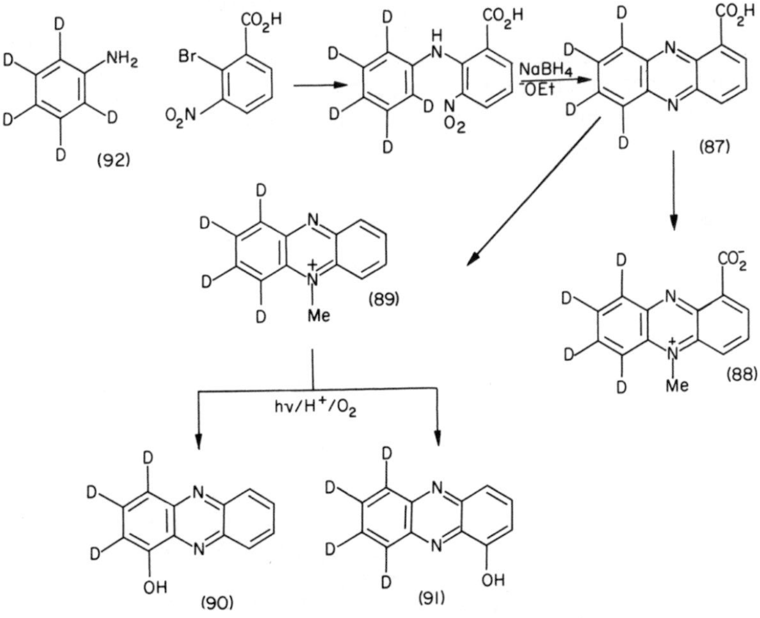

Figure 3.14. Synthesis of deuteriated phenazines[185]

Pyocyanin (74), for which 1-hydroxyphenazine is not a precursor[181], is produced by *Pseudomonas aeruginosa*, alternative strains of which metabolise the aeruginosins A (86) and B (76)[182, 183]. The formation of these three phenazine pigments by different strains of the same organism suggested that their biosynthetic pathways might be similar and in particular that the betaine (85) might be a precursor of pyocyanin (74). These proposals were tested by administration of samples of the deuteriated phenazines (87, 88, 89 and 90+91) to pyocyanin producing strains of *Pseudomonas aeruginosa* over the period of pigment production. Pentadeuteric aniline (92)[184] was used as the starting material to prepare the deuteriated phenazines (87–91) in a novel synthetic sequence[185] (*Figure 3.14*).

After administration of the precursors to the growing cultures of the organism pyocyanin (74) was isolated, dequaternised to give 1-hydroxyphenazine which was examined mass spectrometrically. In agreement with the earlier observation[181], the deuteriated phenazines (90 and 91) were not incorporated into pyocyanin. The quaternary salt (89) was incorporated to a small extent (0.7 per cent) to give pyocyanin containing equal parts of tri- and di-deuteriated species and this is probably due to the fact that the phenazinium ion (89) can act as a casual substrate for hydroxylation. In contrast both (87) and (88) were incorporated into pyocyanin (2–11 per cent) without loss of deuterium and this has led to the proposal that, in the biosynthetic pathway, an oxidative decarboxylation leads from (85) to pyocyanin (74). An arene-oxide intermediate (93) has been proposed for this step.

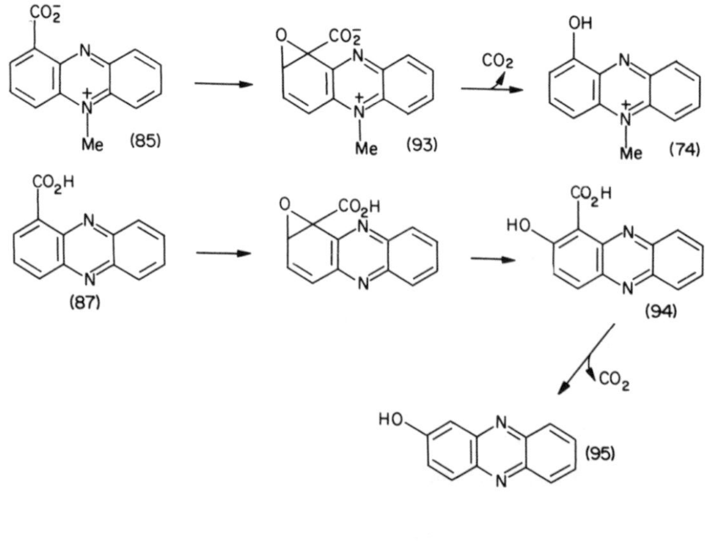

Analogous feeding experiments with *Pseudomonas aureofaciens* established that (87) but not (88) was utilised in the formation of 2-hydroxyphenazine-1-carboxylic acid (94) and 2-hydroxyphenazine (95). These results were interpreted as showing that both of these phenazine metabolites (94 and 95) are metabolites of phenazine-1-carboxylic acid (87) and are formed as outlined above.

3.4.3   BACILYSIN AND ANTICAPSIN

Bacilysin is a hydrophilic substance formed by certain aerobic spore-forming bacteria which causes lysis in cultures of growing staphylococci. Abraham and his collaborators[186–189] have described its production by aerated cultures of a strain of *Bacillus subtilis*, and they showed that on hydrolysis (6M HCl) it gave L-alanine and L-tyrosine, although it did not contain a tyrosine residue. On the basis of physiochemical measurements the unusual dipeptide structure (96) was assigned to bacilysin. In later work[188] a further substance (A A 1), identical with the C-terminal amino acid of bacilysin, was isolated from the culture fluid of *Bacillus subtilis* strains and the same amino acid has been obtained from stirred cultures of *Streptomyces griseoplanus* by Neuss and his collaborators[190] and named anticapsin. Detailed analysis of the products of acid hydrolysis of anticapsin or A A 1 showed them to contain both L-tyrosine and *m*-L-tyrosine in a ratio of 9:2.

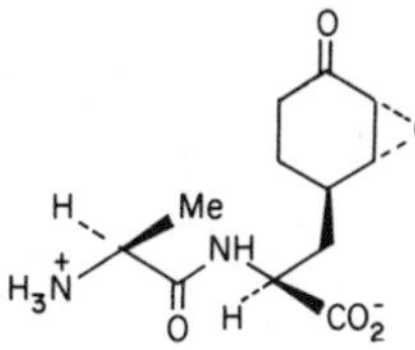

(96)

Observations on the biosynthesis of bacilysin showed[191] that it took place during normal cell growth but that there was no obligatory coupling of protein and antibiotic synthesis. When D L-1-[^{14}C]-alanine was added to a growing culture of *Bacillus subtilis* the isotope was incorporated into bacilysin in the N-terminal L-alanine residue. Under the same conditions virtually no isotopic carbon was incorporated into bacilysin from D L-2-[^{14}C]-tyrosine, L-U-[^{14}C]-tyrosine and 1-[^{14}C]-sodium acetate although these substances were used by the organism for the biosynthesis of other substances. However, the C-terminal amino acid of bacilysin (96) was labelled with ^{14}C when 1,6-[^{14}C$_2$]-($\pm$)-shikimic acid was

employed as a substrate, and the biosynthesis of this unique structure thus appears to involve a diversion from the shikimate pathway
prior to the formation of the aromatic amino acids. Walker and
Abraham suggested that prephenic acid was a possible precursor
but no further experimental data has been presented at this point.

3.4.4 CHLORAMPHENICOL

One of the most useful techniques which may be employed in a study
of the control of protein synthesis is the use of specific inhibitors,
compounds which permit the dissection of the overall sequence
into a number of stages by interfering with the biosynthetic process
at specific points. A classical inhibitor of protein synthesis in
bacteria is chloramphenicol (or chloromycetin, 97)[192] a broad
spectrum antibiotic produced by *Streptomyces venezuelae*[193, 194]
and a concentration of 10–20 micrograms per millilitre will
generally decrease the level of protein synthesis by a factor of ten or
more. Chloramphenicol may be produced synthetically but of the
four stereoisomers only the natural D-threo form (97) shows any
significant activity as an inhibitor of protein synthesis in bacteria.

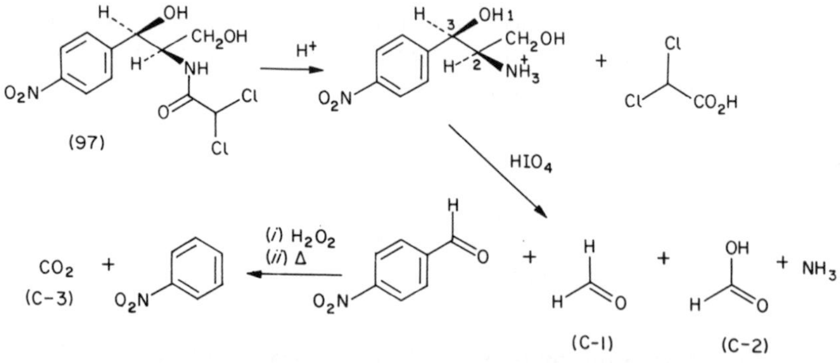

Figure 3.15. Chloramphenicol: chemical degradation[201]

The biosynthesis of chloramphenicol has been studied by a
number of groups including Gottlieb and Carter[195, 196] and
Vining and his collaborators[197–200] and using normal isotopic
tracer techniques they have shown the substance to be formed via
the shikimate pathway. Utilising the extent of incorporation of
individual specifically labelled substrates as a criterion of their
efficiency as precursors of chloramphenicol, Vining and his
group[200] have formulated the final stages in the biosynthesis of the

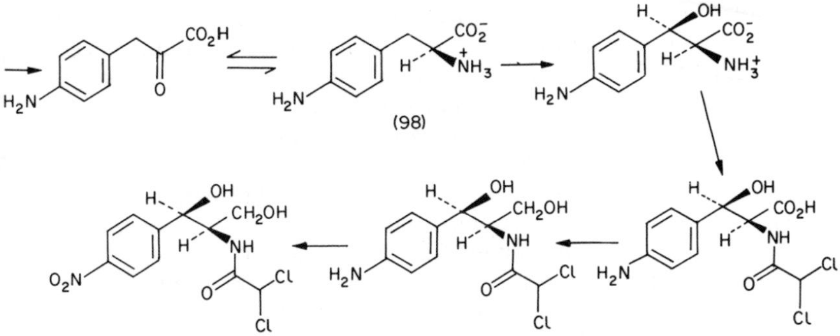

Figure 3.16. Biosynthesis of chloramphenicol: the final stages[200]

antibiotic in the sequence (98 → 97, *Figure 3.16*). Possible intermediates were specifically labelled, normally in the side chain, and
the position of the labelled atom incorporated into chloramphenicol
was determined by chemical degradation[201], *Figure 3.15*. L-*p*-
Aminophenylalanine (98) is regarded as a key intermediate in these
later stages of biosynthesis and it is of interest to note[202] the
occurrence of N-methyl-L-*p*-dimethylaminophenylalanine and N-
methyl-L-*p*-methylaminophenylalanine, N-methylated derivatives

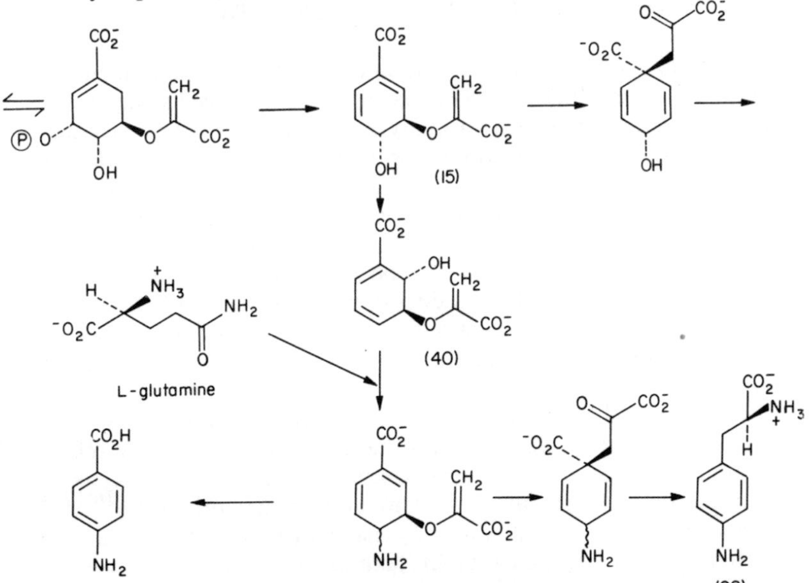

Figure 3.17. Possible route of biosynthesis of L-*p*-aminophenylalanine

of (98), in other natural products such as the vernamycin B group of peptide antibiotics. The functions of the chloramphenicol molecule are introduced in a clearly defined sequence from (98); β-hydroxylation of the side chain, acylation of the aliphatic amino group, reduction of the carboxyl group and finally oxidation of the aryl amino group to nitro.

Experimental work clearly showed that neither L-phenylalanine nor L-tyrosine were incorporated intact into the C_6 . C_3 skeleton of chloramphenicol, and Vining has proposed that the biosynthetic pathway to the antibiotic diverged from the shikimate pathway at a point between 3-enolpyruvylshikimate-5-phosphate and prephenate. Details of the steps prior to the formation of L-p-aminophenylalanine (98) are, however, still shrouded in uncertainty. Based solely on chemical analogy a scheme, such as that shown in *Figure 3.17* in which *iso*chorismate (40) is a key intermediate, would account satisfactorily for the formation of L-p-phenylalanine. The sequence outlined is analogous to that suggested earlier (see page 107) to account for the biosynthesis of p-aminobenzoate (73) from chorismate (15) and as in this case the amino donor is envisaged as the amide amino group of L-glutamine.

3.4.5   D-PHENYLALANINE AND PEPTIDE ANTIBIOTICS

In certain bacteria there is a specific nutritional requirement for D-amino acids which are found as components of cell structures or antimetabolites. Bacteria normally meet this need by the conversion of L-amino acids to D-amino acids and in the case of alanine, methionine and tryptophan the evidence suggests that these reactions are directly catalysed by amino acid racemases which have a cofactor requirement for pyridoxal phosphate[203-207]. An oxidation-reduction cofactor may also be a general feature of racemases of this class. However, the mode of epimerisation of L-phenylalanine to D-phenylalanine necessary for the synthesis of some peptide antibiotics, proceeds in an entirely different way, which as yet has only been partially resolved.

Peptide antibiotics are produced by a variety of micro-organisms and one organism generally produces a family of closely related peptides which are often cyclic and resistant to hydrolysis by peptidases and proteases of plant and animal origin[202, 207, 208]. Their molecular weight is usually less than 3000 and the structures often contain fragments other than L-amino acids, for example hydroxy and fatty acids, imino acids and D-amino acids. D-Phenylalanine occurs in several peptide antibiotics[202, 207, 208] such as the polymyxins (*Bacillus polymyxa*), polypeptin (*Bacillus krzemieniewski*), bacitracin (*Bacillus licheniformis*), the tryocidines and gramicidin S (*Bacillus brevis*). The tyrocidines (99) are cyclic peptides produced by specific strains of *Bacillus brevis* and differ only in analogue type substitutions of aromatic residues (positions 3, 4 and 7). In two positions (1 and 4) D-amino acids are present and at one (9) a non-protein amino acid, L-ornithine. The structure of gramicidin S is closely related to that of the tyrocidines[207, 208].

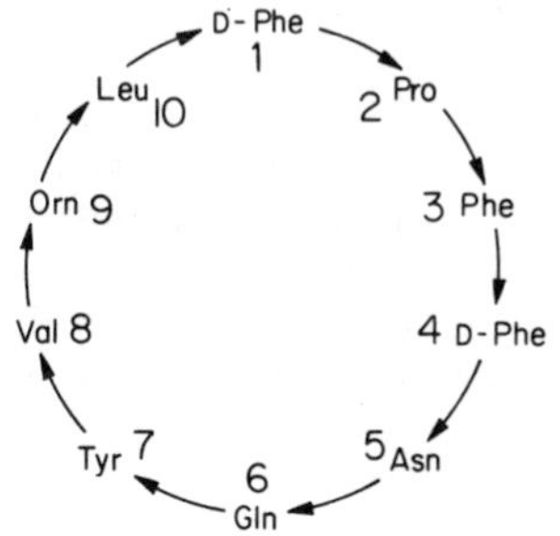

(99)

The development of cell free systems has played an important role in attempts to decipher the mechanism of peptide antibiotic synthesis[209–213]. Most workers agree that this type of synthesis occurs in the complete absence of polynucleotides and that the amino acid sequence is determined by enzyme specificity and organisation in a multi-enzyme complex[214, 215]. The tyrocidines are produced in *Bacillus brevis* by an enzyme system which has been resolved into three complimentary fractions by Sephadex G-200 filtration: a light and an intermediate component (molecular weight 100 000 and 230 000) which activate L-phenylalanine and L-proline respectively and a heavy fraction (460 000) which activates the remaining eight amino acids including L-phenylalanine. Each

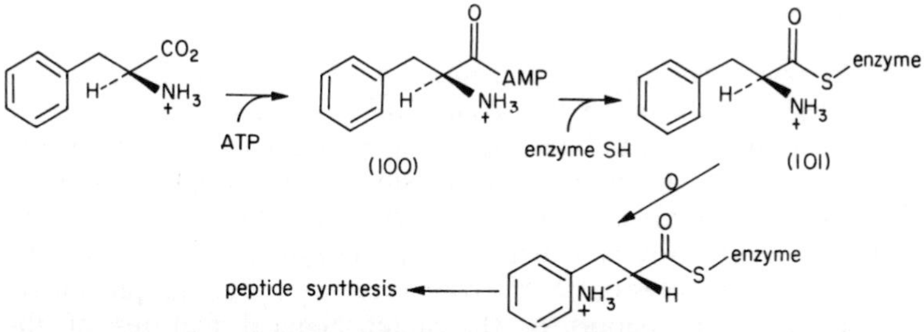

Figure 3.18. Activation and epimerisation of L-phenylalanine[211–213]

enzyme fraction reacts with an amino acid and ATP to form an intermediate amino acyl–adenylate enzyme complex (100) and pyrophosphate ion. The adenylate bound amino acid is then transferred, with the liberation of AMP, to an enzyme thiol group (101) from which it enters the chain elongation process (*Figure 3.18*).

Present data indicates that chain growth in both gramicidin S and the tyrocidines commences at the D-phenylalanine residue adjacent to proline and that the first step in the synthesis is the conversion of L-phenylalanine to D-phenylalanine. In the case of the tyrocidines, the synthesis then proceeds in order from the amino to the carboxyl terminus to form a linear decapeptide ending with a thiol ester linked leucine. The peptide then cyclises relatively slowly to the final product. Yamada and Kurahashi showed[211–213] that the epimerisation of L-phenylalanine in the initial step did not require pyridoxal phosphate or FAD and they suggested that the reaction occurs via the thiol ester enzyme bound form (101), *Figure 3.18*.

3.4.6   β-TYROSINE AND R-β-PHENYL-β-ALANINE

R-β-Phenyl-β-alanine (102) has been reported several times as a component of natural peptides such as islanditoxin from *Penicillium islandicum*[216] and roccanin from the lichen *Roccella canariensis*[217, 218] and some preliminary observations have been made on the biosynthesis of β-tyrosine (103) which occurs in the linear oligopeptide antibiotic edeine in *Bacillus brevis*[219].

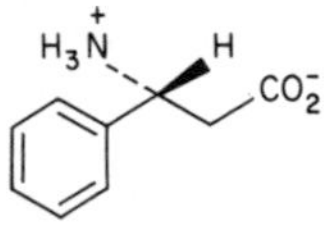

(102) 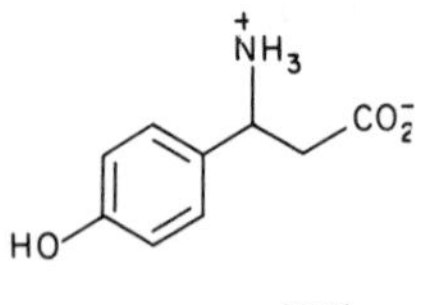   (103)

An enzyme system has been isolated from *Bacillus brevis* which isomerises L-tyrosine to β-tyrosine (103). The enzyme, L-tyrosine α, β-aminomutase, catalyses the direct transfer of the amino group from the α to the β position in L-tyrosine and has a pH optimum of 8.5 and a requirement for ATP as a cofactor. This requirement for ATP contrasts however with similar enzymes involved in the biosynthesis of β-lysine[220, 221] which require pyridoxal phosphate as cofactor. No details of the stereochemical features of the isomerisation of L-tyrosine have been determined.

3.4.7 NEOANTIMYCIN

A further modification of the amino acid L-phenylalanine to which
reference should be made is the structure 3,4-dihydroxy-2,2-
dimethyl-5-phenylvaleric acid (104) which occurs in the cyclic
polyester antibiotic neoantimycin[222] from *Streptoverticillium*

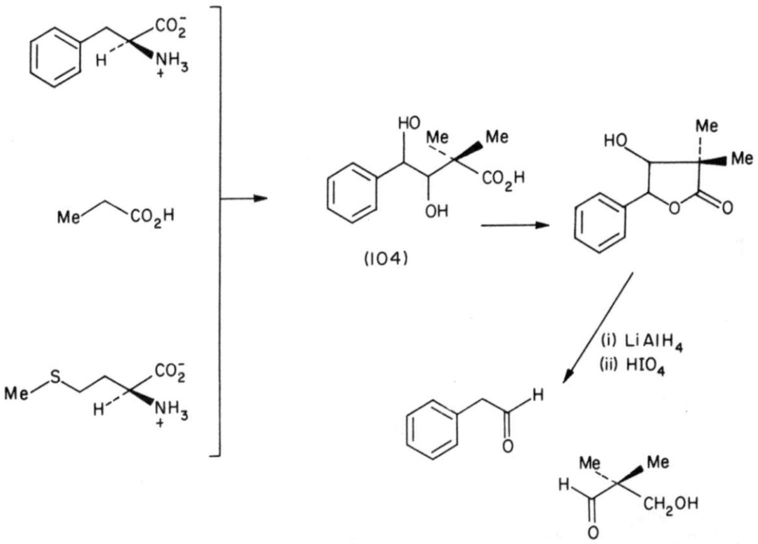

Figure 3.19. Structure, biosynthesis and chemical degradation of 3,4-dihydroxy-
2,2-dimethyl-5-phenylvaleric acid[222, 223]

orinoci. The phenylvaleric acid (104) is formed[223] from L-
phenylalanine, propionic acid which contributes one of the *gem*-
dimethyl groups, and L-methionine which contributes the second.
Isotopic tracer experiments followed by chemical degradation of
(104) permitted this biosynthetic pathway to be formulated[223]
(*Figure 3.19*). This structure represents the first recorded example of
a natural product formed by condensation of a $C_6 . C_3$ and a C_3
unit.

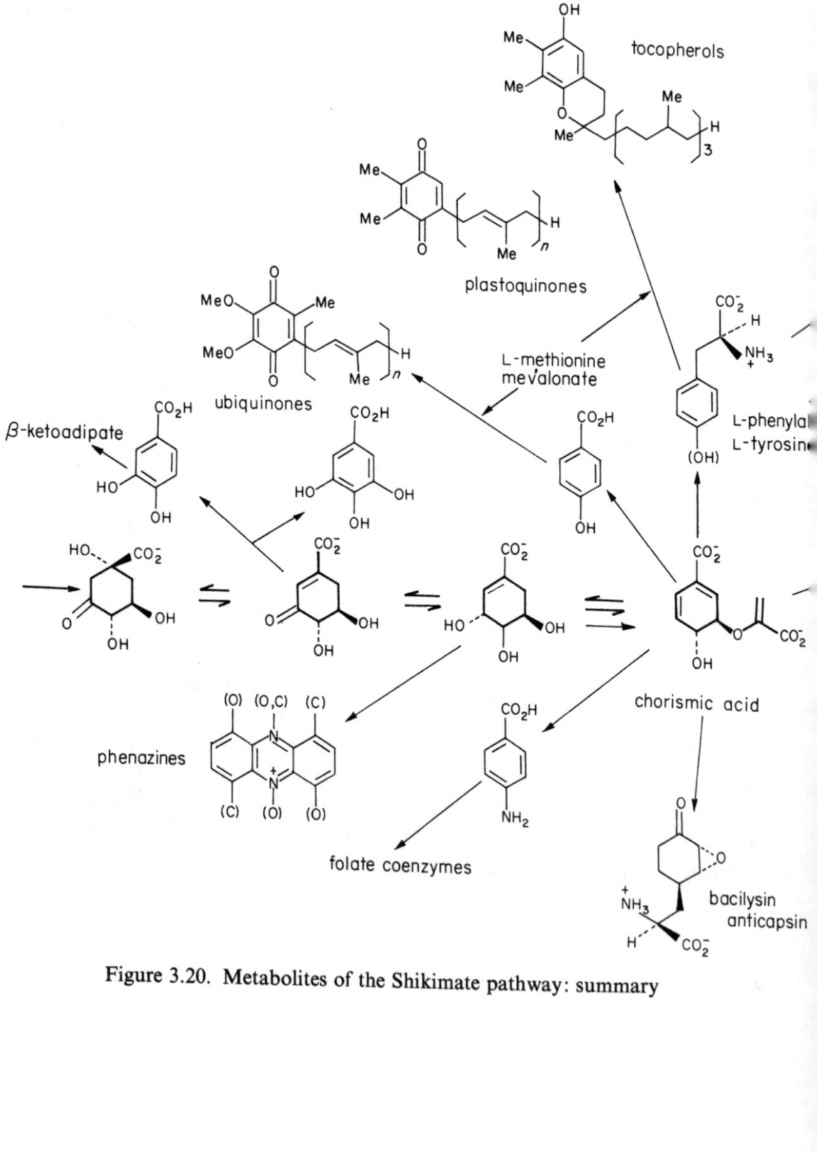

120

Figure 3.20. Metabolites of the Shikimate pathway: summary

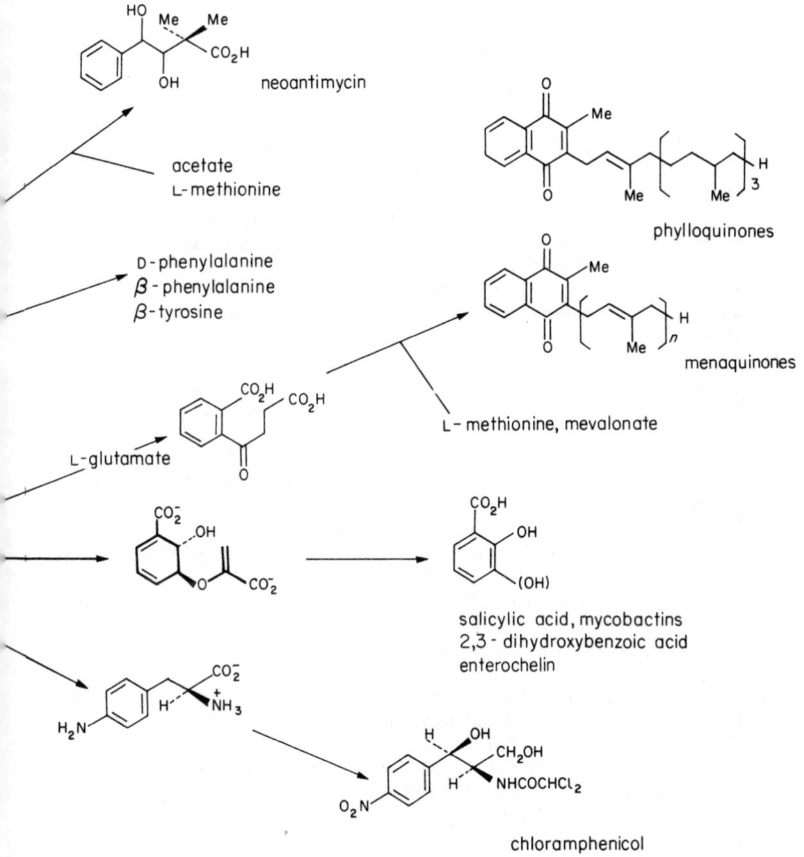

neoantimycin

acetate
L-methionine

phylloquinones

D-phenylalanine
β-phenylalanine
β-tyrosine

menaquinones

L-glutamate

L-methionine, mevalonate

salicylic acid, mycobactins
2,3-dihydroxybenzoic acid
enterochelin

chloramphenicol

REFERENCES

1. Davis, B. D. (1950). *Experentia*, **6**, 41
2. Davis, B. D. (1950). *Nature*, **166**, 1120
3. Davis, B. D. (1949). *Proc. Natl Acad. Sci.*, **35**, 1
4. Davis, B. D. (1951). *J. Biol. Chem.*, **191**, 315
5. Redfearn, E. R. (1961). Ciba Foundation Symp. *Quinones in Electron Transport*, p. 346. Edited by G. E. W. Wolstenholme and C. M. O'Connor. London; Churchill
6. Green, D. E. (1961). Ciba Foundation Symp. *Quinones in Electron Transport*, p. 130. Edited by G. E. W. Wolstenholme and C. M. O'Connor. London; Churchill
7. Clark, V. M., Kirby, G. W., Todd, A. R. and Hutchinson, D. W. (1961). *J. Chem. Soc.*, 715
8. Crane, F. L. (1965). *Biochemistry of Quinones*, p. 183. Edited by R. B. Morton. London; Academic Press
9. Threlfall, D. R. and Whistance, G. R. (1971). *Aspects of Terpenoid Chemistry and Biochemistry*, p. 357. Edited by T. W. Goodwin. London; Academic Press
10. Moore, H. W. and Folkers, K. (1966). *J. Amer. Chem. Soc.*, **88**, 567
11. Pennock, J. F. (1966). *Vitamins and Hormones*, **24**, 307
12. Nilsson, J. L. G., Farley, T. M. and Folkers, K. (1968). *Analyt. Biochem.*, **23**, 422
13. Redfearn, E. R. (1965). *Biochemistry of Quinones*, p. 149. Edited by R. B. Morton. London; Academic Press
14. Trenner, N. R., Arison, B. H., Erickson, R. E., Shunk, C. H., Wolf, D. E. and Folkers, K. (1959). *J. Amer. Chem. Soc.*, **81**, 2026
15. Kofler, M., Langemann, A., Ruegg, R., Jean-dit-Chopard, L. H., Raymond. A. and Isler, O. (1959). *Helv. Chim. Acta*, **42**, 2252
16. Wallwork, J. C. and Pennock, J. F. (1968). *Chem. and Ind.*, 1571
17. Dunphy, P. J., Whittle, K. J. and Pennock, J. F. (1966). *Biochemistry of Chloroplasts*, Volume I, p. 165. Edited by T. W. Goodwin. London; Academic Press
18. Green, J. and McHale, D. (1965). *Biochemistry of Quinones*, p. 261. Edited by R. B. Morton. London; Academic Press
19. Weber, M. M., Matschiner, J. T. and Peck, H. D. (1970). *Biochem. Biophys. Res. Commun.*, **38**, 197
20. Maroc, J., Azerad, R., Kamen, D. and LeGall, J. (1970). *Biochim. Biophys. Acta*, **197**, 87
21. Doisy, E. A. and Matschiner, J. T. (1965). *Biochemistry of Quinones*, p. 317. Edited by R. B. Morton. London; Academic Press
22. Hammond, R. K. and White, D. C. (1969). *J. Chromatog.*, **45**, 446
23. Cornforth, J. W. and Popjak, G. (1966). *Biochem. J.*, **101**, 553
24. Clayton, R. B. (1965). *Quart. Rev.*, **19**, 168, 201
25. Biochemical Society Symp. No. 29. (1970). *Natural Substances formed Biologically from Mevalonic Acid*. Edited by T. W. Goodwin. London; Academic Press
26. Glover, J. (1965). *Biochemistry of Quinones*, p. 207. Edited by R. B. Morton. London; Academic Press
27. Stone, K. J. and Hemming, F. W. (1967). *Biochem. J.*, **104**, 43
28. Dada, O. A., Threlfall, D. R. and Whistance, G. R. (1968). *Eur. J. Biochem.*, **4**, 329
29. Griffiths, W. T., Threlfall, D. R. and Goodwin, T. W. (1968). *Eur. J. Biochem.*, **5**, 124
30. Hemming, F. W., Morton, R. A. and Pennock, J. F. (1963). *Proc. Roy. Soc.*, **158B**, 291

31. Allen, C. M., Alworth, W., MacRae, A. and Block, K. (1967). *J. Biol. Chem.*, **242**, 1895
32. Winrow, M. J. and Rudney, H. (1969). *Biochim. Biophys. Acta*, **37**, 833
33. Raman, T. S., Rudney, H. and Buzzelli, N. K. (1969). *Arch. Biochem. Biophys.*, **130**, 164
34. Momose, R. and Rudney, H. (1972). *J. Biol. Chem.*, **247**, 3930
35. Hamilton, J. A. and Cox, G. B. (1971). *Biochem. J.*, **123**, 435
36. Cox, G. B. and Gibson, F. (1964). *Biochim. Biophys. Acta*, **93**, 204
37. Cox, G. B. and Gibson, F. (1966). *Biochem. J.*, **100**, 1
38. Parsons, W. W. and Rudney, H. (1964). *Proc. Natl Acad. Sci.*, **51**, 444
39. Parsons, W. W. and Rudney, H. (1965). *J. Biol. Chem.*, **240**, 1855
40. Olson, R. E., Bentley, R., Aiyar, A. S., Dialameh, G. H., Gold, P. H., Ramsey, V. G. and Springer, C. M. (1963). *J. Biol. Chem.*, **238**, PC 3146
41. Whistance, G. R., Threlfall, D. R. and Goodwin, T. W. (1966). *Biochem. J.*, **101**, 5P
42. Rudney, H. and Raman, T. S. (1966). *Vitamins and Hormones*, **24**, 531
43. AhLaw, Threlfall, D. R. and Whistance, G. R. (1970). *Biochem. J.*, **123**, 331
44. Spiller, G. H., Threlfall, D. R. and Whistance, G. R. (1968) *Arch. Biochem. Biophys.*, **125**, 786
45. Threlfall, D. R. and Whistance, G. R. (1970). *Phytochemistry*, **9**, 355
46. Powls, R. and Hemming, F. W. (1966). *Phytochemistry*, **5**, 1249
47. Whistance, G. R. and Threlfall, D. R. (1967). *Biochem. Biophys. Res. Commun.*, **28**, 295
48. Gibson, M. I. and Gibson, F. (1962). *Biochim. Biophys. Acta*, **65**, 160
49. Olson, R. E. (1966). *Vitamins and Hormones*, **24**, 251
50. Whistance, G. R., Threlfall, D. R. and Goodwin, T. W. (1967). *Biochem. J.*, **104**, 145
51. Olsen, R. K., Smith, J. L., Daves, G. D., Moore, H. W., Folkers, K., Parson, W. W. and Rudney, H. (1965). *J. Amer. Chem. Soc.*, **87**, 2298
52. Olsen, R. K., Daves, G. D., Moore, H. W., Folkers, K. and Rudney, H. (1966)., *J. Amer. Chem. Soc.*, **88**, 2346
53. Parsons, W. W. and Rudney, H. (1965). *Proc. Natl Acad. Sci.*, **53**, 599
54. Friis, P., Nilsson, J. L. G., Daves, G. D. and Folkers, K. (1967). *Biochem. Biophys. Res. Commun.*, **28**, 324
55. Friis, P., Daves, G. D. and Folkers, K. (1966). *J. Amer. Chem. Soc.*, **88**, 4754
56. Olsen, R. K., Daves, G. D., Moore, H. W., Folkers, K. W., Parson, W. W. and Rudney, H. (1966). *J. Amer. Chem. Soc.*, **88**, 5919
57. Jackman, L. M., O'Brien, I., Cox, G. B. and Gibson, F. (1967). *Biochim. Biophys. Acta*, **141**, 1
58. Threlfall, D. R., Whistance, G. R. and Goodwin, T. W. (1968). *Biochem. J.*, **106**, 107
59. Threlfall, D. R., Whistance, G. R. and Brown, B. S. (1969). *Biochim. Biophys. Acta*, **176**, 895
60. Threlfall, D. R., Whistance, G. R. and Brown, B. S. (1970). *Biochem. J.*, **117**, 119
61. Whistance, G. R., Threlfall, D. R. and Dillon, J. F. (1969). *Biochem. J.*, **111**, 461
62. Nowicki, H. G., Dialameh, G. H. and Olson. R. E. (1972). *Biochemistry*, **11**, 896
63. Cox, G. B., Young, I. G., McCann, L. M. and Gibson, F. (1969). *J. Bact.*, **99**, 450
64. Young, I. G., McCann, L. M., Stroobant, P. and Gibson, F. (1971). *J. Bact.*, **105**, 769
65. Cox, G. B., Newton, N. A., Gibson, F., Snoswell, A. M. and Hamilton, J. A. (1970). *Biochem. J.*, **117**, 551
66. Stroobant, P., Young, I. G. and Gibson, F. (1972). *J. Bact.*, **109**, 134

67. Young, I. G., Leppik, R. A., Hamilton, J. A. and Gibson, F. (1972). *J. Bact.*, **110**, 18
68. Whistance, G. R. and Threlfall, D. R. (1968). *Biochem. J.*, **109**, 482, 577
69. Whistance, G. R. and Threlfall, D. R. (1970). *Biochem. J.*, **117**, 593
70. Evans, W. C. (1963). *J. Gen. Microbiol.*, **32**, 177
71. Whistance, G. R. and Threlfall, D. R. (1971). *Phytochemistry*, **10**, 1533
72. Zenk, M. H. and Leistner, E. (1967). *Z. Naturforsch.*, **22B**, 460
73. Zenk, M. H. and Leistner, E. (1968). *Z. Naturforsch.*, **23B**, 259
74. Leistner, E., Schmitt, J. H. and Zenk, M. H. (1967). *Biochem. Biophys. Res. Commun.*, **28**, 245
75. Campbell, I. M., Coscia, C. J., Kelsey, M. and Bentley, R. (1967). *Biochem. Biophys. Res. Commun.*, **28**, 25
76. Guerin, M., Leduc, M. M. and Azerad, R. G. (1970). *Eur. J. Biochem.*, **15**, 421
77. Campbell, I. M., Robins, D. J., Kelsey, M. and Bentley, R. (1971). *Biochemistry*, **10**, 3069
78. Campbell, I. M., Robins, D. J. and Bentley, R. (1970). *Biochem. Biophys. Res. Commun.*, **39**, 1081
79. Scharf, K. H., Zenk, M. H., Floss, H. G., Onderka, K. D. and Carroll, M. (1971). *Chem. Commun.*, 576
80. Leduc, M. M., Dansette, M. P. and Azerad, R. G. (1971). *Eur. J. Biochem.*, **15**, 428
81. Young, I. G., Batterham, T. J. and Gibson, F. (1968). *Biochim. Biophys. Acta*, **165**, 567
82. Young, I. G., Batterham, T. J. and Gibson, F. (1969). *Biochim. Biophys. Acta*, **177**, 389
83. Richards, J. H. and Hendrickson, J. B. (1964). *The Biosynthesis of Steroids, Terpenes and Acetogenins*, p. 140. New York and Amsterdam; Benjamin
84. Neish, A. C. (1964). *Biochemistry of Phenolic Compounds*, p. 295. Edited by J. B. Harborne. London and New York; Academic Press
85. Corbin, J. L. and Bulen, W. A. (1968). *Biochemistry*, **8**, 757
86. Ito, T. and Neilands, J. B. (1958). *J. Amer. Chem. Soc.*, **80**, 4645
87. Brot, N., Goodwin, J. and Fales, H. (1966). *Biochem. Biophys. Res. Commun.*, **25**, 454
88. Brot, N. and Goodwin, J. (1968). *J. Biol. Chem.*, **243**, 510
89. O'Brien, I. G., Cox, G. B. and Gibson, F. (1970). *Biochim. Biophys. Acta*, **201**, 453
90. O'Brien, I. G. and Gibson, F. (1970). *Biochim. Biophys. Acta*, **215**, 393
91. Dyer, J. R., Heding, H. and Schaffner, C. P. (1964). *J. Org. Chem.*, **29**, 2802
92. Catlin, E. R., Hassall, C. H. and Pratt, B. C. (1968). *Biochim. Biophys. Acta*, **156**, 109
93. Ratledge, C. (1967). *Biochim. Biophys. Acta*, **141**, 55
94. Pollack, J. R. and Nielands, J. B. (1970). *Biochem. Biophys. Res. Commun.*, **38**, 989
95. Young, I. G., Langman, L., Luke, R. K. J. and Gibson, F. (1971). *J. Bact.*, **106**, 51
96. Cox, G. B., Luke, R. K. J., Newton, N. A., Gibson, F., O'Brien, I. G. and Rosenberg, H. (1970). *J. Bact.*, **104**, 219
97. Luke, R. K. J. and Gibson, F. (1971). *J. Bact.*, **107**, 557
98. Young, I. G., Gibson, F. and Cox, G. B. (1967). *Biochim. Biophys. Acta*, **141**, 319
99. Young, I. G., Gibson, F. and Jackman, L. (1967). *Biochim. Biophys. Acta*, **148**, 313
100. Young, I. G., Gibson, F. and Jackman, L. (1967). *Biochim. Biophys. Acta*, **177**, 381

101. Young, I. G. and Gibson, F. (1969). *Biochim. Biophys. Acta*, **177**, 348
102. Young, I. G. and Gibson, F. (1969). *Biochim. Biophys. Acta*, **177**, 401
103. Kjaer, A. and Larsen, P. O. (1963). *Acta Chem. Scand.*, **17**, 2397
104. Larsen, P. O. (1966). *Biochim. Biophys. Acta*, **115**, 529
105. Larsen, P. O. (1967). *Biochim. Biophys. Acta*, **141**, 27
106. Larsen, P. O., Onderka, D. K. and Floss, H. G. (1972). *Chem. Commun.*, 842
107. Hudson, A. T. and Bentley, R. K. (1970). *Tetrahedron Lett.*, 2077
108. Hudson, A. T. and Bentley, R. K. (1970). *Biochemistry*, **9**, 3984
109. Hudson, A. T. (1971). *Phytochemistry*, **10**, 1555
110. Hudson, A. T., Bentley, R. K. and Campbell, I. M. (1970). *Biochemistry*, **9**, 3988
111. Ratledge, C. (1969). *Biochim. Biophys. Acta*, **192**, 148
112. Marshall, B. J. and Ratledge, C. (1972). *Biochim. Biophys. Acta*, **264**, 106
113. Ratledge, C. and Winder, F. G. (1962). *Biochem. J.*, **84**, 501
114. Snow, G. A. (1965). *Biochem. J.*, **97**, 166
115. Snow, G. A. and White, J. A. (1969). *Biochem. J.*, **115**, 1031
116. Snow, G. A. (1970). *Bacteriol. Rev.*, **34**, 99
117. Ornston, L. N. and Stanier, R. Y. (1966). *J. Biol. Chem.*, **241**, 3776
118. Ornston, L. N. (1966). *J. Biol. Chem.*, **241**, 3787, 3795, 3800
119. Hayaishi, O. (1966). *Bacteriol. Rev.*, **30**, 720
120. Hassall, B. E., Darrah, J. A. and Cain, R. B. (1969). *Biochem. J.*, **114**, 75P=
121. Cain, R. B. (1972). *Biochem. J.*, **127**, 15P
122. Rann, D. L. and Cain, R. B. (1969). *Biochem. J.*, **114**, 77P
123. Cain, R. B., Bilton, B. F. and Darrah, J. A. (1968). *Biochem. J.*, **108**, 797
124. Cánovas, J. L. and Stanier, R. Y. (1967). *Eur. J. Biochem.*, **1**, 289
125. Cánovas, J. L., Whellis, M. L. and Stanier, R. Y. (1968). *Eur. J. Biochem*, **3**, 293
126. Tresguerres, M. E. F., de Torrontequi, G. and Cánovas, J. L. (1970). *Arch. Microbiol.*, **70**, 110
127. Tresguerres, M. E. F., de Torrontequi, G., Ingledew, W. M. and Cánovas, J. L. (1970). *Eur. J. Biochem.*, **14**, 445
128. Butkewitsch, W. (1924). *Biochem. Z.*, **145**, 442
129. Butkewitsch, W. (1925). *Biochem. Z.*, **159**, 395
130. Bernauer, K. and Waelsch, H. H. (1932). *Biochem. Z.*, **249**, 223
131. Bernauer, K. and Gorlich, B. (1935). *Biochem. Z.*, **280**, 394
132. Giles, N. H., Partridge, C. W. H., Ahmed. S. I. and Cass, M. E. (1967). *Proc. Natl Acad. Sci.*, **58**, 1930
133. Ahmed, S. I. and Giles, N. H. (1969). *J. Bact.*, **99**, 231
134. Gross, S. R. (1958). *J. Biol. Chem.*, **233**, 1146
135. Mitsuhashi, S. and Davis, B. D. (1954). *Biochim. Biophys. Acta*, **15**, 268
136. Tatum, E. L., Gross, S. R. and Gafford, R. D. (1956). *J. Biol. Chem.*, **219**, 797
137. Scharf, K. H., Zenk, M. H., Onderka, D. K., Carroll, M. and Floss, H. G. (1971). *Chem. Commun.*, 765
138. Sprinson, D. B. (1961). *Adv. Carb. Chem.*, **15**, 235
139. Albrecht, K. and Bernhard, H. (1947). *Helv. Chim. Acta*, **30**, 627
140. Brucker, W. (1954). *Naturwiss*, **41**, 309
141. Brucker, W. and Hashem, M. (1962). *Flora*, **157**, 57
142. Knowles, P. F., Haworth, R. D. and Haslam, E. (1961). *J. Chem. Soc.*, 1854
143. Smith, B. W. and Haslam, E., unpublished observations
144. Hartman, S. C., Levenberg, B. and Buchanan, J. M. (1956). *J. Amer. Chem. Soc.*, **78**, 221, 1057
145. Levenberg, B. and Buchanan, J. M. (1957). *J. Biol. Chem.*, **224**, 1005, 1019
146. Flaks, J. G., Erwin, M. J. and Buchanan, J. M. (1957). *J. Biol. Chem.*, **229**, 603
147. Magasnik, B., Moyed, H. S. and Gehring, L. B. (1957). *J. Biol. Chem.*, **226**, 339

126   Metabolites of the Shikimate Pathway

148. Griffin, M. J. and Brown, G. M. (1964). *J. Biol. Chem.*, **239**, 310, 317, 326
149. Bloch, A. and Coutesogeorgopoulos, C. (1966). *Biochemistry*, **5**, 3345
150. Gibson, F. and Pittard, J. (1968). *Bacteriol. Rev.*, **32**, 465
151. Srinivasan, P. R. and Weiss, B. (1959). *Proc. Natl Acad. Sci.*, **45**, 1491
152. Srinivasan, P. R. and Weiss, B. (1961). *Biochim. Biophys. Acta*, **51**, 597
153. Gibson, F., Gibson, M. and Cox, G. B. (1964). *Biochim. Biophys. Acta*, **82**, 637
154. Hendler, S. and Srinivasan, P. R. (1967). *Biochim. Biophys. Acta*, **141**, 656
155. Altendorf, K. H., Bacher, A. and Lingens, F. (1969). *Z. Naturforsch.*, **24B**, 1602
156. Kane, J. F., Holmes, W. H. and Jensen, R. A. (1972). *J. Biol. Chem.*, **247**, 1587
157. Kane, J. F. and Jensen, R. A. (1970). *Biochem. Biophys. Res. Commun.*, **41**, 328
158. Gerber, N. N. (1966). *Biochemistry*, **5**, 3824
159. Gerber, N. N. (1967). *J. Org. Chem.*, **32**, 4055
160. Gerber, N. N. and Lechevalier, M. P. (1965). *Biochemistry*, **4**, 176
161. Gerber, N. N. and Lechevalier, M. P. (1964). *Biochemistry*, **3**, 598
162. Kluyver, A. J. (1956). *J. Bact.*, **72**, 406
163. Olson, E. S. and Richards, J. H. (1967). *J. Org. Chem.*, **32**, 2887
164. Bentley, R. K. and Holliman, F. G. (1966). *Chem. Commun.*, 312
165. Swan, G. A. and Felton, D. G. I. (1957). *Phenazines*, p. 182. NewYork; Interscience
166. Nakamura, S., Maeda, K. and Umeza, H. (1964). *J. Antibiotics* (Tokyo), **17A**. 33
167. Carter, R. E. and Richards, J. H. (1969). *J. Amer. Chem. Soc.*, **83**, 495
168. Blackwood, A. C. and Neish, A. C. (1957). *Can. J. Microbiol.*, **3**, 165
169. Ingram, J. M. and Blackwood, A. C. (1962). *Can. J. Microbiol.*, **8**, 49
170. Millican, R. C. (1962). *Biochim. Biophys. Acta*, **57**, 407
171. Sheikh, N. M. and McDonald, J. C. (1964). *Can. J. Microbiol.*, **10**, 861
172. Levitch, M. E. and Stadtman, E. R. (1964). *Arch. Biochem. Biophys.*, **106**, 194
173. Levitch, M. E. and Reitz, P. (1966). *Biochemistry*, **5**, 689
174. Podojil, M. and Gerber, N. N. (1967). *Biochemistry*, **6**, 2701
175. Podojil, M. and Gerber, N. N. (1970). *Biochemistry*, **9**, 4616
176. McIlwain, H. (1937). *J. Chem. Soc.*, 1704
177. Holliman, F. G. (1961). *S. African Ind. Chem.*, **15**, 233
178. Flood, M. E., Herbert, R. B. and Holliman, F. G. (1970). *Chem. Commun.*, 1514
179. Flood, M. E., Herbert, R. B. and Holliman, F. G. (1972). *J. Chem. Soc.* (Perkin I), 622
180. Hansford, G. S., Holliman, F. G. and Herbert, R. B. (1972). *J. Chem. Soc.* (Perkin I), 103
181. DeMoss, R. D. and Frank, L. H. (1959). *J. Bact.*, **77**, 776
182. Holliman, F. G. (1969). *J. Chem. Soc.*, (C), 2514
183. Herbert, R. B. and Holliman, F. G. (1969). *J. Chem. Soc.* (C), 2517
184. Fraser, R. R. and Renaud, R. N. (1966). *J. Amer. Chem. Soc.*, **88**, 4365
185. Challand, S. R., Herbert, R. B. and Holliman, F. G. (1970). *Chem. Commun.*, 1423
186. Rogers, H. J., Newton, G. G. F. and Abraham, E. P. (1965). *Biochem. J.*, **97**, 573
187. Rogers, H. J., Lomakina, N. and Abraham, E. P. (1965). *Biochem. J.*, **97**, 579
188. Walker, J. E. and Abraham, E. P. (1970). *Biochem. J.*, **118**, 557
189. Walker, J. E. and Abraham, E. P. (1970). *Biochem. J.*, **118**, 563
190. Neuss, N., Molloy, B. B., Shah, R. and DeLattiguera, N. (1970). *Biochem. J.*, **118**, 571
191. Roscoe, J. and Abraham, E. P. (1966). *Biochem. J.*, **99**, 793

192. Pestka, S. (1971). *Ann. Rev. Microbiol.*, **25**, 487
193. Gottlieb, D., Bhattacharyya, P. K., Anderson, H. W. and Carter, H. E. (1948). *J. Bact.*, **55**, 409
194. Gottlieb, D., Carter, H. E., Legator, M. and Gallicchio, V. (1954). *J. Bact.*, **68**, 243
195. Gottlieb, D., Robbins, P. W. and Carter, H. E. (1956). *J. Bact.*, **72**, 153
196. Gottlieb, D., Carter, H. E., Robbins, P. W. and Burg, R. W. (1962). *J. Bact.*, **84**, 888
197. McGrath, R., Vining, L. C., Sala, F. and Westlake, D. W. S. (1968). *Can. J. Biochem.*, **46**, 587
198. Vining, L. C. and Westlake, D. W. S. (1964). *Can. J. Microbiol.*, **10**, 705
199. Siddiqueullah, M. R., McGrath, R., Vining, L. C., Sala, F. and Westlake, D. W. S. (1967). *Can. J. Biochem.*, **45**, 1881
200. Vining, L. C., Malik, V. S. and Westlake, D. W. S. (1968). *Lloydia*. **31**, 355
201. Rebstock, M. C., Crooks, H. M., Controulis, J. and Bartz, Q. R. (1949). *J. Amer. Chem. Soc.*, **71**, 2458
202. Perlman, D. and Bodanszky, M. (1971). *Ann. Rev. Biochem.*, **40**, 449
203. Free, C. A., Julius, M., Arnow, P. and Barry, G. T. (1967). *Biochim. Biophys. Acta*, **146**, 608
204. Diven, W. F., Scholz, J. J. and Johnston, R. B. (1964). *Biochim. Biophys. Acta*, **85**, 322
205. Glaser, L. (1960). *J. Biol. Chem.*, **235**, 2095
206. Behrmann, E. J. (1962). *Nature*, **196**, 150
207. Bodanszky, M. and Perlman, D. (1968). *Nature*, **218**, 291
208. Katz, E. (1968). *Lloydia*, **31**, 364
209. Roskoski. R., Gevers, W., Kleinkauf, H. and Lipmann, F. (1970). *Biochemistry*, **9**, 4839, 4846
210. Kleinkauf, H. and Gevers, W. (1969). *Cold Spring Harbour Symp. Quant. Biol.*, **34**, 805
211. Kurahashi, K., Yamada, M., Mori, K., Fujikawa, K., Kambe, M., Imae, Y., Sato, E., Takahashi, H. and Sakamoto, Y. (1969). *Cold Spring Harbour Symp. Quart. Biol.*, **34**, 815
212. Yamada, M. and Kurahashi, K. (1968). *J. Biochem.* (Tokyo), **63**, 59
213. Yamada, M. and Kurahashi, K. (1969). *J. Biochem.* (Tokyo), **66**, 529
214. Berg, T. L., Frøholm, L. O. and Laland, S. G. (1966). *Biochem. J.*, **96**, 43
215. Tomino, S., Yamada, M., Itoh, H. and Kurahashi, K. (1967). *Biochemistry*, **6**, 2552
216. Marumo, S. (1959). *Bull. Agr. Chem. Soc.* (Japan), **23**, 428
217. Bohman-Lindgren, G. (1970). *Tetrahedron Lett.*, 3065
218. Bohman-Lindgren, G. (1972). *Tetrahedron*, **28**, 4625
219. Kurylo-Borowska, Z. and Abramsky, T. (1972). *Biochim. Biophys. Acta*, **264**, 1
220. Chirpich, T. P., Zappia, V., Costilow, R. N. and Barker, H. A. (1970). *J. Biol. Chem.*, **245**, 1778
221. Chirpich, T. P., Zappia, V. and Barker, H. A. (1970). *Biochim. Biophys. Acta*, **207**, 505
222. Caglioti, L., Misiti, D., Mondelli, R., Selva, A., Arcamone, F. and Cassinelli, G. (1969). *Tetrahedron*, **25**, 2193
223. Caglioti, L., Cirani, G., Misiti, D., Arcamone, F. and Minghetti, A. (1972). *J. Chem. Soc.* (Perkin I), 1235

4

Metabolism of the Aromatic Amino Acids in Micro-organisms and Higher Organisms

4.1 INTRODUCTION

Protein ingested by a living organism generally undergoes hydrolysis and the liberated amino acids join the various pools of free amino acids in the organism. These pools may be supplemented in appropriate situations by amino acids biosynthesised by the organism itself and from them amino acids are withdrawn for the synthesis of new protein and for various biosynthetic operations. The aromatic amino acids are unique among amino acids for the range of biosynthetic processes in which they engage. They are the parent substances in many organisms for several physiologically important metabolites which are derived by oxidative pathways from the amino acids. Thus L-tryptophan yields 5-hydroxytryptamine (serotonin, 1)—a powerful vasoconstrictor found in mammalian tissue, the plant hormone indole acetic acid, the ommochromes which are insect eye pigments and the amino acid is also a precursor of nicotinic acid and hence of the pyridine ring of some co-enzymes. The hormones adrenalin (2), noradrenalin and thyroxine (3) are similarly derived by oxidative metabolism of L-phenylalanine and L-tyrosine. Under appropriate conditions all the amino acids may, in addition, be metabolised to yield simpler metabolites such as succinate, fumarate, acetoacetate and acetate which can be further metabolised by entering the Krebs cycle or by conversion to fatty acids. In the case of L-phenylalanine and L-tyrosine the recognition of this property gave rise historically to the definition that these two amino acids were ketogenic, i.e. that they gave rise to acetoacetic acid and other ketonic fragments in mammalian systems.

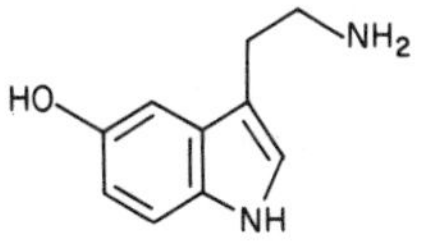

(I, serotonin) 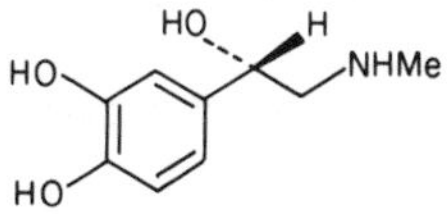   (2, adrenalin) 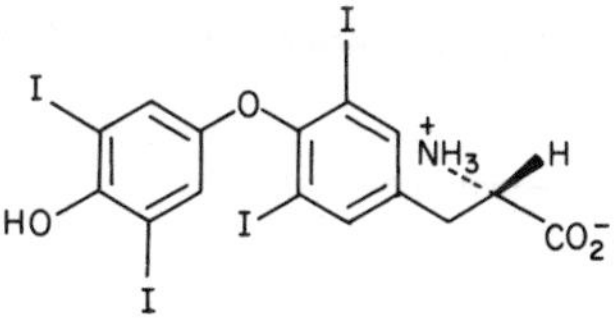   (3, thyroxine)

In this chapter the metabolism, and particularly that which occurs by oxidative pathways, of the aromatic amino acids is discussed with particular reference to higher organisms and micro-organisms. Both L-tryptophan and L-phenylalanine are essential amino acids in higher organisms and they must be provided in the diet. This is a reflection of the absence in these organisms of the enzymes of the shikimate pathway and hence of an inability to synthesise aromatic nuclei from carbohydrate precursors. In higher organisms, L-tyrosine is not normally an essential metabolite since L-phenylalanine is readily transformed to L-tyrosine by the enzyme L-phenylalanine hydroxylase. However, in the congenital biochemical disorder phenylketonuria, in which this conversion is inhibited, L-tyrosine can become essential.

4.2 METABOLISM OF L-PHENYLALANINE AND L-TYROSINE

The oxidative metabolism of L-phenylalanine and L-tyrosine is of intrinsic interest from the medical point of view since a number of diseases resulting from 'inborn errors of metabolism' are associated with this pathway. Some indication of the intermediates in the pathways of oxidative metabolism of L-phenylalanine and L-tyrosine indeed came first from a study of one such disease, alcaptonuria, in which the patient excretes homogentisic acid. Garrod recognised[1] that these errors of metabolism were due to a congenital inability of the organism to carry out a particular enzyme transformation in the normal sequence of events; it is a natural mutant.

4.2.1 CONVERSION OF L-PHENYLALANINE TO L-TYROSINE

Under normal physiological conditions the bulk of L-phenylalanine is oxidatively metabolised by its initial transformation to L-tyrosine[2] which is then oxidised to acetoacetate, fumarate and carbon dioxide (*Figure 4.1*). The aryl hydroxylation reaction probably takes place in all higher animals and occurs in the liver[3, 4]. The same conversion has been detected in some plants[5-7] and bacteria[8, 9] but in most plants and micro-organisms L-phenylalanine and L-tyrosine are synthesised independently by the shikimate pathway and normally interconversion is not observed.

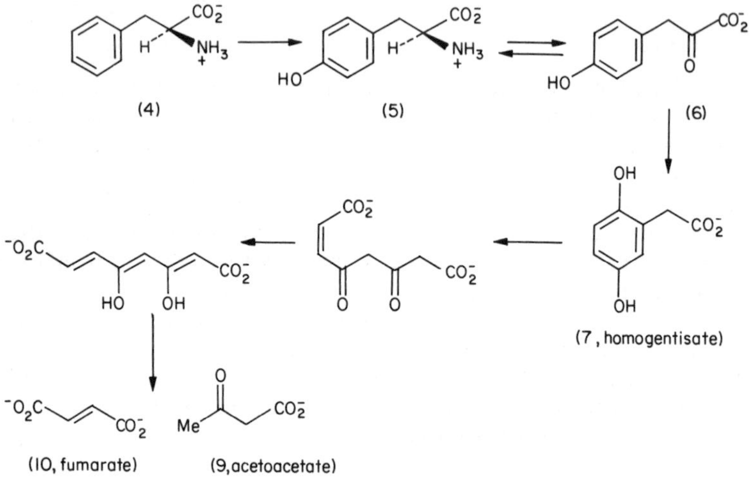

Figure 4.1. Metabolism of L-phenylalanine and L-tyrosine in mammals

The enzyme L-phenylalanine hydroxylase has been isolated and purified (85–95 per cent) from rat liver[10, 11] and its properties have been fully reviewed by Kaufman[12]. It has been shown to exist as two electrophoretically distinguishable iso-enzymes each of which may exist as a monomer (molecular weight 51 000–55 000), dimer or tetramer. Substrates for the enzyme must have an intact alanine side chain but some substitutions and modifications of the phenyl ring are permissible. L-Phenylalanine hydroxylase has also been isolated from Pseudomonas species[8] (ATCC 11299a) and this enzyme shows certain similarities (and some differences) to the enzyme from mammalian tissue. All purified preparations of the bacterial enzyme showed an absolute requirement for ferrous ions and addition of NADH to these preparations increased the production of L-tyrosine twofold. Both mammalian and bacterial enzymes operate as mixed function oxidases utilising a source of molecular oxygen and a tetrahydropteridine as cofactor. Reduced pteridines have been implicated as cofactors in a number of oxygenase and hydroxylation reactions [13, 14], and the rat liver enzyme cofactor has been identified as biopterin (11)[15], and the principal cofactor in the aryl hydroxylase from Pseudomonas species as L-threoneopterin (12)[9].

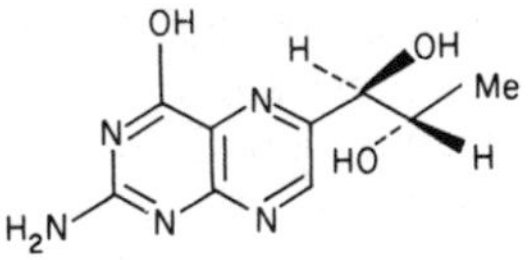

(II, biopterin)   (I2, L-threoneopterin)

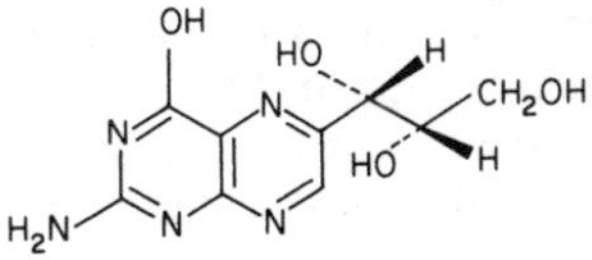

The hydroxylation of aromatic substrates by enzymes such as L-phenylalanine hydroxylase has been extensively studied in the National Institute of Health, Bethesda, USA, and it has been shown[16, 17] that the hydroxylation reaction is frequently accompanied by an intramolecular migration of the group or atom which is being displaced to an adjacent ortho position on the aromatic ring. This effect has been appropriately termed the NIH shift and detailed study of this type of reaction has led to some important conclusions regarding the precise nature of the chemical reactions involved in aromatic hydroxylation. In particular it has led to the suggestion that these enzymic reactions may proceed via arene

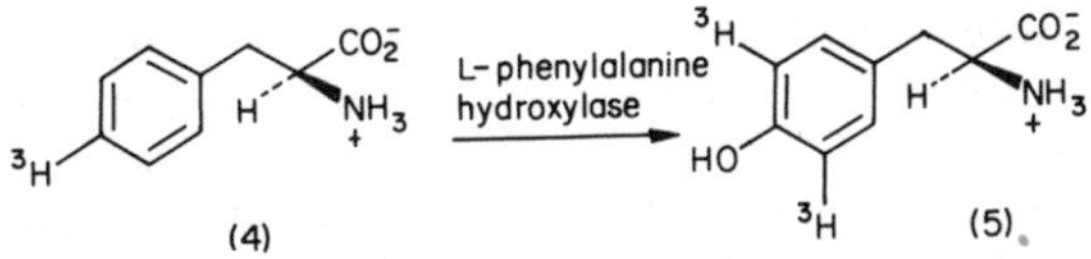

(4)   (5)

oxide intermediates[17-19]. One of the first enzyme controlled reactions of this type to be studied was in fact the conversion of L-phenylalanine to L-tyrosine by L-phenylalanine hydroxylase. Thus using 4'-[³H]-L-phenylalanine (4) as substrate L-tyrosine (5) was isolated which retained over 90 per cent of the tritium in the 3',5' positions[20].

4.2.2 DEGRADATION OF L-TYROSINE

Oxidative degradation of L-tyrosine is initiated by transamination to yield p-hydroxyphenylpyruvate (6) which is then metabolised in a mechanistically novel way to give homogentisate (7), *Figure 4.1*. The unusual nature of this reaction caused an initial reluctance to accept homogentisate (7) as an intermediate in L-tyrosine metabolism.

A single copper containing protein with a requirement for ascorbic acid—p-hydroxyphenylpyruvate oxidase—catalyses the simultaneous hydroxylation of the phenyl ring, migration of the aliphatic side chain to an adjacent position on the ring, and oxidative decarboxylation of the pyruvate fragment[21]. The enzyme has been classified as a mono-oxygenase on the basis of data from the experiments of Mason and his collaborators[22] who showed oxygen to be incorporated into homogentisate both from molecular oxygen and from water. On the basis of these observations mechanisms for the reaction have been discussed by Crandall[23] and Soloway[24]. Subsequently Lindblad, Lindstedt and Lindstedt[25] established that ^{18}O is incorporated into *both* the new phenolic hydroxyl group and the carboxyl group of homogentisate (7) from ^{18}O labelled molecular oxygen. Isotopic oxygen was incorporated into the carboxyl group of more than 95 per cent of the homogentisate molecules. The metabolite, however, contained ^{18}O in the new phenolic hydroxyl group of only 30 per cent of the molecules and this was attributed, on the basis of control experiments with $H_2\ ^{18}O$, to exchange with H_2O of the medium during the reaction. The characteristic feature of the reaction mechanism which these authors proposed for this reaction is the formation of the cyclic peroxide intermediate (13), probably by reaction of the substrate and an oxygen–metal ion–enzyme complex. This then decomposes to give homogentisate (7) by decarboxylation and migration of the carbon side chain in the well-known manner of dienones[26]. The mechanism (*Figure 4.2*), which is very similar to that put forward earlier by Witkop and Cohen[27], accounts for the observed incorporation of molecular oxygen into both the new carboxyl and phenolic hydroxyl groups and exchange with the medium at the carbonyl group of the intermediate (14) may account for the

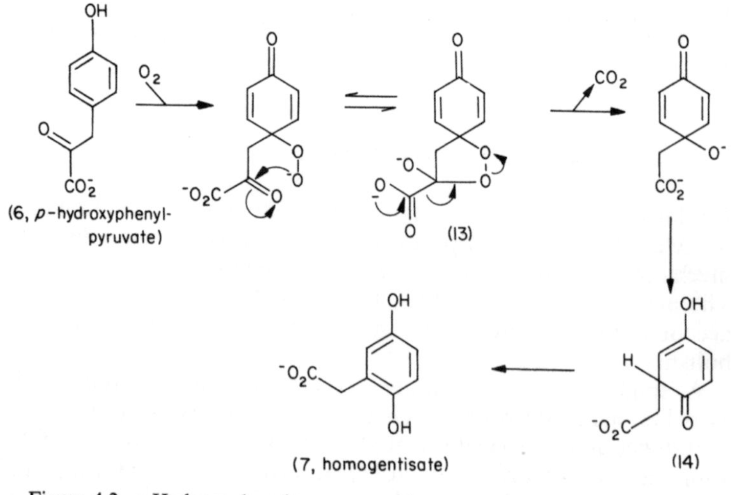

Figure 4.2. p-Hydroxyphenylpyruvate oxidase—a suggested mechanism[25, 27]

apparent loss of some isotopic oxygen from the phenolic hydroxyl group. An analogous cyclic peroxide mechanism has been proposed for the hydroxylation of substrates which are dependent on α-ketoglutarate[28, 29].

The pathway from homogentisate (7) to acetoacetate (9) and fumarate (10) proceeds by oxidative fission of the aromatic ring of homogentisate to give maleyl acetoacetate (8), which is then isomerised to fumaryl acetoacetate[30]. Hydrolysis of this intermediate then yields fumarate and acetoacetate[31], *Figure 4.1*.

4.2.3 CATECHOLAMINE METABOLISM

The catecholamines adrenalin, noradrenalin and dopamine are 3′,4′-dihydroxy derivatives of phenylethylamine. They are widely distributed in the animal kingdom and are found most often in nerve cells. Noradrenalin, for example, is highly localised in peripheral postganglionic sympathetic nerves where its concentration may be as high as 100 microgram per gram of tissue. Catecholamines influence the actions of a wide range of mammalian tissues such as the liver, heart, brain, adipose tissue and vascular smooth muscle. In many cases they appear to act by causing an increase in the activity or amount of a specific enzyme or series of enzymes[32].

The biogenesis of the catecholamines begins with L-tyrosine, *Figure 4.3*. This idea was first suggested by Blaschko[33] in 1939 and was finally confirmed in 1964 with the discovery[14, 34] of the enzyme L-tyrosine hydroxylase (EC 1.14.3) in a number of animal tissues. The enzyme, which catalyses the rate controlling step in catecholamine metabolism—the conversion of L-tyrosine (5) to L-dihydroxyphenylalanine (L-dopa, 15)—was finally isolated and purified from bovine adrenal medulla. It acts as a mixed function oxidase and requires reduced pteridines, ferrous ions and oxygen for maximal activity. It does not hydroxylate D-tyrosine and D L-*m*-tyrosine nor does it hydroxylate tyramine to dopamine. Assay of the enzyme is usually carried out utilising 3′,5′-[^{3}H$_2$]-L-tyrosine as substrate. No NIH shift occurs during this hydroxylation ortho to an existing phenolic hydroxyl group[16, 17], and the tritium directly displaced by the incoming hydroxyl function is separated from the L-tyrosine and L-dopa by ion exchange chromatography[14, 34]. A 'ping-pong' type of mechanism has been postulated for the mode of action of the enzyme in which the substrates add to the enzyme complex in an obligatory sequence[35]. Thus the following steps have been suggested for the enzyme reaction. First a reduction of an oxidised form of the enzyme by 2-amino-4-hydroxy-6,7-dimethyl-tetrahydropteridine which is then followed by dissociation of the oxidised pteridine and aerobic oxidation of L-tyrosine to give L-dopa (15) and the oxidised form of the enzyme.

The formation of dopamine (16) from L-dopa (15) in catecholamine biosynthesis is controlled by the enzyme L-dopa decarboxylase (EC 4.1.1.26) which has been widely found in mammalian tissues. The enzyme utilises pyridoxal phosphate as a cofactor and is specific for L-amino acids. However, it displays a fairly broad specificity in relation to the functional groups which define a particular amino acid. Thus the L-amino acid decarboxylase from guinea-pig kidney catalyses the decarboxylation of L-dopa, 5-hydroxy-L-tryptophan, L-phenylalanine, L-tyrosine, L-tryptophan and L-histidine although L-dopa is the substrate which is most rapidly decarboxylated. A similar lack of precise specificity is also displayed by dopamine-β-hydroxylase (EC 1.14.2.1) which controls the subsequent β-hydroxylation of dopamine (16) to give noradrenalin (17) in adrenalin metabolism. The enzyme, which is a mixed function oxidase, catalyses the conversion of several phenylethylamines, other than dopamine, to their β-hydroxylated analogues[37]. The structural characteristics necessary for substrate activity are an aromatic ring with a side chain of two (or three) carbon atoms terminating in an amino group. The enzyme has been isolated in an essentially homogeneous form from bovine adrenal glands and a molecular weight of 290 000 was estimated for the protein. It contains approximately 2 micromoles of cupric ion per micromole of enzyme and the activity of the enzyme is sensitive to a wide range of chelating agents. Evidence was obtained by Kaufman and Friedman[38] to suggest that the cupric ion in the enzyme undergoes a cyclic reduction and oxidation during the hydroxylation reaction.

Phenylethanolamine-N-methyltransferase (EC 2.1.1.-) catalyses the transfer of the methyl group of S-adenosylmethionine to noradrenalin[39] and this is the final step in the biosynthesis of adrenalin (2), *Figure 4.3*. Although the pathway proposed by Blaschko[33] and described above is the major pathway of catecholamine biosynthesis, the absence of specificity which has been demonstrated for several enzymes in the sequence permits several alternative but minor routes of biosynthesis for these compounds. These have been delineated by Molinoff and Axelrod in their recent review[32].

Subsequent catabolism of the various catecholamines in mammals has been widely investigated and leads to a range of catechol derivatives isolable as urinary metabolites. Two enzymes, catechol-O-methyltransferase (EC 2.1.1.6) and monoamine oxidase (EC 1.4.3.4).

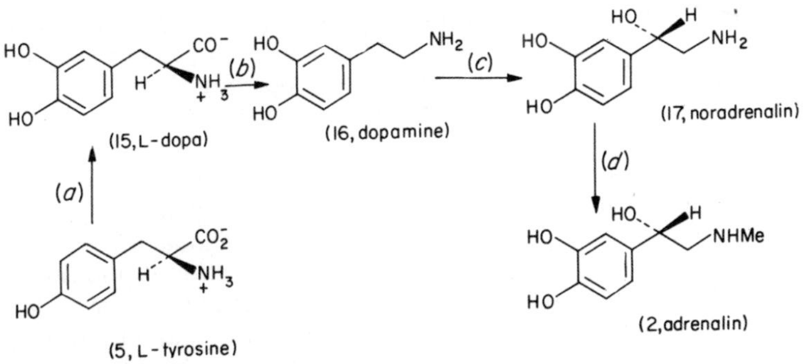

(a) L-tyrosine hydroxylase
(b) L-dopa decarboxylase
(c) dopamine β-hydroxylase
(d) phenylethanolamine – N – methyltransferase

Figure 4.3. Biosynthesis of adrenalin in mammals

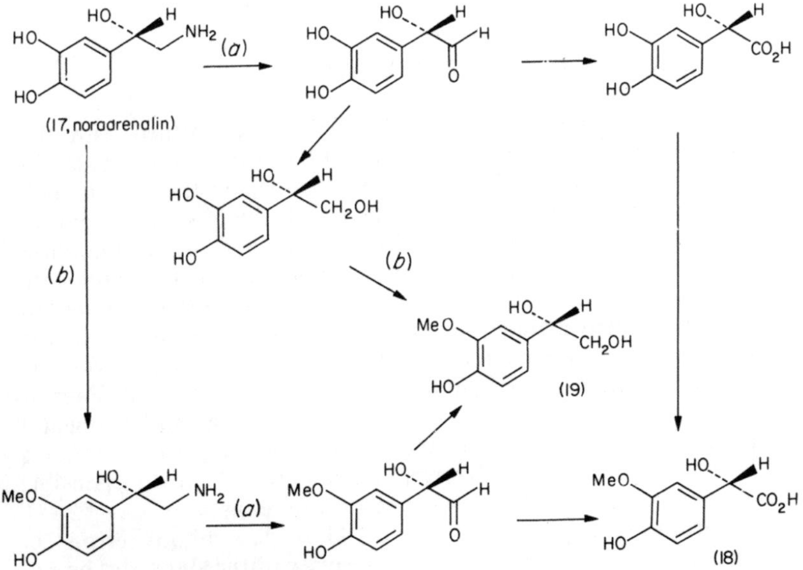

(a) monoamine oxidase
(b) catechol - 0 - methyltransferase

Figure 4.4. Metabolism of noradrenalin

are principally involved in these catabolic processes. Monoamine oxidase deaminates substrates such as adrenalin (2), noradrenalin (17) and dopamine (16) to give the corresponding aldehydes which are then either reduced or oxidised to give respectively an acidic or alcoholic metabolite[40].

Catechol-O-methyltransferase from rat liver[41] requires S-adenosylmethionine as a methyl donor and can methylate catechol but not monohydroxy derivatives of phenylethylamine. *In vivo*, O-methylation occurs exclusively in the meta position to the carbon side chain[42], but with purified preparations of the enzyme *in vitro* methylation can lead to both meta and para methylation. The ratio of products is susceptible.to both the polarity of the substrate and the pH of the medium[43, 44]. In man 3-methoxy-4-hydroxy-mandelic acid (18) comprises about 40 per cent of the total urinary metabolites produced from the catecholamines whilst in other ˜pecies 3-methoxy-4-hydroxyphenylglycol (19), isolated as a sulphate ester, is the predominant breakdown product[45, 46]. A typical metabolic grid which indicates the possible types of pathway leading from noradrenalin (17) to both of these metabolites is shown in *Figure 4.4*; analogous metabolic schemes may be drawn up for both adrenalin and dopamine.

4.2.4 INBORN ERRORS OF METABOLISM

Several 'inborn errors of metabolism' are concerned with the metabolism of L-phenylalanine and L-tyrosine in mammals. In several cases it has been possible to demonstrate that such biochemical disorders are associated with the absence or partial deficiency of a particular enzymatic activity. Phenylketonuria results from the absence of a normal L-phenylalanine hydroxylase activity and individuals suffering from this disease are unable to convert L-phenylalanine to L-tyrosine. Under these conditions the metabolism of the amino acid to phenylpyruvic acid, phenyl-lactic acid and phenyl acetyl glutamine is greatly exaggerated. Phenylketonuria is a severe disorder and results in a marked mental retardation, particularly in children. It is generally assumed that it is the accumulation of abnormal metabolites which is responsible for the mental symptoms associated with the disease.

Alcaptonuria was the first of Garrod's 'inborn errors of metabolism' to be recognised[1] and is caused by the absence of homogentisate oxidase and hence results in an inability to metabolise homogentisic acid further. It is characterised by the excretion of relatively large amounts (up to half a gram daily) of homogentisic acid in the urine, which characteristically becomes black on standing because of the aerial oxidation of the metabolite. Homogentisic acid also occurs in other body fluids and often there is a darkening of various tissues due to the deposition of a pigment formed from the acid.

4.2.5 TYROSINE PHENOL LYASE

Phenol is formed from L-tyrosine and its derivatives in some bacterial cultures not through a stepwise degradation of the molecule but through primary fission of the side chain. The L-tyrosine inducible enzyme L-tyrosine phenol lyase has been prepared in crystalline form from cell extracts of *Escherichia intermedia* A-21 and a molecular weight of 170 000 was estimated[47]. The enzyme catalyses the stoichiometric conversion of L-tyrosine (5) to phenol (20), pyruvate (21) and ammonia in the presence of added pyridoxal phosphate as a cofactor. Brot, Smit and Weissbach[48] have described

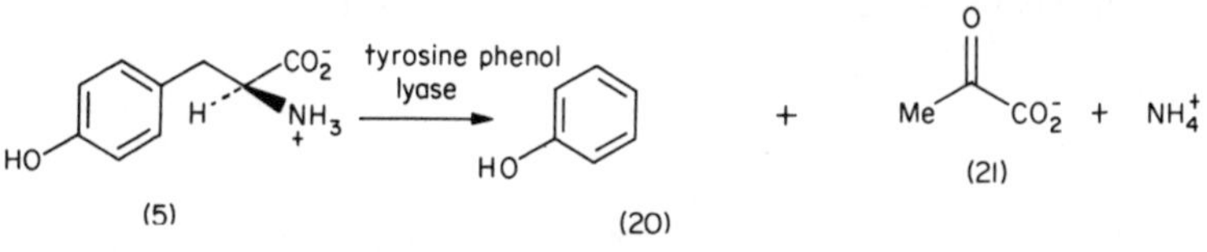

a similar degradative reaction of L-tyrosine using a partially purified enzyme preparation from *Clostridium tetanomorphum*.

Tyrosine phenol lyase, like the tryptophanase of *Escherichia coli* and the cystathionine synthetase of *Salmonella typhimurium*, is an enzyme with a broad substrate specificity and catalyses a whole series of related α, β-elimination, β-replacement and racemisation reactions[49, 50]. In a reversal of the elimination reaction, it is capable of synthesising L-tyrosine from ammonia, pyruvate and phenol[51], and L-tyrosine and L-dopa from serine and phenol or catechol[52].

4.3 METABOLISM OF L-TRYPTOPHAN

Several interesting and important metabolites are derived from L-tryptophan (22) and these include the pyridine ring of nicotinic acid (and hence of the various nicotinanide co-enzymes), serotonin (5'-hydroxytryptamine) a powerful vasoconstrictor found in mammalian nervous tissue, and eye pigments (the ommochromes) of certain insects[53]. From a comparative point of view a distinctive feature of the metabolism of L-tryptophan is the widespread use of a series of common pathways by a variety of organisms. Almost invariably the oxidative breakdown of L-tryptophan leads initially to the formation of kynurenine (24) as the central intermediate from which different pathways radiate yielding different products dependent on the needs of the individual organism[54].

4.3.1   OXIDATIVE PATHWAYS

The first step in the oxidative metabolism of L-tryptophan in both mammals and micro-organisms is mediated by the enzyme L-tryptophan oxygenase (L-tryptophan pyrrolase, EC 1.13.1.12), a dioxygenase enzyme which catalyses the insertion of both atoms of a molecule of oxygen into the pyrrole ring of its substrate L-tryptophan to yield N-formylkynurenine (23)[55]. Hydrolysis of this intermediate by kynurenine formylase then yields the key metabolite kynurenine (24). L-Tryptophan oxygenase has been obtained from both liver[56] and from *Pseudomonas* species[57, 58] and its mode of action has been the subject of some speculation[59, 60]. The enzyme from *Pseudomonas acidovorans* has been highly purified and its molecular weight estimated at around 122 000. The enzyme molecule is composed of four polypeptide chains of essential equivalent mass and associated with one of these chains is the catalytically essential prosthetic group–ferriprotoporphin IX. The intact tetrameric structure of the enzyme is essential for catalytic activity but it may be disrupted by different solvent conditions such as alkaline pH, guanidinium chloride or sodium dodecyl sulphate.

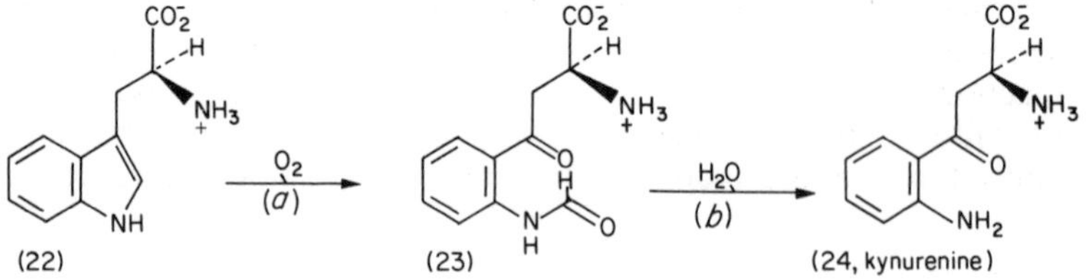

(22)   (23)   (24, kynurenine)

(*a*) L- tryptophan oxygenase   (*b*) kynurenine formylase

Several alternative pathways of L-tryptophan metabolism diverge from kynurenine (24). In mammals the quantitatively major fate of the benzene ring of the amino acid appears[61–63] to be its oxidation to carbon dioxide via 3-hydroxyanthranilic acid (25), *Figure 4.5*. Kynurenine is first hydroxylated by a typical mixed function oxidase and the side chain is then removed, under the influence of the enzyme kynureninase to give L-alanine and 3-hydroxy anthranilate (25). The latter cleavage enzyme is typically dependent on pyridoxal phosphate. Complete oxidation of the aromatic ring then occurs via 2-amino-3-carboxy-*cis, cis*-muconic semi-aldehyde (26) and glutaryl co-enzyme A. Formation of the aldehyde (26) from 3-hydroxyanthranilate (25) occurs by ring fission adjacent to the hydroxyl group and is catalysed by the enzyme 3-hydroxyanthranilic acid oxidase a mixed function oxidase which requires ferrous iron for its action[64, 65].

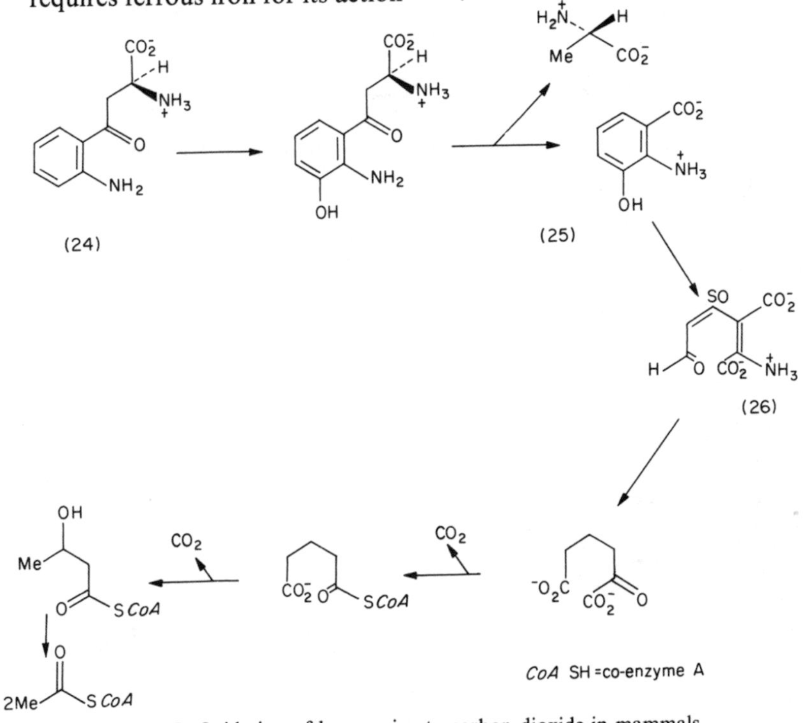

Figure 4.5. Oxidation of kynurenine to carbon dioxide in mammals

A further important pathway of L-tryptophan metabolism also diverges from 2-amino-3-carboxy-*cis, cis* muconic semi-aldehyde (26) the ring fission product of 3-hydroxyanthranilate (25) and this is the biosynthetic pathway which leads to the formation of the pyridine ring of the nicotinamide nucleotides[66-68]. The first step in this pathway appears to be the recyclisation of the oxidation product (26) to give quinolinic acid (27) which is transformed to the pyridine nucleotides, such as NAD^+ (28) in a complex series of reactions, *Figure 4.6.*

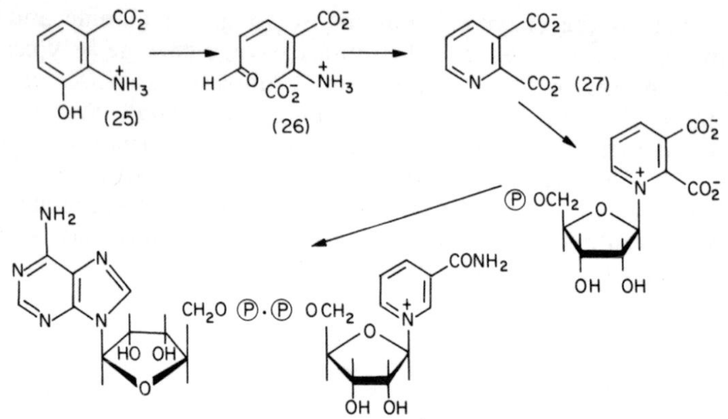

140

Figure 4.6. Biosynthesis of NAD$^+$ from 3-hydroxyanthranilate

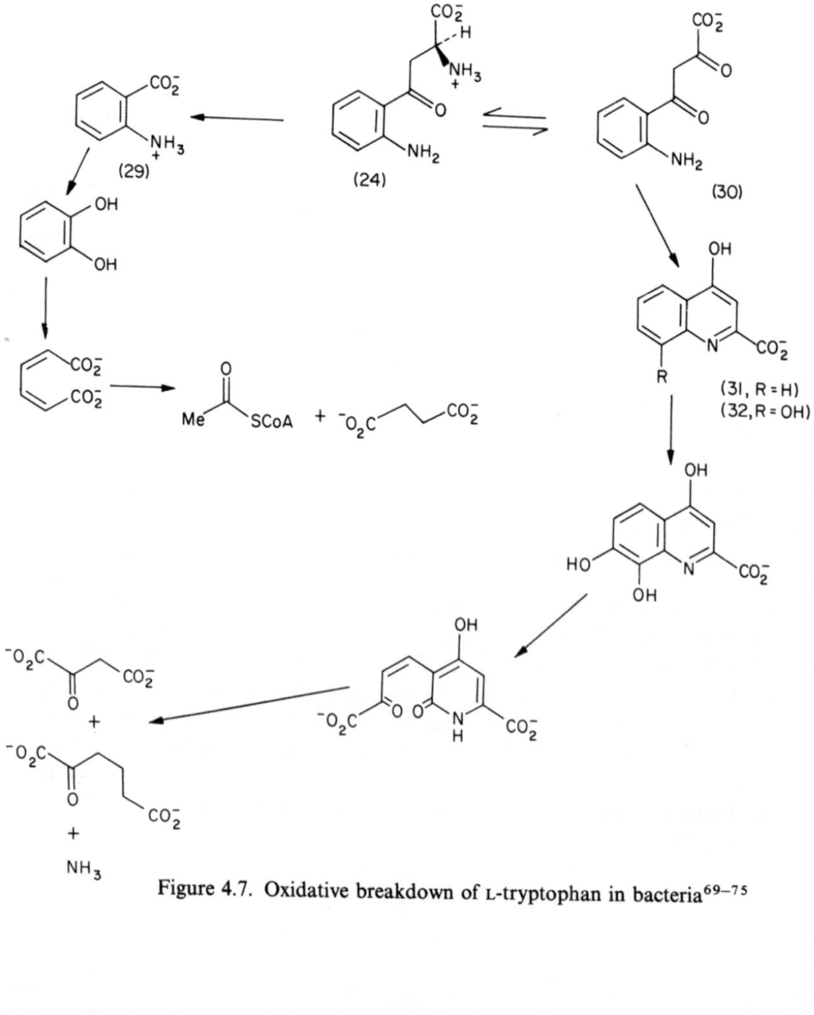

Figure 4.7. Oxidative breakdown of L-tryptophan in bacteria[69-75]

Although the formation of anthranilate and L-alanine by cleavage of kynurenine (24) is of little consequence in mammalian systems it provides a major metabolic route of breakdown in some micro-organisms. Thus several bacteria metabolise L-tryptophan via kynurenine (24) and anthranilic acid (29). The latter is then oxidatively broken down usually via catechol, *cis, cis*-muconate and β-ketoadipate to give finally succinate and acetyl co-enzyme A[67-72], *Figure 4.7*. Transamination of kynurenine (24) by means of the enzyme kynurenine transaminase, which is widely distributed in both the animal kingdom and micro-organisms, gives the corresponding keto acid (30) which spontaneously cyclises to yield kynurenic acid (31). Analogous transformation of hydroxykynurenine gives the corresponding xanthurenic acid (32) which along with kynurenic acid was recognised many years ago as an excretion product of L-tryptophan metabolism in mammals. Soil pseudomonads can also utilise kynurenic acid (31) by means of the oxidative sequence, *Figure 4.7*, to provide α-ketoglutarate, oxaloacetate and nitrogen as ammonia. This pathway[73-75] is often termed the quinoline pathway and is the other major route of oxidative breakdown of L-tryptophan in bacteria.

4.3.2 5'-HYDROXYTRYPTAMINE

Hydroxylation of L-tryptophan (22) in the 5'-position to yield L-5'-hydroxytryptophan (33) is the first and probably the rate-limiting step in the formation of serotonin (1, 5'-hydroxytryptamine) the vasoconstrictor principle in animals[76]. Brain homogenates directly hydroxylate L-tryptophan[77] and the same reaction may be demonstrated *in vivo* by infusing L-tryptophan into the brain. The enzyme responsible is a typical aromatic hydroxylase and requires a reduced pteridine and molecular oxygen in order to function. Analogously the enzyme, like L-phenylalanine hydroxylase, shows the characteristic NIH shift upon hydroxylation of the substrate[16,17]. Thus 5'-[³H]-L-tryptophan is converted by the enzyme to 5'-hydroxy-4'-[³H]-L-tryptophan with greater than 85 per cent retention of the tritium in the 4' position of the amino acid. This experiment constituted the first observation that during aryl hydroxylation of unsymmetrical substrates isotopic hydrogen can selectively migrate and be retained in just one of the two adjacent positions. Serotonin is finally produced from L-5'-hydroxytryptophan by decarboxylation by the enzyme L-5'-hydroxytryptophan decarboxylase[79].

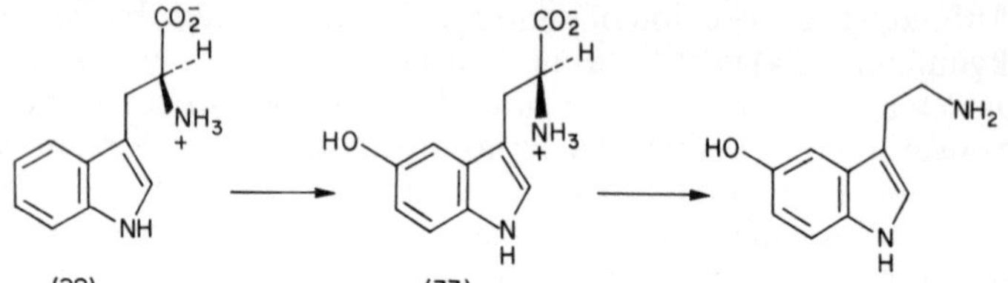

The conversion of L-tryptophan to L-5′-hydroxy-tryptophan has also been observed in *Chromobacterium violaceum*[80]. This organism then utilises L-5′-hydroxy-tryptophan in the formation of a distinctive pigment, violacein[81].

4.3.3　TRYPTOPHANASE

Degradation of L-tryptophan takes place in some bacteria by the tryptophanase reaction in which the amino acid is converted to indole (35), pyruvic acid and ammonia. The reaction was first observed[82] in 1903. Wood and his collaborators first prepared the enzyme tryptophanase which catalyses the change from *Escherichia coli* and showed that pyridoxal phosphate is the co-enzyme involved[83]. Snell and his colleagues have proposed a mechanism for the reaction in which the required cleavage occurs via the intermediacy of a pyridoxal phosphate-tryptophan-metal complex

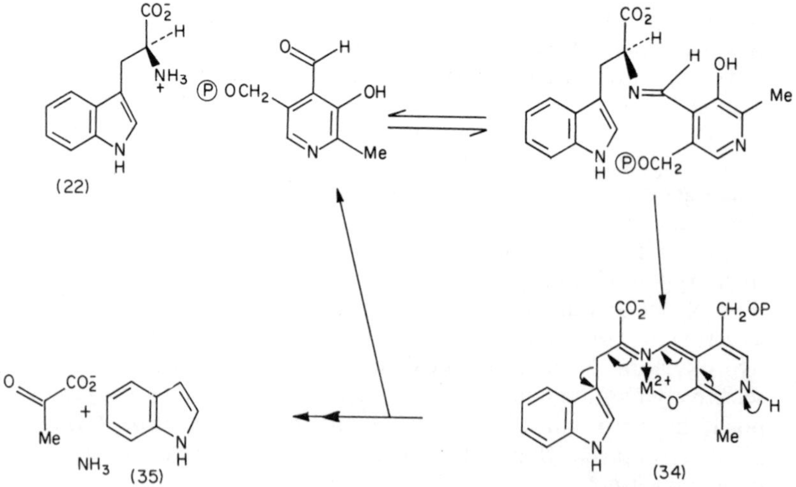

Figure 4.8. Mechanism of the tryptophanase reaction[84]

(34)[84], *Figure 4.8*. An analogous mechanism can be written for the closely related tyrosine phenol lyase reaction.

In subsequent work Newton and Snell[85, 86] reported that the enzyme catalysed a whole series of α β elimination and β replacement reactions. One of these—L-serine+indole → L-tryptophan—is identical with that catalysed by part of the L-tryptophan synthetase complex. Tryptophanases have been isolated from the B and K-12 strains of *Escherichia coli*[87–90] and contrary to earlier reports both enzymes have been shown to have a tetrameric composition. The enzyme from *Escherichia coli*-B/lt7-A contains four identical sub-units of molecular weight around 55 000 with one binding site for pyridoxal phosphate per sub-unit.

Indigo, indirubin, indoxyl and indican which have all been found in mammalian urine[85] are probably formed from indole, itself produced by the action of intestinal bacteria on L-tryptophan.

4.4   OXYGENASES AND BIOLOGICAL OXIDATIONS

The enzymes which participate in biological oxidations may be classified into two major groups. Dehydrogenases catalyse the transfer of hydride ions or electrons from donors to various acceptors; where the acceptor is oxygen the enzyme is termed an oxidase. Oxygenases, in contrast, catalyse the incorporation of molecular oxygen into various organic (and inorganic) substrates. This latter group of enzymes is of particular significance in the metabolism of the aromatic amino acids in higher organisms and micro-organisms since almost all of the oxidative reactions are controlled by oxygenases rather than oxidases and dehydrogenases. Despite their crucial role in metabolism their mechanism of action has not proved easy to study and in the main this is because these enzymes have, until comparatively recently, been difficult to isolate free and not particulate bound[91, 92].

Oxygenases may themselves be divided into two groups: the mono-oxygenases which use one molecule of oxygen per molecule of substrate, and the dioxygenases which incorporate two atoms of oxygen from an oxygen molecule into the substrate. The mono-oxygenases incorporate only one atom of oxygen per molecule of substrate and the other atom of oxygen is reduced to water. Because of their obvious hybrid nature between oxidases and oxygenases these enzymes were classified by Mason[93, 94] as mixed function oxidases.

4.4.1   AROMATIC HYDROXYLASES

An important group of mixed function oxidases which occur in all types of organism and which are critical in the metabolism of the aromatic amino acids and other aromatic substrates are the aromatic hydroxylases. Their mode of action, the overall stoichiometry of which is represented in the sequence below, results in the introduction of a phenolic hydroxyl group in an aromatic ring system and has been the subject of intensive study. Pyridine and flavin nucleotides, cytochromes, metals (Fe, Cu), ascorbate and pteridine derivatives (H_2X) may serve as electron donors. Most, but not all, of these enzyme reactions require transition metal ions for full activity.

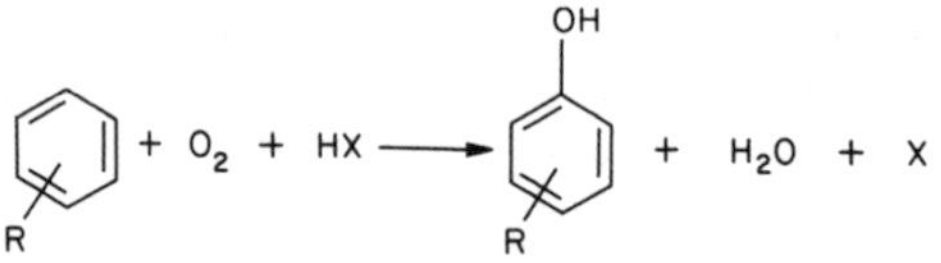

Phenylalanine hydroxylase[12] is typical of this class of enzymes and has perhaps been most widely studied. The conversion of L-phenylalanine to L-tyrosine, which this enzyme catalyses, serves a dual role in the metabolism of higher organisms. It appears not only to be an obligatory step in the catabolism of L-phenylalanine to carbon dioxide but it also provides an endogenous source of the amino acid L-tyrosine and its metabolites. Some insight into the mechanism of this and other enzyme catalysed hydroxylations and the nature of the oxygen species involved has been derived from studies of model reactions and from studies with substrate analogues.

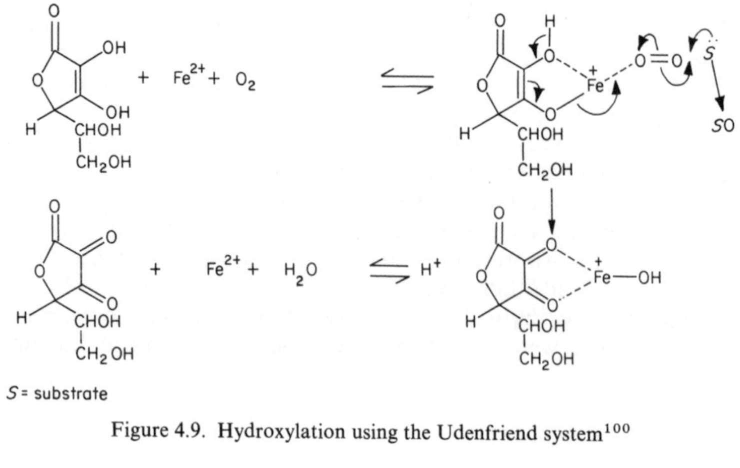

S = substrate

Figure 4.9. Hydroxylation using the Udenfriend system[100]

One model reaction which has been closely scrutinised[95-97] was described by Udenfriend and his collaborators[98]. These workers observed that aromatic compounds are hydroxylated in low yield by oxygen at neutral pH and ambient temperatures in presence of Fe (II) and ascorbic acid. Various species such as H_2O_2, HO· and HOO· have been ruled out as intermediates[95-97, 99] in this reaction and Hamilton[100] has suggested the mechanism shown in *Figure 4.9* for oxidations utilising the Udenfriend system.

Hamilton suggested that the function of the metal ion is to form a complex with the reducing agent and oxygen and to transfer electrons to the oxygen molecule in the transition state of the reaction. Hamilton termed this type of mechanism an 'oxene mechanism' because of its analogies to both carbene and nitrene reactions; the implication is that the oxidising agent which attacks the substrate is an oxygen species—very probably the oxygen atom—which reacts like carbenes and nitrenes. This analogy, Hamilton proposed, extended to the very large number of enzymes of the mixed function oxidase category. Thus, for example, L-phenylalanine hydroxylase is a mixed function oxidase which requires a reduced pteridine as cofactor[9, 15] and Hamilton suggested that this co-enzyme functioned in many ways like the ascorbic acid in the Udenfriend system, *Figure 4.10*. Similar mechanisms could presumably be written to account for the mode of action of L-tyrosine hydroxylase[34] which controls the rate determining step in catecholamine biosynthesis (L-tyrosine → L-dopa) and which bears many similarities to L-phenylalanine hydroxylase including the same requirement for a tetrahydropteridine as a cofactor.

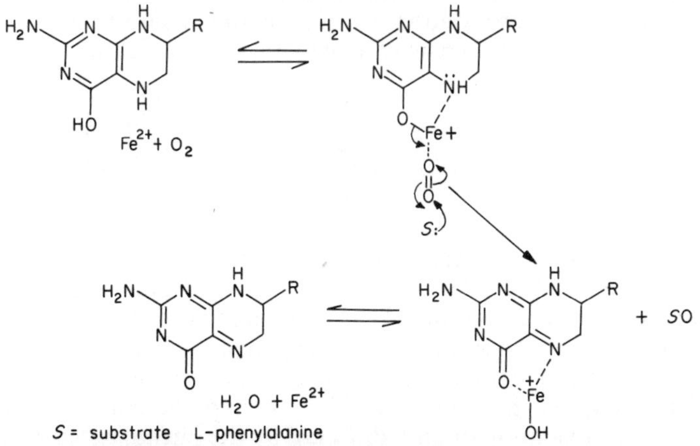

S = substrate L-phenylalanine

Figure 4.10. Suggested mechanism for L-phenylalanine hydroxylase[100]

However, unequivocal proof whether L-phenylalanine hydroxylase from rat liver, for example, is an enzyme containing iron and dependent on this metal for its action is still lacking. Similar doubts surround the bacterial L-phenylalanine hydroxylase since although Guroff and Ito[8] originally reported that it was activated by ferrous ions, more recent studies[9] have shown this to be a quite nonspecific stimulation which can also be promoted by other divalent cations. Kaufman[12] indeed has put forward an alternative view regarding the mode of action of L-phenylalanine hydroxylase. He suggested that from the available evidence oxygen at the level of peroxide, possibly a tetrahydropterin peroxide, is an intermediate in L-phenylalanine hydroxylation.

Tyrosinase is an enzyme complex (phenolase, polyphenol oxidase are other names which have been used for this enzyme), which catalyses of the ortho hydroxylation of monohydric phenols. The enzyme, which should not be confused with L-tyrosine hydroxylase mentioned above[14], contains Cu (I) and catalyses two distinct reactions—the hydroxylation of monohydric phenols to o-diphenols (cresolase activity) and the oxidation of o-diphenols to o-quinones (catecholase or catechol oxidase activity)[94]. Most enzymes of this type, which are widely distributed in both the plant and animal kingdoms, exhibit both catalytic functions. Thus typically, the conversion of L-tyrosine (5) to L-dopa (15) and dopaquinone (36) which occurs in melanin biosynthesis is catalysed by an enzyme of the tyrosinase category. The two activities appear, in the majority of cases, to be functions of the same enzyme. However, certain o-diphenol oxidases such as those from tea[101], sweet potato[102] and tobacco[103] have been reported to show no capacity to catalyse the hydroxylation reaction but this is most probably due to destruction of the cresolase activity during purification.

(5)   (15, L-dopa)   (36, L-dopaquinone)

The consensus of opinion is that the catechol oxidase activity of tyrosinase, and similar enzymes, is an integral part of the hydroxylation mechanism[94, 104, 105] and that the distinctive properties of these enzymes can best be explained in terms of a single enzyme possessing two characteristic activities. In the form proposed by Dawson and his co-workers[106], this hypothesis provides that during the oxidation of an o-diphenol by the phenolase complex a portion of the enzyme becomes momentarily activated or primed so that it can then bring about the hydroxylation of monophenols. Mason[107] later put forward a similar but more detailed mechanism for the mode of action of the enzyme complex on the basis of earlier studies. In the presence of one molecule of the *o*-diphenol, two cupric atoms in the phenolase complex are reduced to the cuprous state[108, 109] and it is this form which Mason suggested exhibits the cresolase activity. He proposed a catalytically active configuration for the enzyme in which two protein bound copper atoms in the cuprous state are linked to one oxygen molecule. The power of the oxygen complex to act in aromatic hydroxylations, he postulated, was dependent on the electron distribution in the complex which in turn is dependent on the spatial disposition of the two copper atoms. A change in the relative positions of these two copper atoms Mason suggested would result in a change in the capacity to bind oxygen and also to carry out hydroxylation but would not seriously affect the ability to directly oxidise catechols to *o*-quinones.

More recently Hamilton[100] has discussed a further mechanism which may be operative in mixed function oxidases of the tyrosinase class, *Figure 4.11*. In this mechanism the reduced enzyme is first

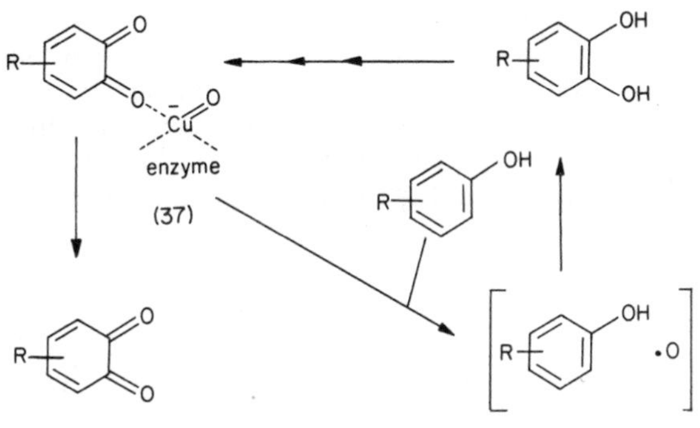

Figure 4.11. Suggested mechanism for the enzyme tyrosinase[100]

oxidised to a form which is then capable of transferring oxygen to the substrate. Hamilton proposed an intermediate of the type (37) as the active form for oxygen transfer; this is formed itself as an intermediate in the oxidation of the resultant *o*-diphenol to the *o*-quinone. As yet no detailed work of a mechanistic type has been carried out with L-tyrosine hydroxylase sufficient to say how this enzyme differs from tyrosinase in its ability to carry out hydroxylation ortho to a phenolic group.

4.4.2   ARENE OXIDES AND THE NIH SHIFT

Regardless of the form of the initial attack on the aromatic substrate there is now strong circumstantial evidence that intermediates of the arene oxide (oxepin) type participate in many of these enzyme catalysed hydroxylations. This evidence derives in the main from work on the NIH shift in which for the first time hydroxylation induced migration of substituents in the aromatic ring was demonstrated[16, 17]. As yet only one arene oxide intermediate has been firmly established as an obligatory intermediate in the conversion of an aromatic substrate to a phenol. This is naphthalene-1,2-oxide (39) which was isolated and identified in 1968 as an intermediate in the hepatic metabolism of naphthalene (38)[18, 110, 111]. It was, moreover, demonstrated that this intermediate is capable of conversion to all the principal metabolites of naphthalene in this system. Thus it rearranges non-enzymatically almost exclusively to the phenolic product 1-naphthol (40) and is hydrated enzymatically to trans-1,2-dihydroxy-1,2-dihydronaphthalene (41) with the correct absolute stereochemistry and similar optical purity as the natural metabolite. In addition soluble liver enzymes catalyse the addition of glutathione (GSH) to naphthalene-1,2-oxide to form a conjugate (42) identical with that obtained during the oxidative breakdown of naphthalene. The nature and distribution of the metabolites which are formed from the intermediate arene oxide is dependent on the balance between a number of factors such as its intrinsic stability in relation to isomerisation to the phenol, its susceptibility to enzymic hydration or conjugation

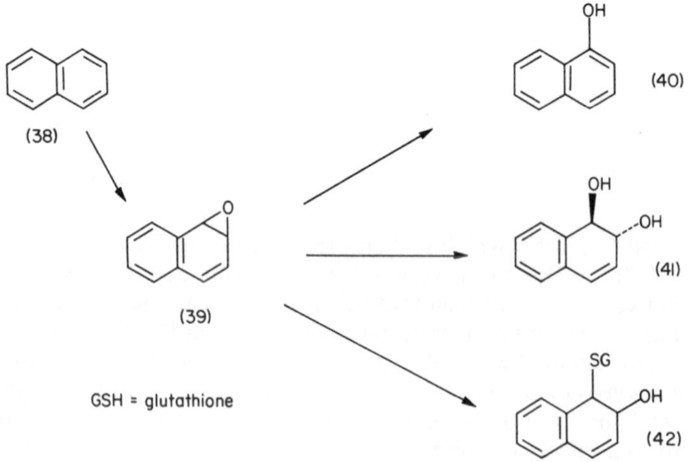

(38)

(39)

GSH = glutathione

(40)

(41)

(42)

with glutathione and to nucleophilic attack by macromolecular components of the tissue.

Since its original discovery the scope and applicability of the NIH shift which accompanies aryl hydroxylation has been widely investigated[17] and some typical examples and observations are discussed below (*Figure 4.12*). L-Phenylalanine hydroxylase reacts with 4'-[^{2}H],-[^{3}H],-Cl and -Br-L-phenylalanine to give products (L-tyrosine derivatives) in which the substituent is retained (80–90 per cent) by migration to the adjacent 3' and 5' positions[20, 112,113]. The source of the enzyme has little effect on the migration of the substituent which depends almost exclusively on the nature of the substrate and hence of the intermediate arene oxide. Thus hydroxylation of 4'-[^{3}H]-L-phenylalanine occurs with greater than 90 per cent retention and migration of the hydrogen isotope with enzymes from soil bacteria[112, 114], *Penicillium* species[115] or liver[112, 114]. Similarly treatment of 3'-[^{3}H]-L-phenylalanine with a microsomal hydroxylase gave L-tyrosine with greater than 95 per cent retention of the hydrogen isotope in the 3' position. Analogously the action of L-tryptophan-5'-hydroxylase on 4' and 5'-[^{3}H]-L-tryptophan gave 4'-[^{3}H]-5'-hydroxy-L-tryptophan with over 85 per cent retention of the tritium. In the case of the 5' substituted tritiated substrate, selective migration of the substituent occurred to the adjacent 4' position. Using a non-specific drug metabolising aryl hydroxylase from liver microsomes it was shown that the NIH shift is not restricted to para-hydroxylation reactions. Thus ortho hydroxylation of 2-deuteriated-anisole (43) and its 4-methyl analogue gave the corresponding catechol derivatives in which the isotope retention and migration was 60 per cent and 53 per cent respectively[116]. On the other hand where hydroxylation occurs ortho or para to an existing hydroxyl group then loss, rather than migration of the displaced substituent occurs. Thus hydroxylation of 3',5'-[^{3}H$_2$]-L-tyrosine by brain or adrenal L-tyrosine hydroxylase leads to L-dopa with the stoichiometric release of one tritium atom as tritiated water[14, 34].

Significantly the non-enzymatic hydroxylation of aromatic substrates by hydroxylating systems which have been favoured as models for enzymatic hydroxylation (e.g., the Udenfriend system[98] —Fe^{2+}, O_2 and an organic reducing agent such as ascorbic acid or a reduced pteridine; Fenton's reagent[95]—Fe^{2+}, hydrogen peroxide; the Hamilton system[99]—Fe^{2+}, hydrogen peroxide and a trace of catechol; horse radish peroxidase[117]) show no measurable retention of the hydrogen isotope displaced during hydroxylation. Witkop[17] has suggested that on the basis of this criterion these model systems have no relevance to the enzymic reactions and probably

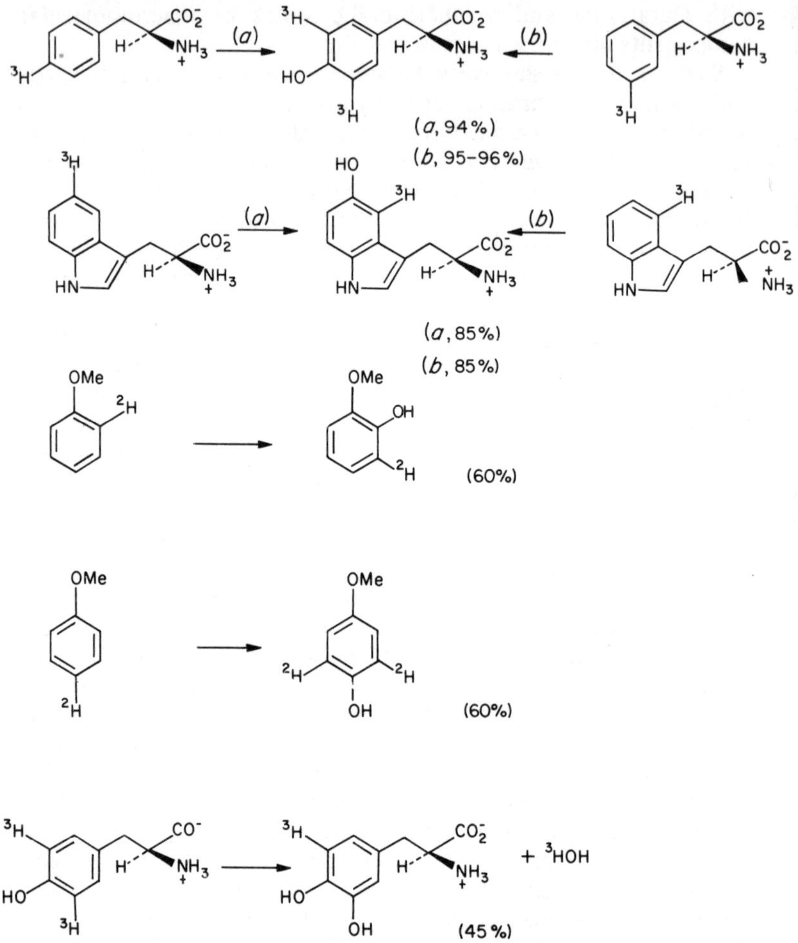

$(a, 94\%)$
$(b, 95-96\%)$

$(a, 85\%)$
$(b, 85\%)$

(60%)

(60%)

$+ \,^3HOH$

(45%)

Figure 4.12. The NIH shift—some examples (Retention of the hydrogen isotope is shown in brackets)[17]

hydroxylate by a free-radical mechanism. However, several peroxy acids hydroxylate a range of deuterium and tritium labelled aromatic substrates with a concomitant NIH shift. Thus trifluorperacetic acid, which probably acts by an oxygen atom transfer mechanism[118], hydroxylates 4-[^{3}H] and 4-[^{2}H]-acetanilide to give the p-hydroxy derivative with 10 per cent retention and migration of the hydrogen isotope[119]. This value may be compared to a retention with migration of 40–65 per cent using a non-specific microsomal hydroxylase[120, 121]. Photo-activated N-oxides also hydroxylate aryl systems with NIH shifts comparable to those of enzymatic hydroxylation although the results are variable and occasionally difficult to interpret due to the photo-isomerisation of the intermediate arene oxides. However, equal amounts of the naphthalene-1,2-oxide and 1-naphthol were obtained[122] from the photo-activated reaction of pyridine N-oxide and naphthalene. Whilst there are still no effective model systems for the oxo-iron species which have been strongly implicated in the actual enzymic hydroxylations several metal ion systems have been described which bring about aryl hydroxylation and also show the NIH shift. These include an $Sn^{2+}-O_2$ system[123], a $Cu^{2+}-O_2$ system[124] and a $Mo(CO)_6-(Me)_3 . C . OOH$ system[122].

Witkop and his collaborators[17] have interpreted the wealth of experimental data on aromatic hydroxylation reactions in model and enzyme systems in terms of an 'oxene' type of mechanism, similar to that proposed by Hamilton[100], in which the initial attack is by a species similar to an oxygen atom to give an arene oxide intermediate (44), *Figure 4.13*. The latter may then rearrange spontaneously (neutral to basic pH) or under acid catalysis to give the phenol and during the course of this rearrangement the characteristic migration of the group undergoing substitution—the NIH shift—occurs. Kinetic measurements suggest that the rate-determining step in the region of spontaneous isomerisation is the ring opening to the zwitterion (45) whilst in acid media the protonated form of (45) is predominant. Two alternate pathways are then proposed and the balance between them is determined by the substituents in the aromatic ring system. Formation of the phenol (46) can occur by direct loss of the substituent (for example a proton)—pathway '*a*'. Alternatively stabilisation of the cation intermediate (45) can take place by rearrangement (NIH shift) to

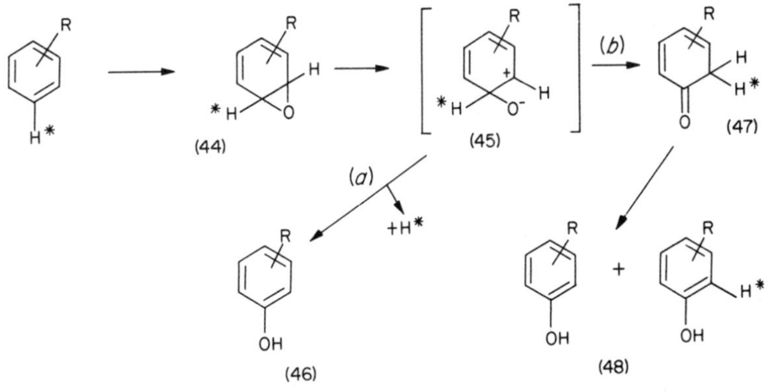

Figure 4.13. Aryl hydroxylation and the NIH shift[17]

give the keto tautomer (47) enolisation of which then gives the phenol (48)—pathway '*b*'.

Evidence in support of this mechanism has been provided by a number of metabolic studies. Thus in an examination[111] of the metabolism of naphthalene to 1-naphthol the rearrangement of synthetic 1-[^{2}H] and 2-[^{2}H]-naphthalene-1,2-oxide (51 and 52) was shown to be strongly pH dependent. In neutral or basic media, both gave 1-naphthol with approximately 80 per cent retention of the deuterium at the 2-position. Microsomal hydroxylation of the deuteriated substrates 1-[^{2}H] and 2-[^{2}H]-naphthalene (49 and 50) gave in both cases 1-naphthol (54) whose retention of deuterium at the 2-position was not statistically different from that observed in the rearrangement of the naphthalene-1,2-oxides (51 and 52). These results were interpreted to suggest that in the metabolism of naphthalene to 1-naphthol, the rate limiting step is the oxidative formation of the naphthalene-1,2-oxide which is followed by a concerted isomerisation of the oxide via the keto form (53) to 1-naphthol. A primary isotope effect of $K_H : K_D = 4.0$ was inferred for the enolisation step. In a similar way rearrangement of the oxide (56) or hydroxylation of 4-[^{2}H]-toluene (55) at the same pH lead to comparable retentions of the deuterium substituent[125].

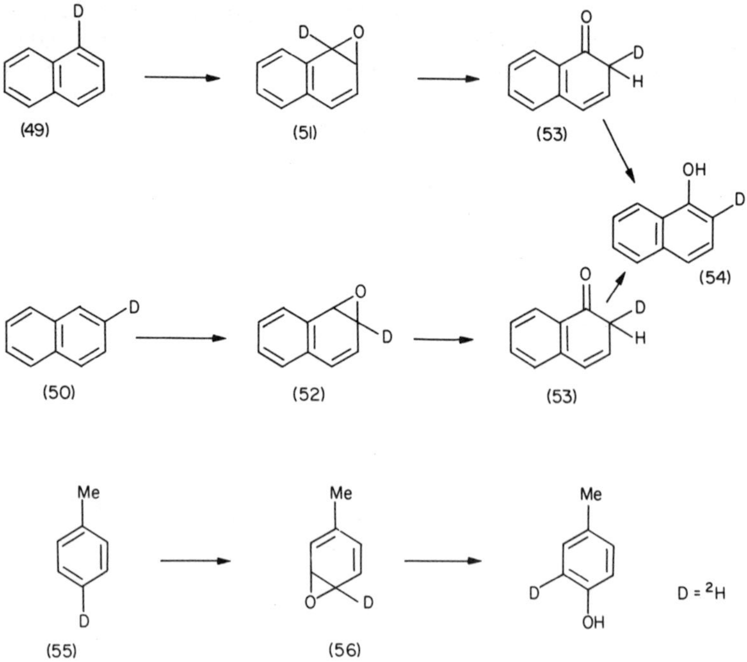

(49)   (51)   (53)

(54)

(50)   (52)   (53)

(55)   (56)

$D = {}^2H$

Similar support for the mechanism (*Figure 4.13*) of aryl hydroxylation was also obtained from studies of the tritium retention and migration which accompanies hydroxylation of 4-[^{3}H]-anisole (57) and 3-[^{2}H]-4-[^{3}H]-anisole (58), and the analogous acetanilide derivatives[120]. Isomerisation of the intermediate keto forms (59 and 60) to 4-hydroxyanisole led to the expected significant primary isotope effect ($K_H > K_D > K_T$) and more tritium was observed to be retained in the phenol derived by hydroxylation of the undeuteriated substrate (57).

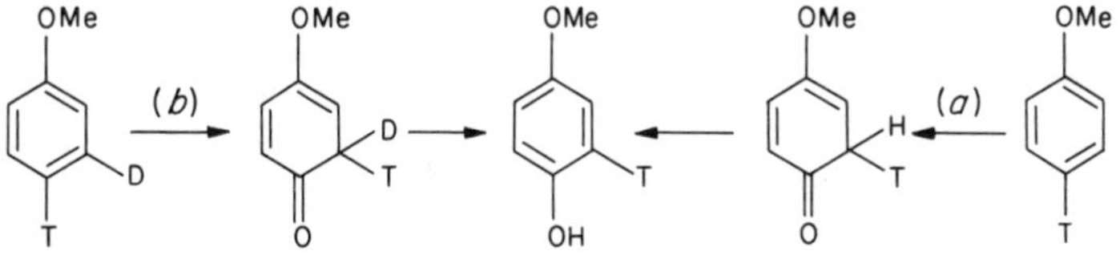

(58) (60) (*a*,77-80%) (*b*,61-64%) (59) (57)

D=^{2}H, T=^{3}H tritium retention

The extensive studies on the migration and retention of deuterium or tritium in aromatic substrates which have been carried out indicate that the magnitude of the retention is considerably influenced by the nature of substituents in the aromatic ring. Several inconsistencies which have been observed are only adequately resolved by assuming that decomposition of the arene oxide (*Figure 4.13*) can take place by both pathways '*a*' and '*b*'. Thus in some cases such as chlorobenzene and anisole (61 and 62) hydroxylation gives the 4-hydroxy derivatives but the retention of deuterium in this product is dependent on whether the deuterium was initially present at the position of substitution or in the adjacent position. In these cases deuterium is preferentially retained when it is originally present in the substrate in the adjacent position to substitution (62)[116, 120], and it must be assumed that both pathways '*a*' and '*b*' are operative in the decomposition of the intermediate arene oxides.

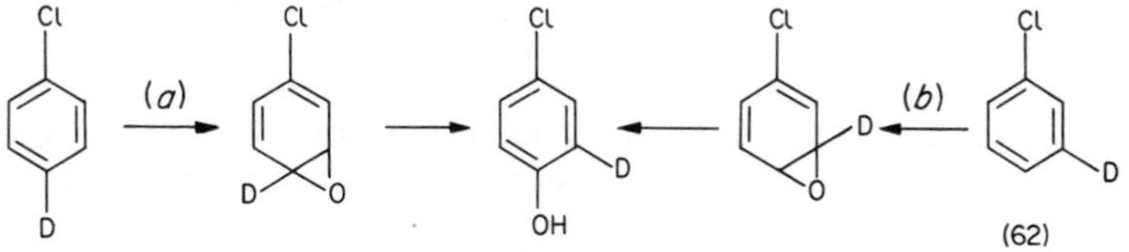

(61) (*a*,54%) (*b*,96%) (62)

D=^{2}H deuterium retention

The stoichiometric loss of tritium (55 per cent) through the introduction of the 3'-hydroxyl group into $3',5'$-[$^3H_2$]-L-tyrosine by L-tyrosine hydroxylase to give L-dopa[14, 34], and various other examples of this type, in which there is tritium or deuterium loss rather than migration following hydroxylation ortho to a preexisting hydroxyl group, are conveniently explained in terms of the arene oxide mechanism (*Figure 4.13*). In these cases loss of the hydrogen isotope is clearly the preferred mode of stabilisation of the appropriate intermediate (63) formed from the arene oxide.

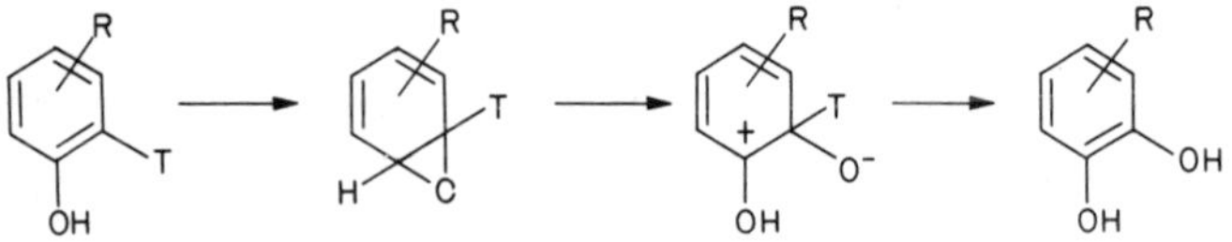

(63)

4.4.3 DIOXYGENASES

The dioxygenases, which incorporate two atoms of oxygen into one molecule of the substrate, are enzymes which are frequently involved in the cleavage of bonds in an aromatic ring. Typical of these are homogentisate oxidase[30] and L-tryptophan oxidase (L-tryptophan pyrrolase)[55] and two bacterial oxygenases pyrocatechase[126] and metapyrocatechase[127].

In bacteria the oxidative degradation of many benzenoid compounds leads to the formation of catechol or protocatechuic acid as the final intermediate possessing an aromatic structure, e.g. *Figure 4.7*. These o-diphenols are then substrates for oxidative ring fission. Pyrocatechase was first isolated from *Pseudomonas* species[126] and incorporates two atoms of oxygen into catechol, breaking the bond *between* the two hydroxyl groups to yield cis, cis-muconate (*Figure 4.7*). Protocatechuate-3,4-oxygenase[128] catalyses a similar oxygenation reaction with protocatechuate as substrate. Metapyrocatechase[127], on the other hand, catalyses an oxygenation reaction with catechol as substrate and incorporates two atoms of oxygen across the bond *adjacent* to one of the hydroxylated carbon atoms to give α-hydroxy muconic semialdehyde. Other enzymes of this class have been isolated but the mechanisms of action of neither group has yet been clearly defined.

4.5 AROMATIC AMINO ACID METABOLITES

4.5.1 MELANIN AND MELANOGENESIS

Melanin is the characteristic pigment of the skin, hair and retina of mammals and is a heterogeneous polymeric material of unknown structure which arises by oxidation of L-tyrosine. The transformations involved in the biogenesis of the melanins were largely discovered and formulated by Raper and his associates[129–130] and confirmed and supported by Mason[131–133]; L-tyrosine (5) is hydroxylated by an enzyme tyrosinase, a copper containing protein, to give L-dopa (15) which is further oxidised to L-dopaquinone (36), *Figure 4.14.* The latter then cyclises spontaneously to give L-cyclodopa (64)[134], which is similarly transformed by oxidation and decarboxylation to five 5,6-dihydroxyindole (66) the immediate precursor of melanin. Early studies on the composition and formation of dopamelanin led to its formulation as a polyindolequinone (67) but no firm conclusions were reached about the ways in which intermonomer linkages were formed. Some workers favoured 4,7 linked structures[135, 136] and others, such as Bu'Lock and Harley-Mason[137], suggested 3,7 linked polymers.

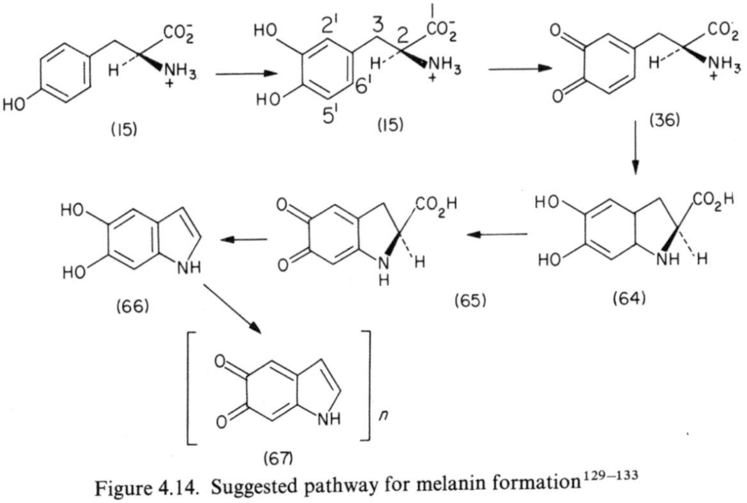

Figure 4.14. Suggested pathway for melanin formation[129–133]

The over-simplifications inherent in the Raper–Mason scheme for melanin formation (*Figure 4.14*) have been emphasised by degradative and biogenetic studies from a number of different laboratories.

This work, some aspects of which are outlined below, has served to show that melanins are in fact highly irregular polymers in which several different types of monomer unit are linked in a variety of ways. From the results of an extensive series of experiments in which D L-dopa deuteriated separately at the 1, 2, 2', 5' and 6' positions and D L-1-[^{14}C]-dopa were converted into dopa-melanin autoxidatively or by enzymic oxidation in the presence of mushroom polyphenol oxidase, Swan and his collaborators[138–141] have concluded that the principal structural fragments in the polymer are probably (*i*) uncyclised units of dopa (68, 0.1), (*ii*) indoline carboxylic acid units (69, 0.1), (*iii*) indole units (70, 0.65) and (*iv*) pyrrole carboxylic acid units (71, 0.15). The indoline (69) and indole (70) units are presumed to exist in the polymer half in the phenolic and half in the quinonoid forms. It was also suggested that the pyrrole carboxylic unit (71) is probably derived by oxidative fission, during the synthesis of the dopa-melanin, of a benzenoid unit such as (70).

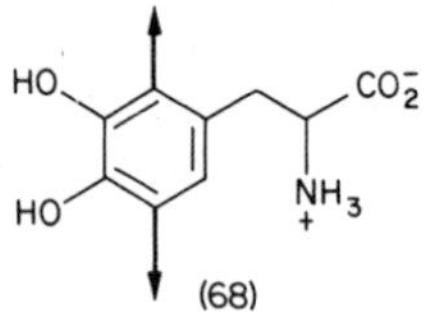

(68) 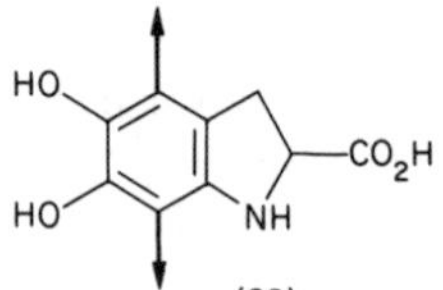   (69) 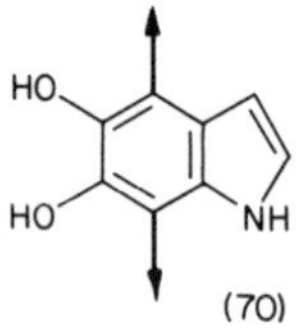   (70)

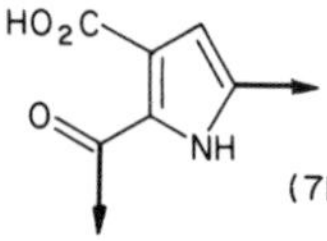

(71)

Kirby and Ogunkoya[142] have similarly demonstrated that dopa-melanin probably contains some uncyclised amino acid chains such as (68) in the polymer structure. Specimens of D L-dopa labelled singly with tritium at the 2, 3, 2', 5' and 6' positions and triply at the 2', 5' and 6' positions were each mixed with D L-2-[^{14}C]-dopa and then converted into melanin by mushroom tyrosinase. Since there is negligible loss of C-2 of dopa during its conversion to dopa-melanin[143], Kirby and Ogunkoya[142] argued that the measured changes of ^{3}H to ^{14}C observed in proceeding from substrate to product were a reflection of the loss of particular hydrogen atoms of dopa in the polymerisation. As expected a considerable loss of tritium was observed from the 6' position but the significant quantity retained (13 per cent) was interpreted in terms of the presence of uncyclised amino acid units (68) in the polymer or that in some units cyclisation had occurred on to the 2' rather than the 6' position of dopaquinone. The results of Kirby and Ogunkoya[142] also demonstrated unequivocally the irregular nature of the dopa-melanin polymeric structure and they showed that the regular $3 \rightarrow 7$ and $4 \rightarrow 7$ linked structures suggested by earlier workers were not compatible with the experimental evidence. Hempel[144], by a similar though probably less exact experimental procedure, studied melanogenesis in melanomas in mice. He concluded that some 70 per cent of the polymer units retained the carboxyl groups of the original amino acid and that many of these units were probably uncyclised.

Nicolaus and his collaborators[145, 146] have made a study of the melanin from the ink sac of *Sepia officinalis*, the so called sepiomelanin. Close structural similarities were noticed between the natural pigment and those derived by autoxidation of D L-dopa and 5,6-dihydroxyindole (66), and Nicolaus concluded therefore that the natural pigment contained units of the type (72) and (64) in a state varying from the fully reduced, through the semiquinonoid, to the quinonoid. Nicolaus also showed that during melanogenesis there was good evidence for the oxidative breakdown of indolequinone units to give pyrrole carboxylic acid units of the type (73) and (74). To account for the high percentage of carboxyl groups in sepiomelanin Nicolaus suggested that the molecule of dopachrome (65) was also involved in the polymerisation process.

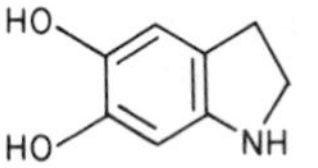

(72) 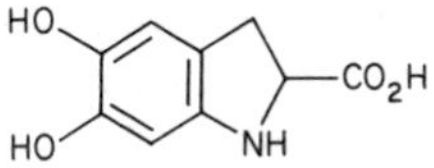   (64) 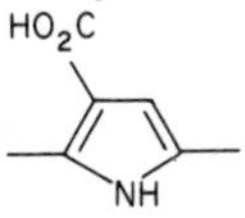   (73) 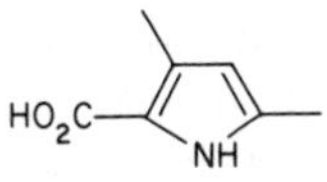   (74)

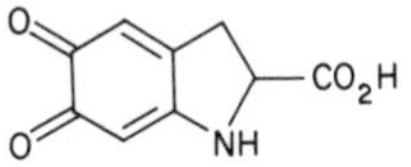

(65)

The work of Swan and his collaborators[147] also demonstrated, contrary to the expectations of the Raper-Mason hypothesis, that the melanin pigment derived from dopamine (16) by autoxidation differs substantially from a similarly derived dopa-melanin. In particular more uncyclised chains and intermonomer carbon-nitrogen linkages were shown to be present in the dopamine pigment.

The only enzyme which is apparently concerned with the conversion of L-tyrosine (4) to the polymer melanin is tyrosinase which catalyses the initial transformation of the amino acid to

L-dopaquinone (36). *Figure 4.14.* The conversion of L-dopaquinone (36) to dopa-melanin is then assumed to be spontaneous. Albinism, which is characterised in mammals by a complete or partial lack of melanin pigments in the skin and hair, results from a deficiency in tyrosinase activity. It represents a further example of a disease which has resulted from an 'inborn error of metabolism'.

4.5.2 THYROXINE AND THE THYROID HORMONES

Thyroglobulin is a glycoprotein of approximately 660 000 molecular weight which comprises the bulk of the colloid in the thyroid follicles. It serves as a storage vehicle for the two thyroid hormones thyroxine (80) and 3,3',5-tri-iodothyronine (76) and provides a matrix for their synthesis. The molecule of thyroglobulin contains approximately 120 residues of L-tyrosine and under normal conditions some 10–25 of these are present in iodinated forms and up to five in the form of thyroxine[148].

Numerous L-tyrosine containing peptides and proteins may be utilised to synthesise thyroxine (80) by iodination and without enzymic mediation[149–154]. Iodine reacts with the L-tyrosine residues in the peptide to form mono- and di-iodo-L-tyrosine residues. An oxidative coupling of two, appropriately placed, di-iodo-L-tyrosyl residues then occurs to give thyroxine (80). Considerable evidence has been accumulated[155] to suggest that thyroxine biosynthesis follows the same pathway *in vivo*, but other mechanisms have been suggested and examined[156–162]. The exact role of the protein thyroglobulin in thyroxine biosynthesis is, however, not clear for although L-tyrosine residues of other proteins may be readily iodinated *in vitro* only thyroglobulin is known to make thyroxine *in vivo*.

The study *in vitro* of various model reactions has contributed considerably to this area of investigation and to the formulation of alternative proposals regarding the mode of biogenesis of thyroxine. Thus it has been suggested that thyroxine may also be formed by oxidative coupling of protein bound di-iodo-L-tyrosine and free di-iodo-L-tyrosine or di-iodo-4-hydroxyphenylpyruvic acid. Nearly fifty years ago Harrington and Barger[163] suggested that 3', 5'-di-iodo-L-tyrosine (75) was the precursor of thyroxine *in vivo* and this idea was supported by the results of several studies on model systems[151–156]. Johnson and Tewkesbury[156] proposed that oxidation of the substrate (75) gave the phenoxyradical (77) which coupled with its mesomeric form (78) to give the intermediate dimer (79). From this they postulated that loss of aminoacrylic

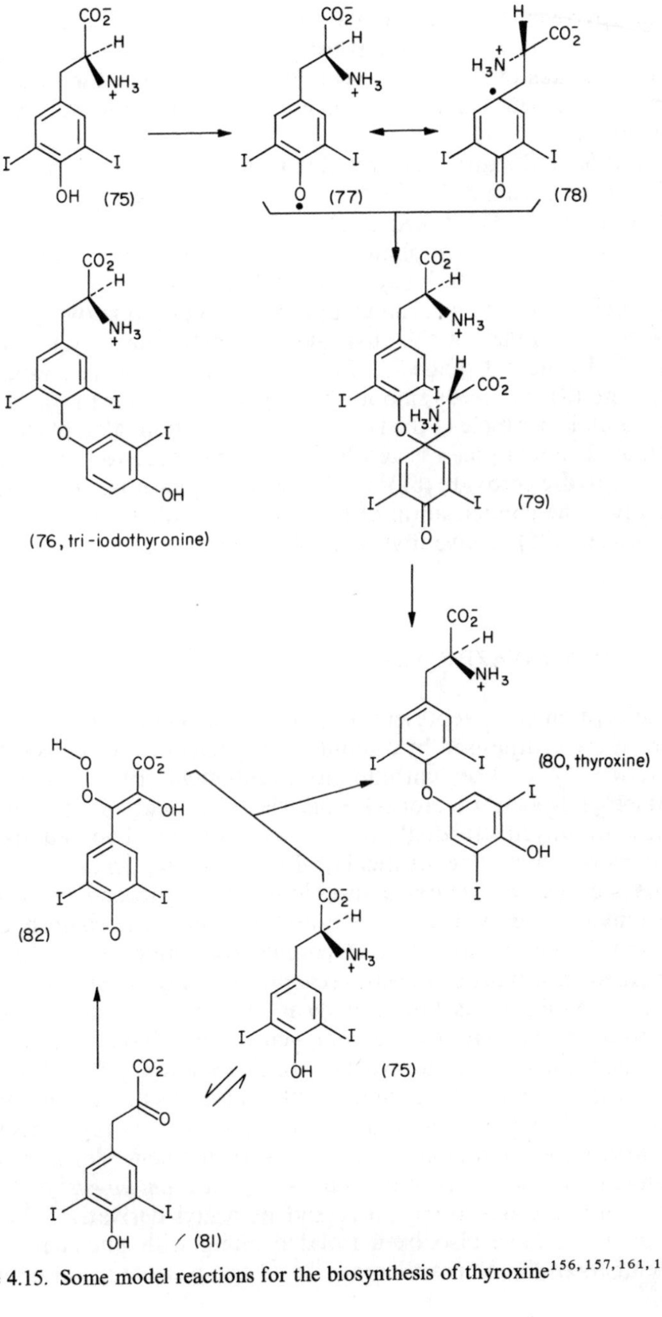

Figure 4.15. Some model reactions for the biosynthesis of thyroxine[156, 157, 161, 162]

acid (pyruvic acid+ammonia on hydrolysis) occurred to give thyroxine (80), *Figure 4.15*. Subsequent work[164, 165] has sought to clarify details of this mechanism and in particular the fate of the amino acid side chain which is eliminated, but this still remains uncertain

In 1956, Hillman[157] suggested that the first step in the formation of thyroxine from di-iodo-L-tyrosine (75) might be the deamination of the amino acid to its keto analogue (81), and later work has shown that the oxidative coupling of (75) and (81) is an efficient non-enzymic model for the biosynthesis of thyroxine[158, 159, 161, 162]. The keto acid (81) also reacts under appropriate conditions with di-iodo-L-tyrosyl residues in the thyroglobulin molecule to give thyroxine[160]. In the pH range 7.2–7.6, the *in vitro* oxidative coupling of (75) and (81) proceeds smoothly and gives thyroxine in yields of up to 0.4 mole per mole of the keto acid. The reaction takes place in two stages. The first phase is aerobic[161, 162] and leads to the formation of the hydroperoxide (82)[161]. The second phase is anaerobic and involves the condensation of the hydroperoxide (82) with di-iodo-L-tyrosine (75) to give thyroxine (80), *Figure 4.15*.

4.5.3   PHENOXAZINONES

The actinomycins represent a family of chromopeptide antibiotics which are synthesised by a number of species of the genus *Streptomyces*[166–169]. The chromophore actinocin (91, 2-amino-4,6-dimethyl-3-phenoxazinone-1,9-dicarboxylic acid) is attached in the actinomycins through its carboxyl groups to two pentapeptide chains which may be identical or differ in amino acid composition. The peptides are cyclic structures in which the carboxyl group of the terminal N-methyl-L-valine is esterified to the side chain hydroxyl group of the amino acid L-threonine. An actinomycin producing organism normally synthesises several actinomycins simultaneously and these substances differ usually at only one amino acid site on the peptide chains. The two actinomycins which have received most detailed study are actinomycin C_1 (or D, 83) and C_3. Phenoxazinone structures similar to actinocin (91) also occur in the ommochromes[170] (e.g. xanthommatin, 84), a group of eye pigments found in arthropods, cinnabarin (85), cinnabarinic acid (86) and tramesanguin (87) pigments of the fungus *Coriolus sanguineus*[171–174] and the antibiotic questiomycin A and its acetyl derivative[175]. Phenoxazinones have also been isolated along with phenazines from members of the Norcardiaceae[176–178] and a structural fragment of the antibiotic anthramycin[179] is identical to 3-hydroxy-4-methyl-anthranilic acid which may be regarded as the biological precursor of actinocin the chromophore of actinomycin.

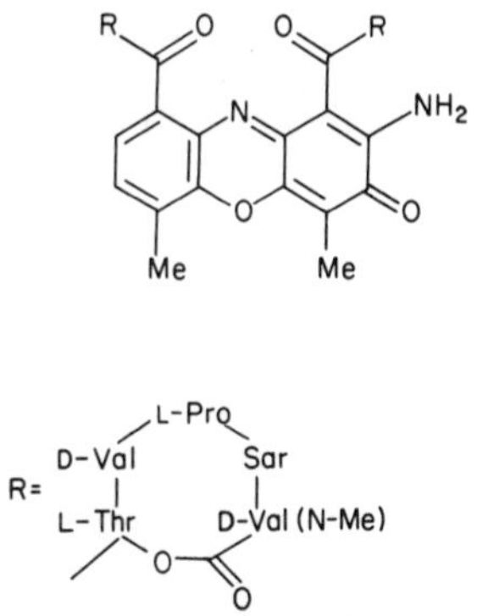

(84, xanthommatin)

(83, actinomycin C_1)

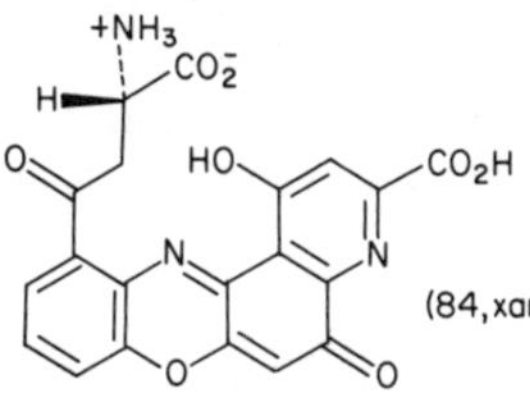

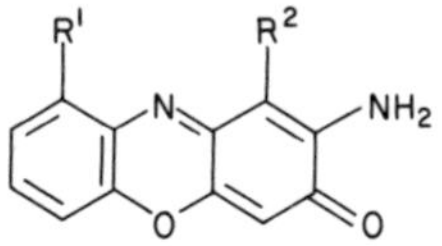

(85, R^1 = CH_2OH, R^2 = CO_2H, cinnabarin)

(86, R^1 = R^2 = CO_2H, cinnabarinic acid)

(87, R^1 = CO_2H, R^2 = CHO, tramesanguin)

There is now firm evidence from genetic, biochemical and chemical investigations that products of L-tryptophan oxidative metabolism play a central role in phenoxazinone biosynthesis. Thus 3-hydroxyanthranilic acid (25) is readily dimerised oxidatively to give cinnabarinic acid (86) and enzymes having cinnabarinate synthetase activity have been isolated from mammalian mito-chondria[180], liver[181, 182], plants[183] and micro-organisms[184, 185]. The same oxidative dimerisation has been effected by a range of inorganic oxidants[178] and may also occur on tlc plates on exposure to air, *Figure 4.16*.

Butenandt[170] has similarly shown that the ommochrome eye pigments found in *Drosophila* species were not formed in some mutants unless kynurenine (24) or 3-hydroxykynurenine (88) was provided in the diet. In addition xanthommatin (84), rhodommatin and ommatin C, ommochromes synthesised by the blowfly *Callio-phora erythrocephala*, were demonstrated to possess radioactivity after injection of either D L-3-[14C]-tryptophan or D L-5-[14C]-kynurenine[186, 187]. Based on these observations Butenandt pro-posed[170] that xanthommatin (84) was biosynthesised by an oxidative condensation of two molecules of 3-hydroxykynurenine (88) and a biogenetically patterned synthesis of (84) was achieved by treatment of 3-hydroxykynurenine (88) with potassium ferri-cyanide. The synthesis proceeds probably by oxidation to the *o*-quinoneimine, condensation to the phenoxazinone (89) and finally cyclisation of the alanine side chain to form a quinoline ring, *Figure 4.16*. The enzymology of the initial steps of L-tryptophan metabolism and the synthesis of ommochromes in *Drosophila melanogaster* has been well defined[188-190] and more recently it was demonstrated that a unique enzyme system exists in the insect for the final oxidative sequence from 3-hydroxykynurenine (88) to the pigment[191].

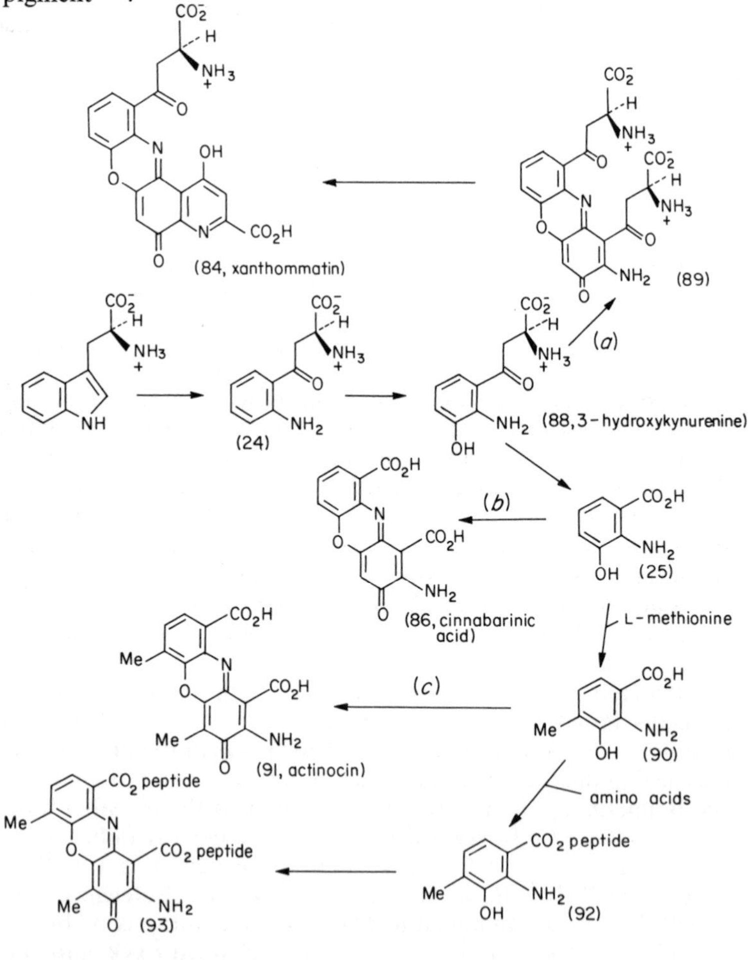

(a) *Drosophila melanogaster*
(b) cinnabarinate synthetase
(c) phenoxazinone synthetase

Figure 4.16. The oxidative transformation of some L-tryptophan metabolites

Actinocin (91) and its derivatives, the actinomycins (83), can be analogously prepared chemically in almost quantitative yield from 3-hydroxy-4-methylanthranilic acid (90) and it has been postulated that this compound is the immediate precursor of the actinomycin chromophore *in vivo*[192, 193]. Direct biochemical evidence for this hypothesis has been obtained by Katz and Weissbach and their collaborators. Actinomycin synthesis was initially studied[194] with washed mycleia of *Streptomyces antibioticus* and several important results were obtained. Stimulation of ^{14}C labelled amino acid incorporation (L-valine, L-methionine, D L-tryptophan and L-proline) was achieved by addition of 3-hydroxy-4-methylanthranilic acid (90), 3-hydroxyanthranilic acid (25) or 3-hydroxykynurenine (88—precursors of the antibiotic chromophore—and from incubations with ^{14}C-labelled D L-tryptophan or L-methionine (methyl-^{14}C), radioactively labelled 3-hydroxy-4-methylanthranilic acid (90) was isolated. In addition, utilising 5′-[^{3}H]-D L-tryptophan as substrate a stoichiometric loss of one tritium atom was observed during the conversion of two molecules of the amino acid to actinomycin. In later work the enzymic synthesis of actinocin (91) from 3-hydroxy-4-methylanthranilic acid (90) was achieved[195, 196] employing a cell free system from *Streptomyces antibioticus*. The metabolic origin of 3-hydroxyanthranilic acid (25) from L-tryptophan is well established (*Figure 4.5*) and this same pathway has been observed in *Streptomyces antibioticus*[197–199]. Isotopic labelling experiments have also established that the methyl groups of actinocin are derived from L-methionine[199, 200] but the point at which methylation is effected is not known precisely, although it seems probable that this occurs before the oxidative condensation. In a similar way the point at which the peptide chains are attached is not known with certainty. Brockmann[192, 201] has suggested on the basis of chemical models that the actinomycins (93) are derived from 3-hydroxy-4-methylanthranilic acid peptides (92). Support for this particular route of biosynthesis came from the observation that such peptide derivatives are smoothly transformed to actinomycin analogues by the phenoxazinone synthetase of *Streptomyces antibioticus*[202].

(88)

(94, questiomycin A)

L-Tryptophan metabolites have also been implicated in the biosynthesis of the two antibiotics questiomycin A (94) and questiomycin B. Thus when ring ^{14}C labelled 3-hydroxykynurenine (88) was administered to *Streptomyces chromofuscus*, the antibiotic (94) was found to incorporate the label to the same extent as *o*-aminophenol (questiomycin B)[203]. This result was interpreted to suggest that the side chain of 3-hydroxykynurenine (88) was eliminated prior to the oxidative condensation of two molecules to give questiomycin A (94).

4.5.4 DIKETOPIPERAZINES AND RELATED PRODUCTS

The diketopiperazines and their derivatives form a restricted group of microbial metabolites and biosynthetic studies indicate that they are formed, as expected, from the corresponding natural L-amino acids. Thus echinulin (95) has been shown[204-207] to be derived from L-tryptophan and L-alanine. Mycelianamide (100), from *Penicillium griseofulvum*, which can be considered as a diketopiperazine in which the ring has suffered oxidation, is similarly formed[208, 209] from L-tyrosine and L-alanine. In the case of echinulin (95), cyclo-L-alanyl-L-tryptophan (96) has been shown to be a precursor of the metabolite[210] and is a substrate for the subsequent isoprenylation reactions. Thus Allen[211] has shown that ammonium sulphate fractions of cell free extracts of *Aspergillus amstelodami*, catalyse the transfer of an isoprenoid unit from isopentenylpyrophosphate to cyclo-L-alanyl-L-tryptophan to form (97) a further intermediate in the biosynthesis of echinulin (95). Bycroft and Landon[212] have suggested, mainly on the basis of chemical analogy, that the γγ-dimethylallyl group may be introduced into the L-tryptophan nucleus of echinulin and other metabolites by formation of an enzyme thiol addition product (98) followed by a sulphonium ylide rearrangement via the indolenine (99).

The assignment of the structures of the brevianamides A–E, five closely related diketopiperazine metabolites isolated from the mould *Penicillium brevicompactum*, was based partly on the results of biosynthetic studies and biogenetic considerations[213, 214]. The presence of an indole unit and a diketopiperazine ring in the brevianamides suggested the possibility of the formation of these metabolites from L-tryptophan and L-proline (101) with the additional C_5 unit in the molecules originating in mevalonate (102). Biosynthetic evidence to support this conclusion was obtained in the case of brevianamide A (105). Thus D L-3-[14C]-tryptophan and L-U-[14C]-proline were both readily incorporated into the

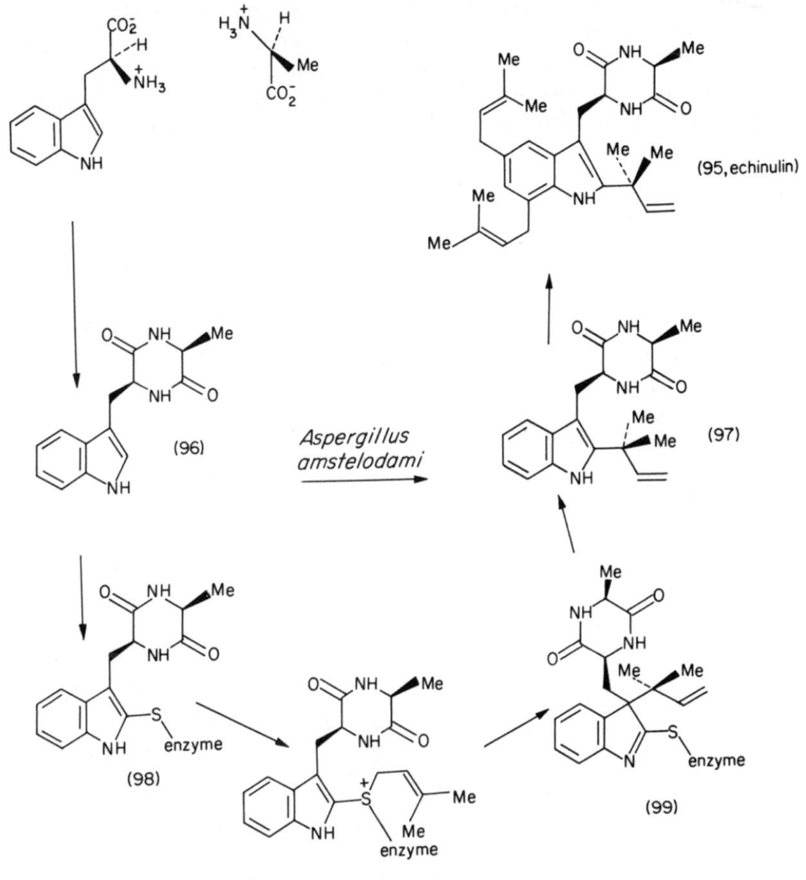

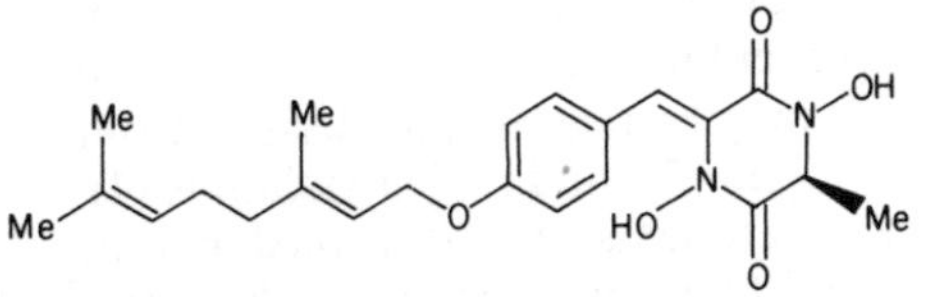

(95, echinulin)

(96)

Aspergillus amstelodami

(97)

(98)

(99)

(100, mycelianamide)

metabolite in typical feeding experiments. Birch and Wright[213] tentatively postulated that the isoprenylated cyclo-L-tryptophanyl-L-proline (104) might act as a biogenetic precursor for the various brevianamides, e.g. A (105) and E (106). Support for this hypothesis was obtained by Steyn[215] who isolated both (104) and austamide a product of alternative ring closure of (104), from *Aspergillus ustus.* Cyclo-L-tryptophanyl-L-proline, brevianamide F (103), was also isolated from *Penicillium brevicompactum* by Birch and Russell[214] but it has not yet been unequivocally proved whether either (103) or (104) or both are true biosynthetic intermediates to the various other brevianamides as shown in *Figure 4.17.*

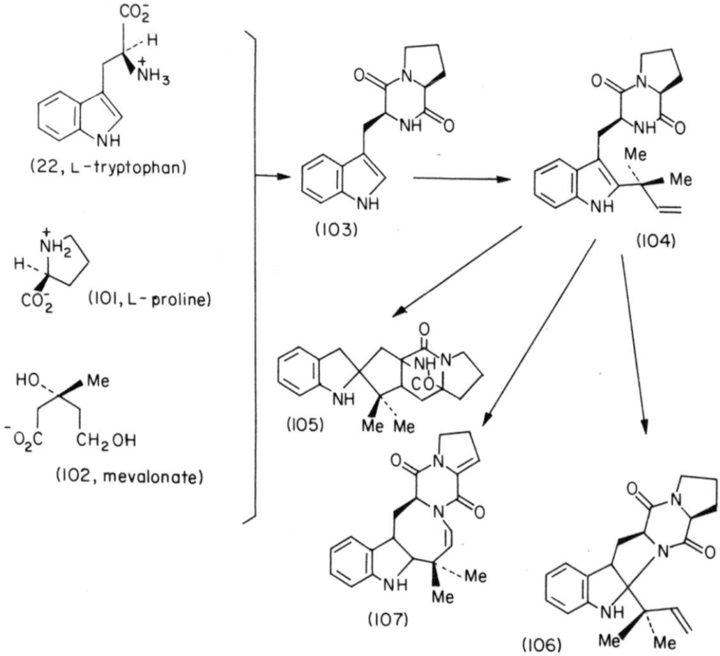

Figure 4.17. Biogenesis of the brevianamides

Extensive biosynthetic studies have been carried out[216,217] on the benzodiazepin alkaloids (—)-cyclopenin and (—)-cyclopenol (108 and 109) isolated from the culture medium of *Penicillium cyclopium*[218, 219]. Both alkaloids may be regarded as derived from the cyclic dipeptide of anthranilic acid (29) and L-phenylalanine (4), and typical biosynthetic feeding experiments have confirmed that the carbon skeleton of both (108) and (109) originate from these two amino acids. These experiments have also shown that the N-methyl group derives from L-methionine, that the two nitrogen atoms result directly from the amino groups of (4) and (29) and that the introduction of the hydroxyl function into the aromatic ring occurs via a mixed function oxidase as one of the later stages of the biosynthetic pathway. *m*-Tyrosine significantly was only incorporated into the alkaloids in an unspecific fashion.

Cyclopenase, an enzyme isolable from the mycelia of *Penicillium viridicatum* and *Penicillium cyclopium*, catalyses the transformation of the benzodiazepins (108 and 109) to the quinoline derivatives viridicatin (110) and viridicatol (111)[220,221]. This rearrangement, which may also be accomplished with acids *in vitro*, is accompanied by the elimination of molar quantities of carbon dioxide and methylamine. 5-[^{14}C]-Cyclopenin, biosynthesised from carboxyl labelled [^{14}C]-anthranilic acid (29), gave viridicatin (110) which was devoid of radioactivity. All the radioactivity was found to be present in the carbon dioxide evolved. When the enzymatic transformation was carried out in ^2H_2O or $H_2{}^{18}O$ there was no significant incorporation of deuterium or ^{18}O into the products. These observations have enabled a mechanism for this complex biosynthetic rearrangement to be formulated[221] in which carbon dioxide is eliminated from C-5 of (108 or 109) and the epoxide ring is opened to allow the formation of the hydroxyl group on the quinoline ring and the eventual union of C-5a and C-10.

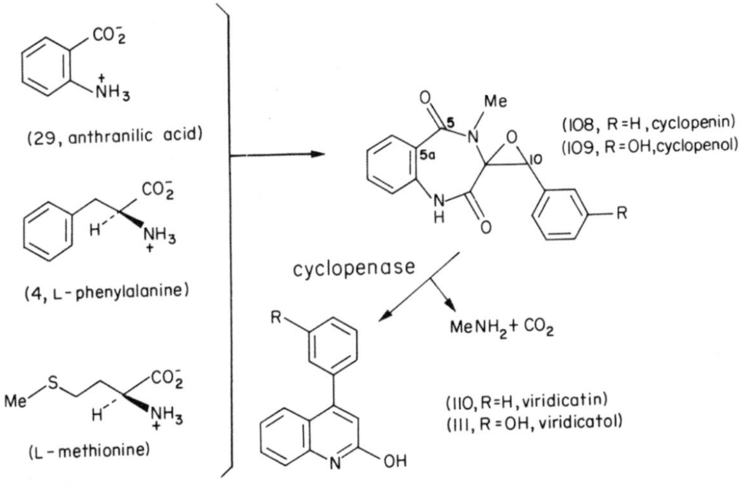

(29, anthranilic acid)

(4, L-phenylalanine)

(L-methionine)

cyclopenase

(108, R=H, cyclopenin)
(109, R=OH, cyclopenol)

$MeNH_2 + CO_2$

(110, R=H, viridicatin)
(111, R=OH, viridicatol)

Gliotoxin (112) was the first member of a series of microbial metabolites containing the unusual disulphide bridged diketo-piperazine ring system to be isolated and characterised[222,223]. More recently additional members of this class of compound such as arantoin (113)[224], verticillin A (114)[225] and the closely related chaetocin (115)[226], and the various sporidesmins (e.g. 116)[227-229] have been found. Biogenetic considerations have lent considerable assistance to their structural formulation and comparative CD measurements show that the configuration of the two asymmetric centres of the disulphide bridge in verticillin A (114) and chaetocin (115) are antipodal to those in gliotoxin (112), arantoin (113) and the sporidesmins (116).

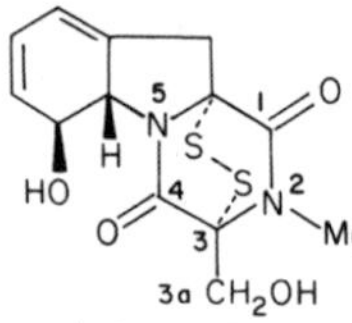

(112, gliotoxin)

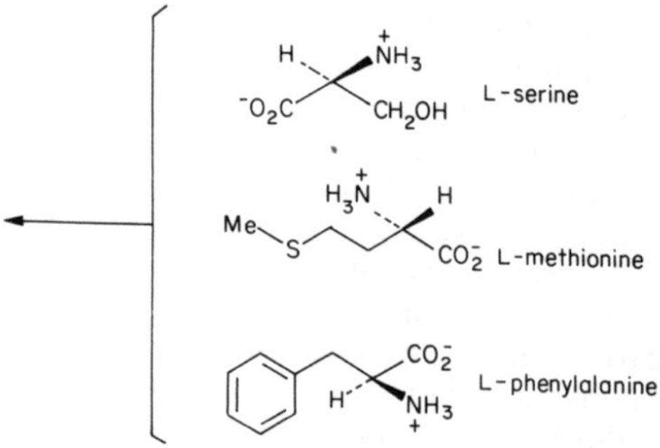

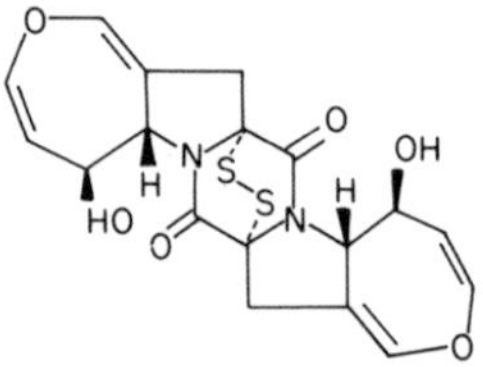

(113, arantoin)

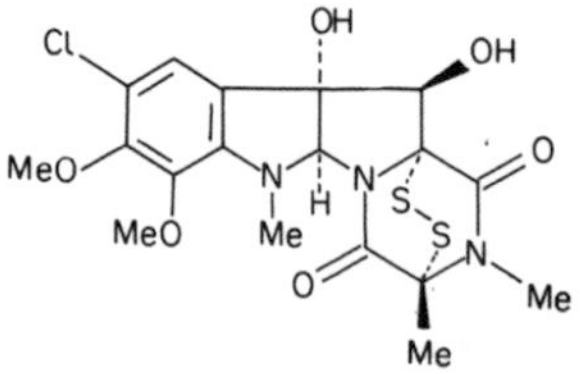

(116, sporidesmin)

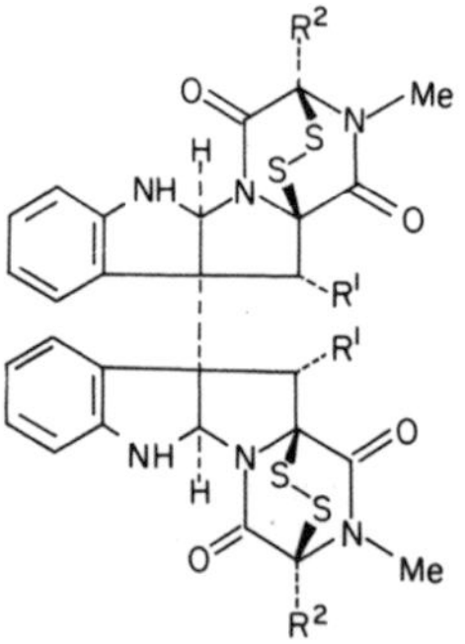

(114, R^1=OH , R^2= Me , verticillin A)
(115, R^1=H , R^2 = CH$_2$OH, chaetocin)

L-serine

L-methionine

L-phenylalanine

Considerable attention has been focused on the biosynthesis of gliotoxin (112) in *Trichoderma viride*. Vigorous aeration is required for maximum gliotoxin production when the organism is grown on a D-glucose-salts medium at 27° and pH 3.5. Suhadolnik and his collaborators showed by conventional methods that L-phenylalanine was a precursor of the indole portion of the gliotoxin molecule and contributes, without rearrangement, all its nine carbon atoms to the structure[230-232]. L-Methionine was shown to serve as a precursor of the N-methyl group and L-serine as a source of the carbon atoms 3, 3a and 4. Winstead and Suhadolnik[231] also demonstrated that D L-m-tyrosine-[^{3}H] was readily incorporated in high yield (88 per cent—assuming only the natural isomer was utilised) into the metabolite. Suhadolnik therefore implicated L-m-tyrosine as an intermediate on the biosynthetic pathway to gliotoxin and suggested that meta hydroxylation of L-phenylalanine occurred to give L-m-tyrosine before cyclisation of the alanine side chain. Examination of gliotoxin biosynthesised from ^{15}N-L-phenylalanine indicated[232] that only N-5 was labelled but using ^{15}N-glycine randomisation of the nitrogen label was observed. This was attributed to the metabolism of glycine to 'active' formate and the release of the amino-group to the nitrogen pool. The authors suggested[232] that this nitrogen pool was used for the formation of substantial quantities of L-phenylalanine by transamination. Using double labelled intermediates it was, however, later shown[233] that L-phenylalanine and N-methyl-L-serine (or L-serine) readily participate in the nitrogen pool but that the exchange of the amino-group between L-phenylalanine and L-serine is limited.

Subsequent work by Bu'Lock and Kirby and their collaborators[234–236] has led to definitive proposals regarding the biosynthesis of gliotoxin and in particular to a reconsideration of the status of L-m-tyrosine as an intermediate on the pathway. Bu'Lock and Ryles[234] demonstrated that 2-[^{14}C]-L-m-tyrosine and 2′,4′, 6′-[^{3}H$_3$]-L-m-tyrosine and similarly labelled L-o-tyrosine and L-2′, 3′-dihydroxyphenylalanine were not incorporated into gliotoxin. On the other hand 2′, 3′, 4′, 5′, 6′-[^{2}H$_5$]-L-phenylalanine was incorporated intact into gliotoxin, with the retention of all of its aromatic hydrogen atoms. This observation, along with a consideration of the stereochemistry of the amino alcohol system of gliotoxin (112), led to the postulate that the indole ring of gliotoxin could originate by interaction of the amino nitrogen of L-phenylalanine and an epoxide group in the aromatic ring (119). Kirby and Johns[235] similarly concluded that L-m-tyrosine is not an obligatory intermediate in the conversion of L-phenylalanine to gliotoxin. Thus D L-2,4,6-[^{3}H$_3$]-m-tyrosine and D L-1-[^{14}C]-phenylalanine fed in admixture [^{3}H:^{14}C = 11] to *Trichoderma viride* gave gliotoxin [^{3}H:^{14}C = 0.1], although good incorporation of the isotopic carbon (4.9 per cent) had occurred. Further evidence to support the intervention of a benzene oxide type intermediate in the biosynthesis of gliotoxin was also obtained by Kirby and Johns. D L-3′,5′-[^{3}H$_2$]-phenylalanine and D L-1-[^{14}C]-phenylalanine were fed in admixture [^{3}H:^{14}C = 7.07] to *Trichoderma viride*. The gliotoxin [^{3}H:^{14}C = 7.51] isolated showed that no loss of tritium had occurred in the transformation. Dehydration and desulphurisation of the metabolite gave (117, $^3H:^{14}C = 7.51$] and dehydrogenation yielded dehydro-gliotoxin (118, $^3H:^{14}C = 3.91$). This evidence showed that hydroxylation of L-phenylalanine to give L-*m*-tyrosine in a process which typically produces a loss or migration (NIH shift)[17] of the tritium from the site of attack had not occurred during the biosynthesis of gliotoxin.

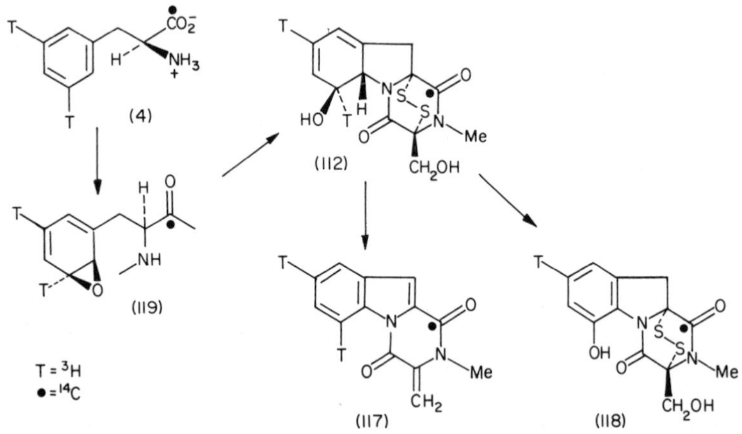

In later work, Bu'Lock, Kirby, Johns and Ryles[236] demonstrated that the methylene group of L-phenylalanine contributed directly to the benzylic methylene group of gliotoxin (112). The interpretation of these results was, however, complicated by the observation that under the conditions of feeding, some process, quite independent of gliotoxin synthesis, caused a stereospecific loss of *one* of the β-methylene protons of L-phenylalanine.

Two reactions dominate the chemistry of arene oxides; the first which has been discussed above is their facile conversion to phenols and the second is their valence bond tautomerism to oxepins. Circumstantial evidence for the role of an aromatic epoxide type of intermediate (119) in the biosynthesis is provided by the location of two oxepin structural units in the related metabolite arantoin (113). The origin of these rings has been plausibly attributed to valence bond tautomerism of an L-phenylalanine epoxide[224] (*Figure 4.18*) and some experimental data to support this suggestion has been obtained by Brannon, Mabe, Molloy and Day[237]. These workers showed that $[^2H_8]$-L-phenylalanine was readily incorporated into acetylarantoin in *Aspergillus terreus* and contributed seven (or fourteen) hydrogen atoms to the molecule. In this case, in

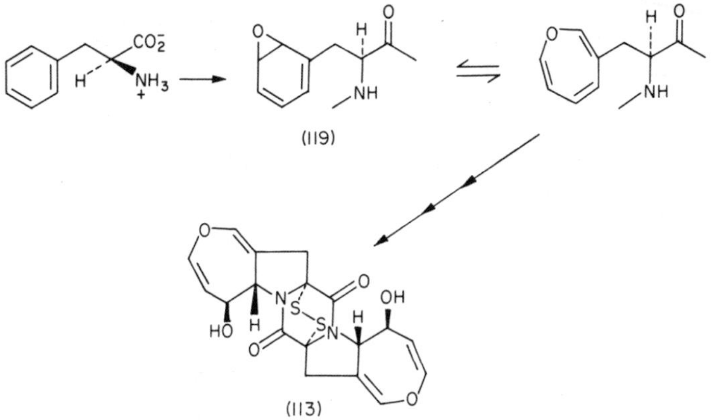

Figure 4.18. Origin of the oxygen ring systems of arantoin[17, 224]

contrast to that of gliotoxin, the β-methylene group of the amino acid was assimilated intact into the metabolite.

As yet the stages at which cyclisation of the amino nitrogen and the introduction of the disulphide bridges occur have not been elucidated, nor has the origin of the sulphur of the disulphide bridge been ascertained. Similarly, although no experimental work has been reported on the sporidesmins (116), verticillin-A (114) and chaetocin (115), it can by analogy be predicted that these metabolites are derived from L-tryptophan by similar processes to those described for gliotoxin.

4.5.5   NOVOBIOCIN

Novobiocin is a fungal metabolite with antibiotic properties produced by *Streptomyces niveus*. The structure of novobiocin (120) contains some unique features including the sugar noviose containing a carbamyl group and the unusual 3-amino-4-hydroxy coumarin fragment[238, 239].

Particular attention has focused on one aspect of the biosynthesis of novobiocin—the formation of the coumarin ring. Kenner and his

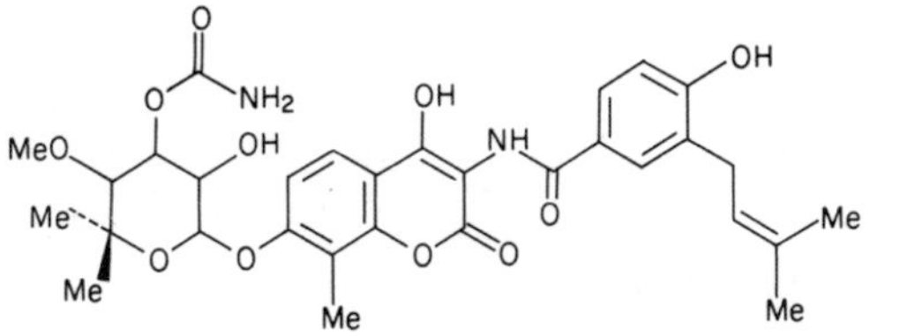

(120, novobiocin)

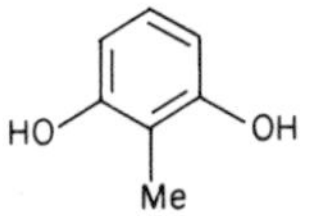

(121)

associates[240] have shown that, in contrast to the synthesis of the coumarin ring in plants, the synthesis in novobiocin takes place by an oxidative cyclisation reaction. L-U-[^{14}C]-Tyrosine was incorporated into the coumarin fragment of novobiocin and also the *p*-hydroxybenzoyl grouping. The relative activities of the two segments of structure showed that L-tyrosine was equally effective as a precursor of both aromatic nuclei. When *Streptomyces niveus* was grown in the presence of carboxyl-[^{18}O]-L-tyrosine (prepared

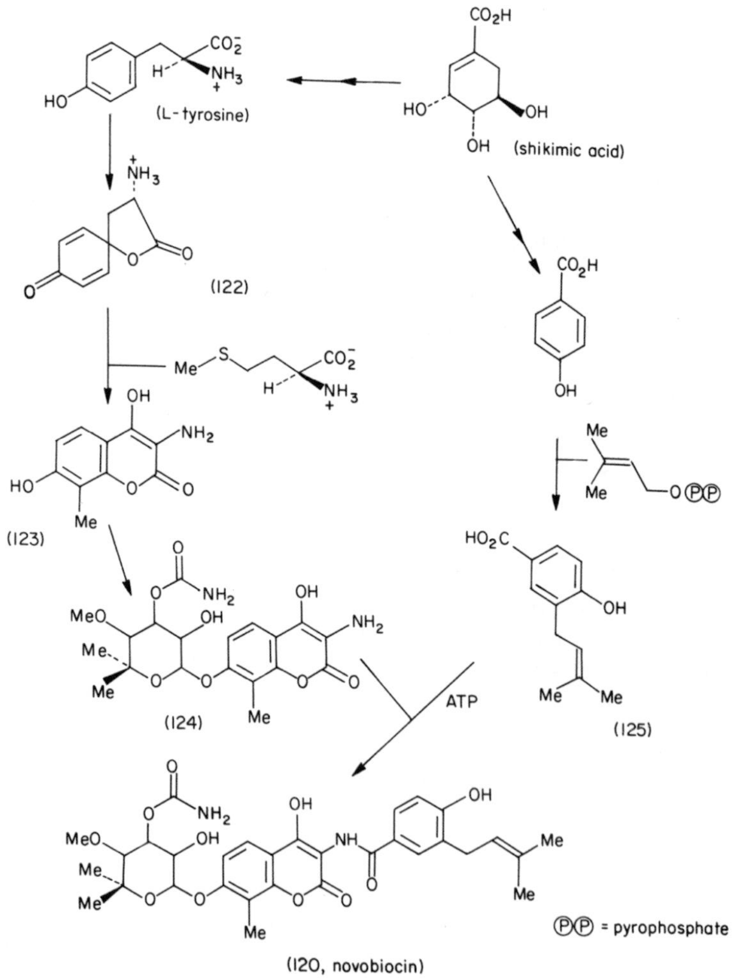

Figure 4.19. Suggested pathway of novobiocin biogenesis[200, 240, 241]

from the amino acid by hydrogen bromide catalysed exchange with $H_2^{18}O$) then novobiocin labelled with ^{18}O was isolated. The metabolite was degraded to give 2-methyl resorcinol (121) from the coumarin ring and this product showed a small but significant atom excess of ^{18}O consistent with the inclusion of ^{18}O from the carboxylate group of L-tyrosine in the heterocyclic ring of novobiocin. This experiment supported the conclusion that a mode of oxidative cyclisation is involved in the biogenesis of the coumarin ring of novobiocin. Kenner has suggested that this could occur via a spirodienone type of intermediate (e.g. 122) followed by rearrangement.

Fermentation of *Streptomyces niveus* in the presence of Me-[^{14}C]-L-methionine yielded novobiocin with an incorporation of 10 per cent of the added isotope[200]. The radioactivity was distributed between the O-methyl and gem-dimethyl groups of the sugar noviose (66 per cent) and the C-methyl group of the coumarin fragment (123, 35 per cent) and thus indicated that the coumarin methyl group was formed by a process of C-methylation. The timing of this step is, however, not known. Enzyme systems have been isolated[241] from *Streptomyces niveus* which are capable of rapid amide bond formation between the isoprenyl 4-hydroxybenzoic acid (125) and the coumarin (123) to give novobiocic acid. The reaction requires ATP and presumably involves some form of activation of the p-hydroxybenzoyl carboxyl function.

On the basis of this and the other biosynthetic evidence, Kominek[241] has identified the rate limiting steps in the biosynthesis of novobiocin as the formation of the coumarin (123) and p-hydroxybenzoic acid from (−)-shikimic acid. He proposed a pathway of biosynthesis (*Figure 4.19*) in which the final step is condensation of the noviosyl coumarin (124) and the p-hydroxybenzoic acid derivative (125).

4.5.6   MISCELLANEOUS METABOLITES

The terphenyl quinones are a unique group of natural products confined, with the exception of volucrisporin (128), to the Basidiomycetes and certain lichens. They are all derivatives of 2,5-diphenyl p-benzoquinone and have a varying pattern of hydroxylation in the aromatic ring. Typical members of this class are atromentin (126) and the bronze pigment polyporic acid (127) which constitutes over 20 per cent of the dry weight of *Polyporus rutilans*. The red pigment volucrisporin (128) is the only quinone of this group found in a lower fungus, *Volucrisporia aurantiaca*, and differs from the other terphenyl quinones in the absence of hydroxyl groups attached to the quinone ring and the *m*-hydroxylation pattern in the phenyl groups.

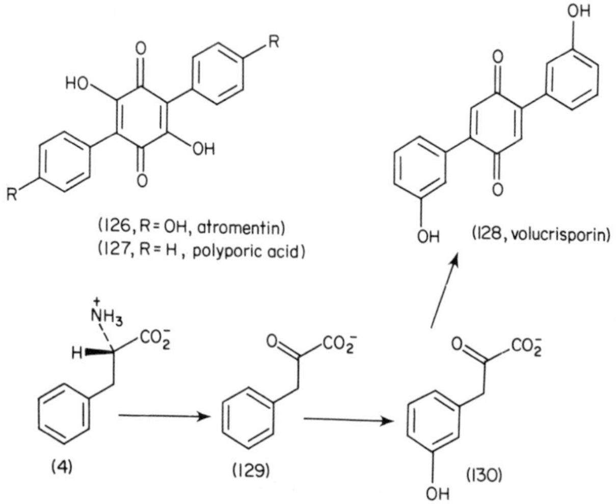

(126, R = OH, atromentin)
(127, R = H, polyporic acid)

(128, volucrisporin)

(4)   (129)   (130)

Feeding experiments with ^{14}C labelled precursors showed that shikimic acid, L-phenylalanine and L-*m*-tyrosine were readily incorporated into volucrisporin (128) in *Volucrisporia aurantiaca*, and it was suggested[242, 243] that biogenesis most probably proceeds via phenylpyruvic acid (129) and *m*-hydroxyphenylpyruvic acid (130) before condensation to give the quinone pigment. An analogous pathway has recently been confirmed[244] in a study of the biosynthesis of phlebiarubrone (131) in *Phlebia strigosozonata* using ^{13}C labelled precursors. These experiments showed that the molecule was assembled from two molecules of L-phenylalanine and one of formic acid. It seems very probable that other terphenyl quinones are biosynthesised by similar condensation of appropriate $C_6 . C_3$ precursors.

The terphenyl quinones are closely related to the pulvinic acid pigments (e.g. 132, 133) which are confined to lichens and when they

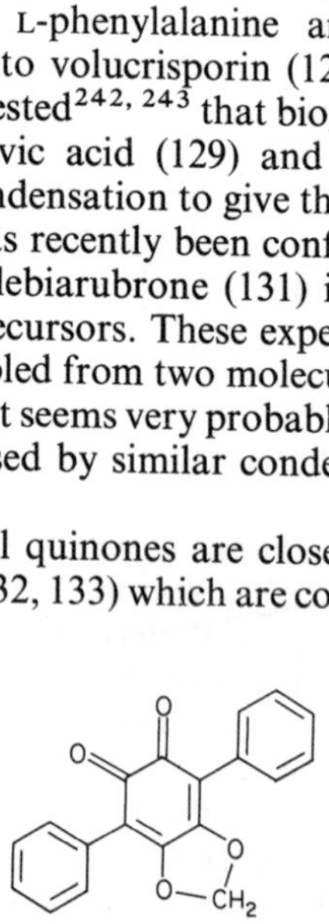

(131, phlebiarubrone)

occur together it is conceivable that the quinones are the biogenetic precursors of the pulvinic acid compounds. Indeed such conversions can be carried out oxidatively in the laboratory and tracer experiments in *Pseudocyphellaria crocata* have indicated that polyporic acid (127) can act as a precursor of pulvinic lactone (132)[245].

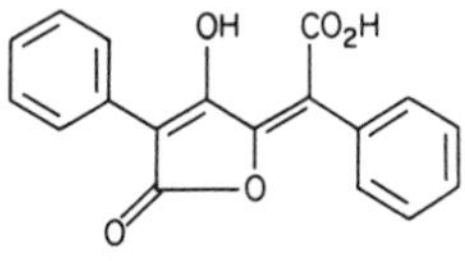

(133, pulvinic acid) 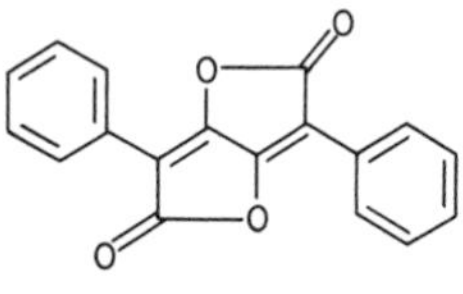   (132, pulvinic lactone)

Xanthocillin (134) is a yellow antibiotic pigment produced by cultures of *Penicillium notatum*. A novel structural feature[246], which is of considerable interest from the biogenetic point of view, is the presence of the isonitrile grouping. Because of its symmetrical structure, it has been assumed that the xanthocillin molecule is formed by union of two biogenetically equivalent precursors and tracer experiments performed by Achenbach and Grisebach[247] showed that the $C_6 . C_2$ unit was most probably derived from tyrosine. Thus D L-2-[^{14}C]-tyrosine was incorporated into the antibiotic in high yield (25 per cent) and the position of the isotopic label determined by degradation to the α-diketone (135) and anisic

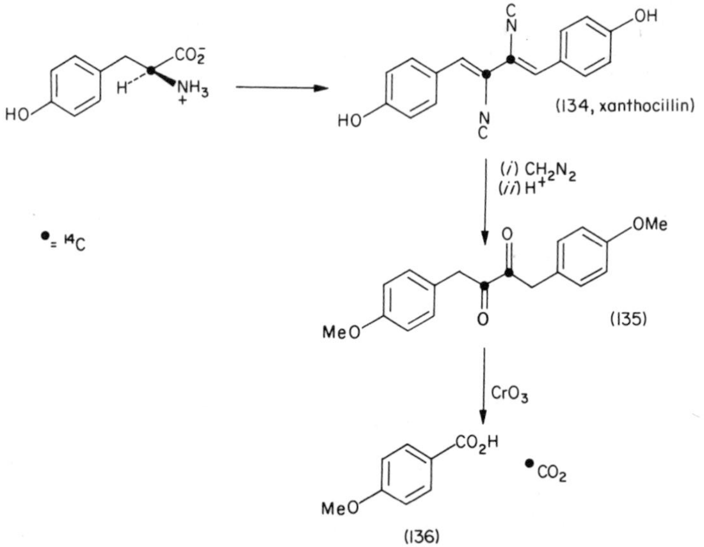

(134, xanthocillin)

(i) CH$_2$N$_2$
(ii) H$^+$

(135)

CrO$_3$

(136)

$\bullet = {}^{14}C$

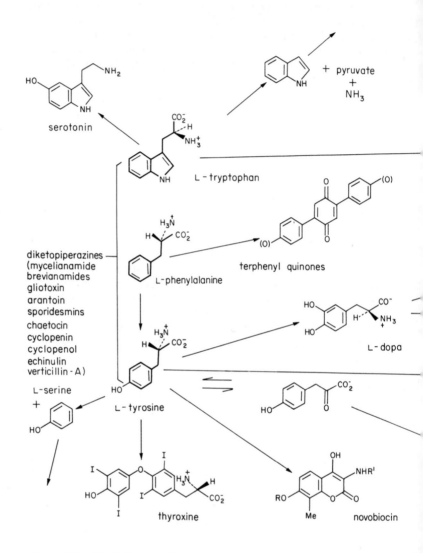

Figure 4.20. Metabolism of the aromatic amino acids in higher organisms: summary of pathways

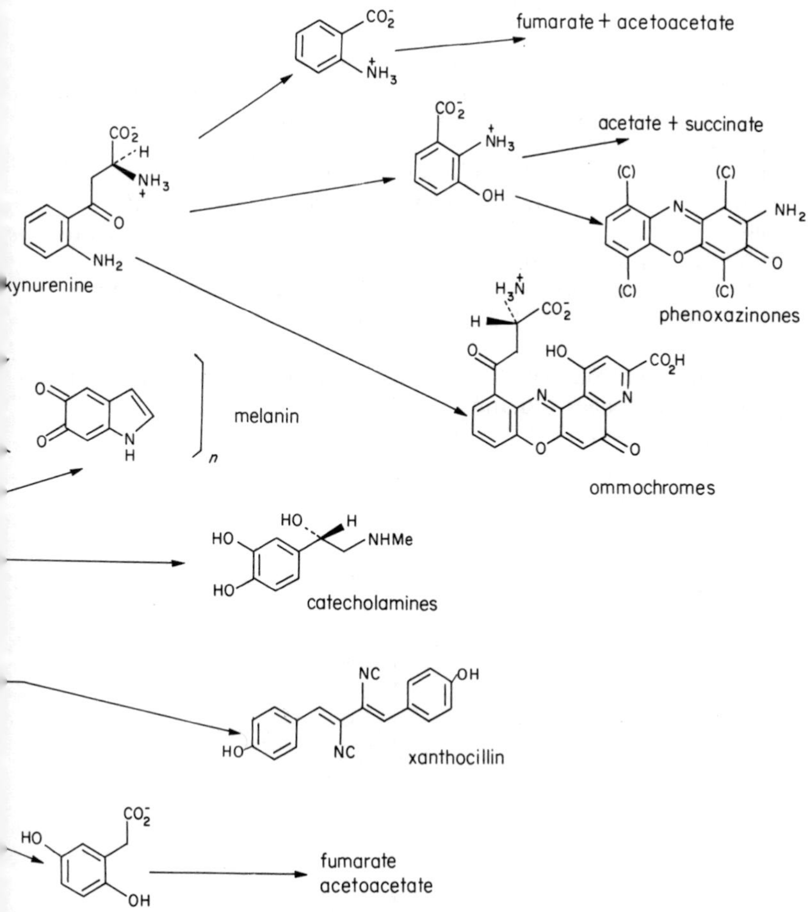

kynurenine

fumarate + acetoacetate

acetate + succinate

phenoxazinones

melanin

ommochromes

catecholamines

xanthocillin

fumarate
acetoacetate

acid (136). During the biogenesis from tyrosine, the carboxyl group
of the amino acid must be lost and this was confirmed by the
negligible incorporation of radioactivity from D L-1-[^{14}C]-tyrosine
into xanthocillin. Although a number of possible precursors of the
isonitrile function (acetate, formate, L-methionine) were examined,
no firm conclusions regarding its biogenesis were reached but the
suggestion that it is formed by dehydration of an N-formyl group
appears improbable.

Pyrrolnitrin (137) and 3-chloroindole (138) are two products of
tryptophan metabolism in *Pseudomonas pyrrocinia* and *Pseudomonas aureofaciens*[248-250]. The occurrence of phenylpyrrole compounds in nature is rare and it has been suggested[250] that the
antifungal compound pyrrolnitrin (137) is produced from tryptophan in a reaction initiated by a chloroperoxidase system (*Figure
4.21*).

A summary of the pathways discussed in this chapter is given in
Figure 4.20 on pages 176 and 177.

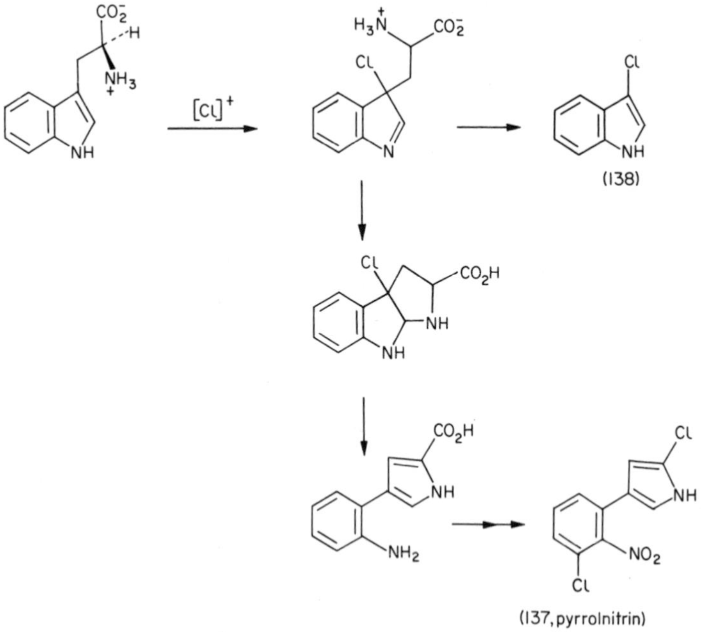

Figure 4.21. Suggested pathway of pyrrolnitrin biogenesis[250]

REFERENCES

1. Garrod, A. E. (1923). *Inborn Errors of Metabolism.* London; Froude, Holder and Houghton
2. Kaufman, S. (1963). *The Enzymes,* vol. 8, p. 373. Edited by P. D. Boyer, H. Lardy and K. Myrback. Second Edition. London and New York; Academic Press
3. Kaufman, S. (1957). *J. Biol. Chem.,* **226**, 511
4. Kaufman, S. (1959). *J. Biol. Chem.,* **234**, 2677
5. Nair, P. M. and Vining, L. C. (1965). *Phytochemistry,* **4**, 401
6. Rosenberg, H., McLaughlin, J. L. and Paul, A. G. (1967). *Lloydia,* **30**, 100
7. Leete, E., Bowman, R. M. and Manuel, M. F. (1971). *Phytochemistry,* **10**, 3029
8. Guroff, G. and Ito, T. (1965). *J. Biol. Chem.,* **240**, 1175
9. Guroff, G. and Rhoads, C. A. (1969). *J. Biol. Chem.,* **244**, 142
10. Kaufman, S. (1964). *J. Biol. Chem.,* **239**, 232
11. Kaufman, S. and Fischer, D. B. (1970). *J. Biol. Chem.,* **245**, 4745
12. Kaufman, S. (1970). *Adv. Enzymol.,* **35**, 245
13. Hayaishi, O. (1962). *Oxygenases.* New York; Academic Press
14. Nagatsu, T., Levitt, M. and Udenfriend, S. (1964). *J. Biol. Chem.,* **239**, 2910
15. Kaufman, S. (1963). *Proc. Natl Acad. Sci.,* **50**, 1085
16. Guroff, G., Daly, J. W., Jerina, D. M., Renson, J., Witkop, B. and Udenfriend, S. (1967). *Science,* **158**, 1524
17. Daly, J. W., Jerina, D. M. and Witkop, B. (1972). *Experentia,* **28**, 1129
18. Jerina, D. M., Daly, J. W., Witkop, B., Zaltzmann-Nirenberg, P. and Udenfriend, S. (1968). *Arch. Biochem. Biophys.,* **128**, 176; *J. Amer. Chem. Soc.,* **90**, 6525
19. Daly, J. W., Jerina, D. M., Ziffer, H., Witkop, B., Klarner, G. F. and Vogel, E. (1970). *J. Amer. Chem. Soc.,* **92**, 702
20. Guroff, G., Levitt, M., Daly, J. W. and Udenfriend, S. (1966). *Biochem. Biophys. Res. Comm.,* **25**, 253
21. Hager, S. E., Gregerman, R. I. and Knox, W. E. (1957). *J. Biol. Chem.,* **225**, 935
22. Yasonubu, Y., Tanaka, T., Knox, W. E. and Mason, H. S. (1958). *Fed. Proc. Amer. Soc. Expl. Biol.,* **17**, 340
23. Crandall, D. J. (1965). *Oxidases and related Redox systems,* p. 263. Edited by T. E. King, H. S. Mason and M. Morrison. New York; Wiley
24. Soloway, A. H. (1966). *J. Theoret. Biol.,* **13**, 100
25. Lindblad, B., Lindstedt, G. and Lindstedt, S. (1970). *J. Amer. Chem. Soc.,* **92**, 7446
26. Witkop, B. and Goodwin, S. (1952). *Experentia,* **8**, 377
27. Witkop, B. and Cohen, L. (1963). *Angew. Chem.,* **73**, 253
28. Holme, E., Lindstedt, G., Tofft, M. and Lindstedt, S. (1968). *Fed. Eur. Biochem. Soc. Lett.,* **2**, 29
29. Lindblad, B., Lindstedt, G. and Lindstedt, S. (1969). *J. Amer. Chem. Soc.,* **91**, 4604
30. Knox, W. E. and Edwards, S. W. (1955). *J. Biol. Chem.,* **216**, 489
31. Ravdin, R. G. and Crandall, D. I. (1951). *J. Biol. Chem.,* **189**, 137
32. Molinoff, P. B. and Axelrod, J. (1971). *Ann. Rev. Biochem.,* **40**, 465
33. Blaschko, H. (1939). *J. Physiol.,* **96**, 50P
34. Nagatsu, T., Levitt, M. and Udenfriend, S. (1964). *Anal. Biochem.,* **9**, 122
35. Ikeda, M., Fahien, L. A. and Udenfriend, S. (1966). *J. Biol. Chem.,* **241**, 4452
36. Lovenberg, W., Weissbach, H. and Udenfriend, S. (1962). *J. Biol. Chem.,* **237**, 89
37. Levin, E. Y. and Kaufman, S. (1961). *J. Biol. Chem.,* **236**, 2043
38. Kaufman, S. and Friedman, S. (1965). *J. Biol. Chem.,* **240**, PC 552
39. Kirshner, N. and Goodall, McC. (1957). *Biochim. Biophys. Acta,* **24**, 658

40. Blaschko, H. (1952). *Pharm. Rev.*, **4**, 415
41. Axelrod, J. and Tomchick, R. (1968). *J. Biol. Chem.*, **233**, 702
42. Axelrod, J., Senoh, S. and Witkop, B. (1968). *J. Biol. Chem.*, **233**, 697
43. Senoh, S., Daly, J. W., Axelrod, J. and Witkop, B. (1959). *J. Amer. Chem. Soc.*, **81**, 6240
44. Creveling, C. R., Dalgard, M., Shimizu, H. and Daly, J. W. (1970). *Mol. Pharmacol.*, **6**, 691
45. Armstrong, M. D. and McMillan, A. (1959). *Pharmacol. Rev.*, **11**, 394
46. Axelrod, J., Kopin, I. J. and Mann, J. D. (1959). *Biochim. Biophys. Acta*, **36**, 576
47. Kumagai, H., Yamada, H., Matsui, H., Ohkishi, H. and Ogata, K. (1970). *J. Biol. Chem.*, **245**, 1767, 1773
48. Brot, N., Smit, Z. and Weissbach, H. (1965). *Arch. Biochem. Biophys.*, **112**, 1
49. Kumagai, H., Matsui, H., Ohkishi, H., Ogata, K., Yamada, H., Uneo, T. and Fukami, H. (1969). *Biochem. Biophys. Res. Commun.*, **34**, 266
50. Kumagai, H., Kashima, N. and Yamada, H. (1970). *Biochem. Biophys. Res. Commun.*, **39**, 796
51. Yamada, H., Kumagai, H., Kushima, N., Torrii, H., Enei, H. and Okumura, S. (1972). *Biochem. Biophys. Res. Commun.*, **46**, 370
52. Enei, H., Matsui, H., Okmura, S. and Yamada, H. (1971). *Biochem. Biophys. Res. Commun.*, **43**, 1345
53. Dalgliesh, C. E. (1955). *Adv. Protein Chem.*, **10**, 33
54. Henderson, L. M., Gholson, R. K. and Dalgliesh, C. E. (1962). *Comparative Biochemistry*, p. 246. Edited by M. Florkin and H. S. Mason. New York: Academic Press
55. Kotake, Y. and Masayama, T. (1936). *Z. Physiol. Chem.*, **243**, 237
56. Greengard, O. and Feigelson, P. (1962). *J. Biol. Chem.*, **237**, 1903
57. Koike, K., Poillon, W. N. and Feigelson, P. (1969). *J. Biol. Chem.*, **244**, 3457
58. Maeno, H. Koike, K., Poillon, W. N. and Feigelson, P. (1969). *J. Biol. Chem.*, **244**, 3447
59. Maeno, H. and Feigelson, P. (1967). *J. Biol. Chem.*, **242**, 596
60. Ishimura, Y., Nozaki, M., Hayaishi, O., Nakamura, Y., Tamura, M. and Yamazaki, I. (1970). *J. Biol. Chem.*, **245**, 3593
61. Nishizuka, Y., Ichiyama, A., Gholson, R. K. and Hayaishi, O. (1965). *J. Biol. Chem.*, **240**, 733
62. Ichiyama, A., Nakamura, S., Kawai, H., Honjo, T., Nishizuka, Y., Hayaishi, O. and Senoh, S. (1965). *J. Biol. Chem.* **240**, 740
63. Lan, J. J. and Gholson, R. K. (1965). *J. Biol. Chem.*, **240**, 3934
64. Hayaishi, O., Katagiri, M. and Rothberg, S. (1957). *J. Biol. Chem.*, **229**, 905
65. Stevens, C. O. and Henderson, L. M. (1959). *J. Biol. Chem.*, **234**, 1188
66. Nishizuka, Y. and Hayaishi, O. (1963). *J. Biol. Chem.*, **238**, 3369, PC 483
67. Gholson, R. K., Ueda, I., Ogasawara, N. and Henderson, L. M. (1964). *J. Biol. Chem.*, **239**, 1208
68. Ikeda, M., Tsuj, H., Nakamura, S., Ichiyama, A., Nishizuka, Y. and Hayaishi, O. (1965). *J. Biol. Chem.*, **240**, 1395
69. Ichihara, A., Adachi, K., Hosokawa, K. and Takeda, Y. (1962). *J. Biol. Chem.*, **237**, 2296
70. Taniuchi, H., Hatanaka, M., Kuno, S., Hayaishi, O., Nakajima, M., and Kurihara, N. (1964). *J. Biol. Chem*, **239**, 2204
71. Stanier, R. Y. and Hayaishi, O. (1951). *Science*, **114**, 326
72. Subba Rao, P. V., Sreeleda, N. S., Premakumar, R. and Vaidyanathan, C. S. (1971). *J. Bact.*, **107**, 100
73. Hayaishi, O., Taniuchi, H., Tashiro, M. and Kuno, S. (1961). *J. Biol. Chem.*, **236**, 2492

74. Horibata, K., Taniuchi, H., Tashiro, M., Kuno, S. and Hayaishi, O. (1961). *J. Biol. Chem.*, **236**, 2991
75. Dagley, S. and Johnson, P. A. (1963). *Biochim. Biophys. Acta*, **78**, 577
76. Page, I. H. (1968). *Serotonin*. Chicago; Year Book Medical Publishers
77. Graeme-Smith, D. G. (1964). *Biochem. J.*, **92**, 52
78. Renson, J., Daly, J., Weissbach, H., Witkop, B. and Udenfriend, S. (1966). *Biochem. Biophys. Res. Commun.*, **25**, 504
79. Udenfriend, S., Clark, C. T. and Titus, E. (1953). *J. Amer. Chem. Soc.*, **75**, 501
80. Mitoma, C., Weissbach, H. and Udenfriend, S. (1955). *Nature*, **175**, 994
81. Beer, R. J. S., Jennings, B. E. and Robertson, A. (1954). *J. Chem. Soc.*, 2679
82. Hopkins, F. G. and Cole, S. W. (1903). *J. Physiol.*, **29**, 451
83. Wood, W. A., Gunsalus, I. C. and Umbreit, W. W. (1947). *J. Biol. Chem.*, **170**, 313
84. Metzler, D. A., Ikawa, M. and Snell, E. E. (1954). *J. Amer. Chem. Soc.*, **76**, 648
85. Newton, W. A. and Snell, E. E. (1964). *Proc. Natl Acad. Sci.*, **51**, 382
86. Newton, W. A., Morino, Y. and Snell, E. E. (1965). *J. Biol. Chem.*, **240**, 1211
87. Morino, Y. and Snell, E. E. (1967). *J. Biol. Chem.*, **242**, 2793, 5591, 5602
88. Kagamiyama, H., Wade, H., Matsubura, H. and Snell, E. E. (1972). *J. Biol. Chem.*, **247**, 1571
89. Kagamiyama, H., Matsubura, H. and Snell, E. E. (1972). *J. Biol. Chem.*, **247**, 1576
90. London, J. and Goldberg, M. E. (1972). *J. Biol. Chem.*, **247**, 1566
91. Hayaishi, O. (1966). *Bacteriol. Rev.*, **30**, 720
92. Hayaishi, O. (1969). *Ann. Rev. Biochem.*, **38**, 21
93. Mason, H. S. (1957). *Science*, **125**, 1185
94. Mason, H. S. (1957). *Adv. Enzymol.*, **19**, 79
95. Norman, R. O. C. and Smith, J. R. L. (1965). *Oxidases and Related Redox Systems*, p. 131. Edited by T. E. King, H. S. Mason and M. Morrison. New York; Wiley
96. Norman, R. O. C. and Radda, G. K. (1962). *Proc. Chem. Soc.*, 138
97. Hamilton, G. A., Workman, R. J. and Woo, L. J. (1964). *J. Amer. Chem. Soc.*, **86**, 3390
98. Udenfriend, S., Clark, C. T., Axelrod, J. and Brodie, B. B. (1954). *J. Biol. Chem.*, **208**, 731
99. Hamilton, G. A. and Friedman, J. P. (1964). *J. Amer. Chem. Soc.*, **86**, 3391
100. Hamilton, G. A. (1969). *Adv. Enzymol.*, **32**, 55
101. Gregory, R. P. F. and Bendall, D. S. (1966). *Biochem. J.*, **101**, 569
102. Eiger, I. Z. and Dawson, C. R. (1949). *Arch. Biochem. Biophys.*, **21**, 194
103. Clayton, R. A. (1959). *Arch. Biochem. Biophys.*, **81**, 404
104. Vaughan, P. F. T. and Butt, V. S. (1969). *Biochem. J.*, **113**, 109
105. Vaughan, P. F. T. and Butt, V. S. (1970). *Biochem. J.*, **119**, 89
106. Dawson, C. R. and Tarpley, W. B. (1951). *The Enzymes*, Vol. 2, p. 454. Edited by J. B. Sumner and K. Myrback. New York; Academic Press
107. Mason, H. S. (1956). *Nature*, **177**, 79
108. Kubowitz, F. (1938). *Biochem. Z.*, **296**, 443
109. Kubowitz, F. (1938). *Biochem. Z.*, **299**, 32
110. Jerina, D. M., Daly, J. W., Witkop, B., Zaltzman-Nirenberg, P. and Udenfriend S. (1970). *Biochemistry*, **9**, 147
111. Boyd, D. R., Daly, J. W. and Jerina, D. M. (1972). *Biochemistry*, **11**, 1961
112. Guroff, G., Reifsnyder, A. and Daly, J. W. (1966). *Biochem. Biophys. Res. Comm.*, **24**, 720
113. Guroff, G., Kondo, K. and Daly, J. W. (1966). *Biochem. Biophys. Res. Comm.*, **25**, 622
114. Guroff, G. and Daly, J. W. (1967). *Arch. Biochem. Biophys.*, **122**, 212

115. Nover, L. and Luckner, M. (1969). *Fed. Eur. Biochem. Soc. Lett.*, **3**, 292
116. Daly, J. W. and Jerina, D. M. (1969). *Arch. Biochem. Biophys.*, **134**, 266
117. Daly, J. W. and Jerina, D. M. (1970). *Biochim. Biophys. Acta*, **208**, 340
118. Henbest, H. B. (1965). *Chemical Soc. Special Publications*, **19**, 83
119. Jerina, D. M., Daly, J. W., Landis, W., Witkop, B. and Udenfriend, S. (1967). *J. Amer. Chem. Soc.*, **89**, 3347
120. Daly, J. W., Jerina, D. M. and Witkop, B. (1968). *Arch. Biochem. Biophys.*, **128**, 517
121. Daly, J. W., Jerina, D. M., Farnsworth, J. and Guroff, G. (1969). *Arch. Biochem. Biophys.*, **131**, 238
122. Jerina, D. M., Boyd, D. R. and Daly, J. W. (1970). *Tetrahedron Lett.*, 457
123. Ullrich, V. and Staudinger, H. (1969). *Z. Naturforsch.*, **24B**, 583
124. Butte, W. A. and Price, C. C. (1962). *J. Amer. Chem. Soc.*, **84**, 3567
125. Jerina, D. M., Daly, J. W. and Witkop, B. (1968). *J. Amer. Chem. Soc.*, **90**, 6523
126. Hayaishi, O. and Hashimoto, K. (1950). *J. Biochem.* (Tokyo), **37**, 37
127. Kojima, Y., Itada, N. and Hayaishi, O. (1961). *J. Biol. Chem.*, **236**, 2223
128. Stanier, R. Y. and Ingraham, J. L. (1954). *J. Biol. Chem.*, **210**, 799
129. Raper, H. S. (1927). *Biochem. J.*, **21**, 89
130. Raper, H. S. (1938). *J. Chem. Soc.*, 125
131. Mason, H. S. (1947). *J. Biol. Chem.*, **168**, 433
132. Mason, H. S. (1948). *J. Biol. Chem.*, **172**, 83
133. Mason, H. S. (1967). *Advances in Biology of Skin*, Vol. 3, p. 293. Oxford; Pergamon
134. Wyler, H. and Chiovini, J. (1968). *Helv. Chim. Acta*, **51**, 1476
135. Mason, H. S. (1959). *Pigment Cell Biology*, p. 563. Edited by M. Gordon. New York; Academic Press
136. Beer, R. J. S., Broadhurst, T. and Robertson, A. (1954). *J. Chem. Soc.*, 1947
137. Bu'Lock, J. D. and Harley-Mason, J. (1951). *J. Chem. Soc.*, 703, 2248
138. Chapman, R. F. and Swan, G. A. (1970). *J. Chem. Soc.* (C), 865
139. Binns, F., Chapman, R. F., Robson, N. C., Swan, G. A. and Waggott, A. (1970). *J. Chem. Soc.* (C), 1128
140. Swan, G. A. and Waggott, A. (1970). *J. Chem. Soc.* (C), 1409
141. Swan, G. A., King, J. A. G., Percival, A. and Robson, N. C. (1970). *J. Chem. Soc.* (C), 1419
142. Kirby, G. W. and Ogunkoya, L. (1965). *Chem. Commun.*, 546
143. Swan, G. A. and Wright, D. (1965). *J. Chem. Soc.*, 1549
144. Hempel, K. (1967). *Z. Naturforsch*, **22B**, 173
145. Piatelli, M. and Nicolaus, R. A. (1961). *Tetrahedron*, **15**, 66
146. Piatelli, M., Fattorusso, E., Magno, S. and Nicolaus, R. A. (1962). *Tetrahedron*, **18**, 941
147. Binns, F., King, J. A. G., Mishra, S. N., Percival, A., Robson, N. C., Waggott, A. and Swan, G. A. (1970). *J. Chem. Soc.* (C), 2063
148. Dunn, J. T. (1970). *J. Biol. Chem.*, **245**, 5954
149. van Zyl, A. and Edelhoch, H. (1967). *J. Biol. Chem.*, **242**, 2423
150. Taurog, A. and Howells, E. M. (1966). *J. Biol. Chem.*, **241**, 1329
151. von Mutzenbecher, P. (1939). *Z. Physiol. Chem.*, **261**, 253
152. Harrington, C. R. and Pitt-Rivers, R. V. (1945). *Biochem. J.*, **39**, 157
153. Pitt-Rivers, R. V. (1948). *Nature*, **161**, 308
154. Pitt-Rivers, R. V. (1948). *Biochem. J.*, **43**, 223
155. Pitt-Rivers, R. V. and Cavalieri, R. R. (1964). *The Thyroid Gland*, Vol. 1. Edited by R. V. Pitt-Rivers and W. R. Trotter. London; Butterworths
156. Johnson, T. B. and Tewkesbury, L. B. (1942). *Proc. Natl Acad. Sci.*, **28**, 73
157. Hillman, G. (1956). *Z. Naturforsch.*, **11B**, 424

158. Meltzer, R. I. and Stanaback, R. J. (1961). *J. Org. Chem.*, **26**, 1977
159. Shiba, T. and Cahnmann, H. J. (1962). *J. Org. Chem.*, **27**, 1773
160. Toi, K., Salvatore, G. and Cahnmann, H. J. (1965). *Biochim. Biophys. Acta*, **97**, 523
161. Nishinaga, A., Cohnmann, H. J., Kon, H. and Matsuura, T. (1968). *Biochemistry*, **7**, 388
162. Blasi, F., Fragomelle, F. and Covelli, I. (1968). *Eur. J. Biochem.*, **5**, 215
163. Harrington, C. R. and Barger, G. (1927). *Biochem. J.*, **21**, 169
164. Matsuura, T. and Cahnmann, H. J. (1959). *J. Amer. Chem. Soc.*, **81**, 871
165. Matsuura, T. and Cahnmann, H. J. (1960). *J. Amer. Chem. Soc.*, **82**, 2050, 2055
166. Katz, E. (1968). *Actinomycin*, p. 45. Edited by S. A. Waksman. New York; Interscience
167. Katz, E. (1967). *Antibiotics, II, Biosynthesis*, p. 276. Edited by D. Gottlieb and P. D. Shaw. Berlin, Heidelberg and New York; Springer Verlag
168. Johnson, A. W. (1968). *Actinomycin*, p. 33. Edited by S. A. Waksman. New York; Interscience
169. Brockmann, H. (1964). *Angew. Chem.*, **76**, 1
170. Butenandt, A. (1957). *Angew. Chem.*, **69**, 16
171. Cavill, G. W. K., Ralph, B. J., Tetar, J. R. and Werner, R. L. (1953). *J. Chem. Soc.*, 525
172. Gripenberg, J. (1951). *Acta Chem. Scand.*, **5**, 590
173. Gripenberg, J. (1963). *Acta Chem. Scand.*, **17**, 703
174. Gripenberg, J., Honkanen, E. and Patoharjn, O. (1957). *Acta Chem. Scand.*, **11**, 1485
175. Anzai, K., Isono, K., Okuma, K. and Suzuki, S. (1960). *J. Antibiotics* (Japan), **13A**, 125
176. Gerber, N. N. and Lechevalier, M. P. (1964). *Biochemistry*, **3**, 598
177. Gerber, N. N. (1966). *Biochemistry*, **5**, 3824
178. Gerber, N. N. (1968). *Can. J. Chem.*, **46**, 790
179. Leimgruber, W., Batcho, A. D. and Schenker, F. (1965). *J. Amer. Chem. Soc.*, **87**, 5793
180. Nagasawa, H. T., Gutmann, H. R. and Morgan, M. A. (1959). *J. Biol. Chem.*, **234**, 1600
181. Morgan, L. R., Weimorts, D. M. and Aubert, C. C. (1965). *Biochim. Biophys. Acta*, **100**, 393
182. Subba-Rao, P. V., Jegannathan, N. S. and Vaidyanathan, C. S. (1965). *Biochem. J.*, **95**, 628
183. Subba-Rao, P. V. and Vaidyanathan, C. S. (1966). *Arch. Biochem. Biophys.*, **115**, 27
184. Nair, P. M. and Vining, L. C. (1964). *Can. J. Biochem.*, **42**, 1515
185. Katz, E. and Weissbach, H. (1967). *Develop. Ind. Microbiol.*, **8**, 67
186. Butenandt, A. and Beckmann, R. (1955). *Z. Physiol. Chem.*, **301**, 115
187. Butenandt, A. and Newbert, G. (1955). *Z. Physiol. Chem.*, **301**, 109
188. Glassman, E. (1956). *Genetics*, **41**. 556
189. Baglioni, C. (1959). *Nature*, **184**, 1085
190. Ghosh, D. and Forrest, H. S. (1967). *Genetics*, **55**, 423
191. Phillips, J. P. and Forrest, H. S. (1970). *Biochem. Genetics*, **4**, 489
192. Brockmann, H. and Lackner, H. (1967). *Chem. Ber.*, **100**, 353
193. Johnson, A. W. (1958). *Ciba Foundation Symp. on Amino acids and peptides with Antimetabolic Activity*, p. 143. Edited by G. E. W. Wolstenholme and C. M. O'Connor
194. Weissbach, H., Redfield, B. G., Beaven, V. and Katz, E. (1965). *J. Biol. Chem.*, **240**, 4377
195. Katz, E. and Weissbach, H. (1963). *J. Biol. Chem.*, **238**, 666

184 Metabolism of the Aromatic Amino Acids

196. Marshall, R., Redfield, B. G., Katz, E. and Weissbach, H. (1968). *Arch. Biochem. Biophys.*, **123**, 317
197. Sivak, A. and Katz, E. (1959). *Bact. Proc.*, 124
198. Katz, E. and Weissbach, H. (1962). *J. Biol. Chem.*, **237**, 882
199. Sivak, A., Meloni, M. L., Nobili, F. and Katz, E. (1962). *Biochim. Biophys. Acta*, **57**, 283
200. Birch, A. J., Cameron, D. W., Holloway, P. W. and Rickards, R. W. (1960). *Tetrahedron Lett.*, (25), 26
201. Brockmann, H. and Manegold, J. (1964). *Naturwiss*, **51**, 383
202. Salzmann, Z., Weissbach, H. and Katz, E. (1969). *Arch. Biochem. Biophys.*, **130**, 536
203. Hollstein, U., Burton, R. A. and White, J. A. (1966). *Experentia*, **22**, 210
204. Birch, A. J. and Farrer, K. R. (1963). *J. Chem. Soc.*, 4277
205. Birch, A. J., Blance, G. E., David, S. and Smith, H. (1961). *J. Chem. Soc.*, 3128
206. McDonald, J. C. and Slater, G. P. (1966). *Can. J. Microbiol.*, **12**, 455
207. Casnati, G., Cavalleri, R., Piozzi, F. and Quilico, A. (1962). *Gazzetta*, **92**, 105
208. Birch, A. J. and Smith, H. (1958). *Ciba Foundation Symp. on Amino Acids and Peptides with Antimetabolic Activity*, p. 247. Edited by G. E. Wolstenholme and C. M. O'Connor. London; Churchill
209. Birch, A. J., Massy-Westrop, R. A. and Rickards, R. W. (1956). *J. Chem. Soc.*, 3717
210. Slater, G. P., MacDonald, J. C. and Nakashima, R. (1970). *Biochemistry*, **9**, 2886
211. Allen, C. M. (1972). *Biochemistry*, **11**, 2154
212. Bycroft, B. W. and Landon, W. (1970). *Chem. Commun.*, 967
213. Birch, A. J. and Wright, J. J. (1970). *Tetrahedron*, **26**, 2329
214. Birch, A. J. and Russell, R. A. (1972). *Tetrahedron*, **28**, 2999
215. Steyn, P. S. (1971). *Tetrahedron Lett.*, 3331
216. Nover, L. and Luckner, M. (1969). *Fed. Eur. Biochem. Soc. Lett.*, **3**, 292
217. Nover, L. and Luckner, M. (1969). *Eur. J. Biochem.*, **10**, 268
218. Bracken, A., Pocker, A. and Raistrick, H. (1954). *Biochem. J.*, **57**, 587
219. Birkinshaw, J. H., Luckner, M., Mohamed, Y. S., Mothes, K. and Stickings, C. E. (1963). *Biochem. J.*, **89**, 196
220. Luckner, M. and Mothes, K. (1962). *Tetrahedron Lett.*, 1035
221. Luckner, M., Winter, K. and Reisch, J. (1969). *Eur. J. Biochem.*, **7**, 380
222. Bell, M. R., Johnson, J. R., Wildi, B. S. and Woodward, R. B. (1958). *J. Amer. Chem. Soc.*, **80**, 1001
223. Buchan, A. F., Friedrichsons, J. and McL. Mathieson, A. (1966). *Tetrahedron Lett.*, 3131
224. Nagarajan, N., Neuss, N. and Marsh, M. M. (1968). *J. Amer. Chem. Soc.*, **90**, 6519
225. Minato, H., Matsumoto, M. and Katayama, T. (1971). *Chem. Commun.*, 44
226. Hauser, D., Weber, H. P. and Sigg, H. D. (1970). *Helv. Chim. Acta*, **53**, 1061
227. Fridrichsons, J. and McL. Mathieson, A. (1962). *Tetrahedron Lett.*, 1265
228. Jamieson, W. D., Rahman, R. and Taylor, A. (1970). *J. Chem. Soc.* (C), 1564
229. Rahman, R., Safe, S. and Taylor, A. (1970). *J. Chem. Soc.* (C), 1665
230. Suhadolnik, R. J. and Chenoweth, R. G. (1958). *J. Amer. Chem. Soc.*, **80**, 4391
231. Suhadolnik, R. J. and Winstead, J. A. (1960). *J. Amer. Chem. Soc.*, **82**, 1644
232. Bose, A. K., Das, K. G., Funke, P. T., Kugachovsky, I., Sukla, O. P., Kanchandani, K. S. and Suhadolnik, R. J. (1968). *J. Amer. Chem. Soc.*, **90**, 1038
233. Bose, A. K., Kanchandani, K. S., Tavares, R. and Funke, P. T. (1968). *J. Amer. Chem. Soc.*, **90**, 3593
234. Bu'Lock, J. D. and Ryles, A. P. (1970). *Chem. Commun.*, 1404
235. Johns, N. and Kirby, G. W. (1971). *Chem. Commun.*, 163

Metabolism of the Aromatic Amino Acids   185

236. Bu'Lock, J. D., Ryles, A. P., Kirby, G. W. and Johns, N. (1972). *Chem. Commun.*, 100
237. Brannon, D. R., Mabe, J. A., Molloy, B. B. and Day, B. A. (1971). *Biochem. Biophys. Res. Comm.*, **43**, 588
238. Hinman, J. W., Caron, E. L. and Hoeksema, H. (1957). *J. Amer. Chem. Soc.*, **79**, 3789
239. Stammer, C. H., Walton, E., Wilson, A. N., Walker, R. W., Trenner, N. R. Holly, F. W. and Folkers, K. W. (1958). *J. Amer. Chem. Soc.*, **80**, 137
240. Bunton, C. A., Kenner, G. W., Robinson, M. J. T. and Webster, B. R. (1963). *Tetrahedron*, **19**, 1001
241. Kominek, L. A. (1967). *Antibiotics, II, Biosynthesis*. Edited by D. Gottlieb and P. D. Shaw. Berlin, Heidelberg, New York; Springer Verlag
242. Read, G., Vining, L. C. and Haskins, R. H. (1962). *Can. J. Chem.*, **40**, 2357
243. Chandra, P., Read, G. and Vining, L. C. (1966). *Can. J. Biochem.*, **44**, 403
244. Bose, A. K., Khanchandani, K. S., Funke, P. T. and Anchel, M. (1969). *Chem. Commun.*, 1347
245. Maass, W. S. G. and Neish, A. C. (1967). *Can. J. Bot.*, **45**, 59
246. Hagedorn, I. and Tonjes, H. (1957). *Pharmazie*, **12**, 567
247. Achenbach, H. and Grisebach, H. (1965). *Z. Naturforsch*, **20B**, 137
248. Arima, K., Imanaka, H., Kousaka, M., Fukuda, A. and Tamura, G. (1964). *Agr. Biol. Chem.*, **28**, 575
249. Arima, K., Imanaka, H., Kousaka, M., Fukuda, A. and Tamura, G. (1965). *J. Antibiotics* (Japan), **18A**, 201
250. Gorman, M. and Lively, D. H. (1967). *Antibiotics, II, Biosynthesis*, p. 433. Edited by D. Gottlieb and P. D. Shaw. Berlin, Heidelberg, New York; Springer Verlag

5

The Shikimate Pathway in Higher Plants: Phenylpropanoid Compounds and their Derivatives

5.1 INTRODUCTION

Organic chemists have long been familiar with the wealth of substances of diverse structural type which are derived from plants or micro-organisms and to which no general function in the overall metabolism can be assigned. These secondary metabolites[1], or natural products as they are frequently designated[2], have provided organic chemistry with a seemingly inexhaustible supply of intriguing experimental subjects for the past ten or more decades. As, however, the purely chemical problems relating to structure have been solved, attention has rightly switched to become focused upon more biochemical considerations such as their mode of biosynthesis and physiological function in the plant or organism. Secondary metabolites have also become of particular interest to plant taxonomists and geneticists since it is clear that they represent a distinctive chemical manifestation of the individuality of species[1]. Knowledge of their distribution thus assists in elucidating lines of evolution and plant classifications. One distinction of this type which has already been noted on several occasions is that between 'woody' and 'non-woody' plants. This distinction is sufficiently pronounced for it to be possible to speak of a typically 'woody' pattern of secondary metabolites in the leaves of some plants[3].

Higher plants, ferns, mosses and liverworts contain a range of secondary metabolites—alkaloids, essential oils, pigments, phenolic glycosides and tannins—and many of these are derived biosynthetically from intermediates in, or the amino acid end-products of, the shikimate pathway[4]. Predominant amongst these substances are the various plant phenols[5] whose structural complexity varies from that of hydroquinone, usually encountered as its β-D-glucoside arbutin (1), to that of the unique structural polymer lignin. Rather more restricted in their distribution in higher plants are the alkaloids and their phenolic amine precursors which are based on the aromatic amino acids[6]. Discussion of this group of secondary metabolites is, however, reserved for a companion volume in this series.

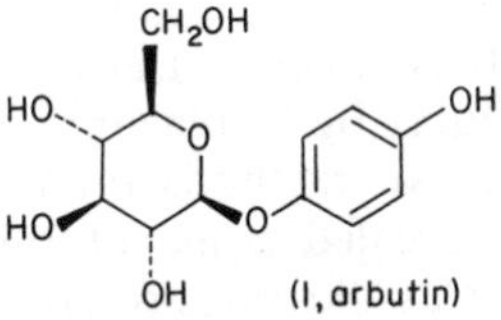

(I, arbutin)

With the important exception of lignin, the function of the majority of these phenolic plant secondary metabolites has remained obscure despite much conjecture and rather less definitive experimentation. The restricted form of their distribution in the plant kingdom cannot fit them for a general role in the basic metabolism of plants. Suggestions[4, 7–9] as to their specific roles have varied from ones of storage and transport, the attraction and repulsion of insects to the idea that they protect the plant against natural predators or infection by disease. It has also been suggested[10] that they are merely interesting by-products of metabolism, but Geissman and Crout[2] have rejected such a conclusion arguing that simply because no exact role has yet been discovered, it is therefore incorrect to assume that none exists.

In contrast the function of lignin in the plant is clear. It has long been recognised by botanists and plant physiologists that an important change—lignification—occurs early in the development of the cell walls of the woody tissues of vascular plants. In this process a group of closely related polymers (lignin) is laid down between the cellulose microfibrils already present in the cell wall and these act to resist compression forces on the cell. The lignified tissue no longer plays an active metabolic role in the life of the plant but it serves as a supporting skeletal structure for the whole developing organism.

5.2   PLANT PHENOLS

5.2.1   CLASSIFICATIONS

The organic chemist's early intuitive approach to the elucidation of biosynthetic relationships amongst secondary metabolites was based on the study of their comparative anatomy[11]. Certain patterns of atoms were found to occur frequently and these were made the basis of biogenetic classifications. The most familiar of these is the isoprene rule, recognised by Wallach and formulated by Ruzicka for use in terpene chemistry[12], and other theories of biosynthesis which evolved along these lines concerned alkaloids and phenols and were advanced notably by Robinson[11], Schopf[13] and Collie[14]. Whilst the development of methods for the direct study of biosynthetic routes has revealed some of the limitations of these empirical suggestions, classifications of groups of natural products on the basis of structural comparisons still remains useful from the biogenetic point of view. The division into groups of plant phenols (*Table 5.1*) is based both on structural grounds and upon information gained subsequently from the study of biosynthetic pathways.

Table 5.1   PLANT PHENOLS

Carbon skeleton	Plant phenol
Group A $C_6 . C_3$	Cinnamic acids, Cinnamyl alcohols, Coumarins, Phenylpropenes
$[C_6 . C_3]_2$	Lignans, Betacyanins
$[C_6 . C_3]_n$	Lignins
Group B $C_6 . C_2$	Phenylacetic acids, β-Phenylethylamines, Acetophenones
$C_6 . C_1$	Benzoic acids, Benzaldehydes, Benzyl alcohols
C_6	Phenols
Group C $C_6 . C_3 + C_2$	Phenylalkylketones, Cinnamylidene acetic acids, Cinnamyl alkyl ketones, Phenyl α-pyrones
$C_6 . C_3 + (C_2)_2$	Cinnamyl α-pyrones
$C_6 . C_3 + (C_2)_3$	Stilbenes, Flavonoids, Isoflavonoids
$[C_6 . C_3 + (C_2)_3]_2$	Biflavonoids, Proanthocyanidins
$(C_6 . C_3)_2 + C_2$	Diarylheptanoids, Phenalenones
Group D $C_6 . C_1 + C_4$	Naphthaquinones
$C_6 . C_1 + C_4 + C_5$	Anthraquinones
$C_{10} + C_5$	Anthraquinones
$C_6 . C_1 + (C_2)_3$	Xanthones, Benzophenones

All the compounds listed in *Table 5.1* contain at least one aromatic nucleus which is derived from the shikimate pathway. Key compounds in this classification and members of the first group are those with a $C_6 . C_3$ skeleton which play a central role in plant phenol biosynthesis. A further two classes can be envisaged as being derived from this group, either by shortening of the aliphatic C_3 side chain (group B) or lengthening of the C_3 chain by the addition of other carbon atoms (group C). Although the net structural result of the latter may be the addition of only one carbon atom this is recorded (*Table 5.1*) as a two carbon extension ($C_6 . C_3 . C_2$) to accord with biosynthetic considerations which suggest that this process occurs by condensation of acetate units (or their biological equivalent—malonate units). Thus for example it has been proposed[15] that the $C_6 . C_4$ carbon skeleton of zingerone (2) is formed by condensation of a $C_6 . C_3$ and a C_2 unit followed by loss (decarboxylation) of the intermediate. A fourth category (group D) lists miscellaneous types of phenol which appear in general to be derived by elaboration of a $C_6 . C_1$ building unit.

(R = −S CoA)   (2, zingerone)

The basic carbon skeletons upon which the classification (*Table 5.1*) is based are found additionally in structurally modified forms. Very often, for example, this results from the addition of one or more branched C_5 units derived from isopentenyl pyrophosphate or γ,γ-dimethylallylpyrophosphate[16, 17]. Subsequent modifications of the isoprenoid addendum may lead to structural changes which mask the underlying biogenetic relationships. Thus osthenol (4), columbianetin (5), angelicin (7), archangelicin (6) and isobergapten (8) are a typical group of coumarins from the Umbelliferae. Biosynthetic studies by Brown and his associates[18] have shown that the biogenetic relationship, illustrated in *Figure 5.1*, probably exists between these metabolites and their precursor, umbelliferone (3), and that the angular furan ring is formed from the γ,γ-dimethylallyl residue of osthenol. Similar correlations have been made by other workers[19–21]. However, for the purpose of this review it is intended to limit the biosynthetic discussion, except in specific cases, to the biological derivation of the fundamental $C_6 . C_3$ carbon skeleton of plant phenols, its modification by degradation

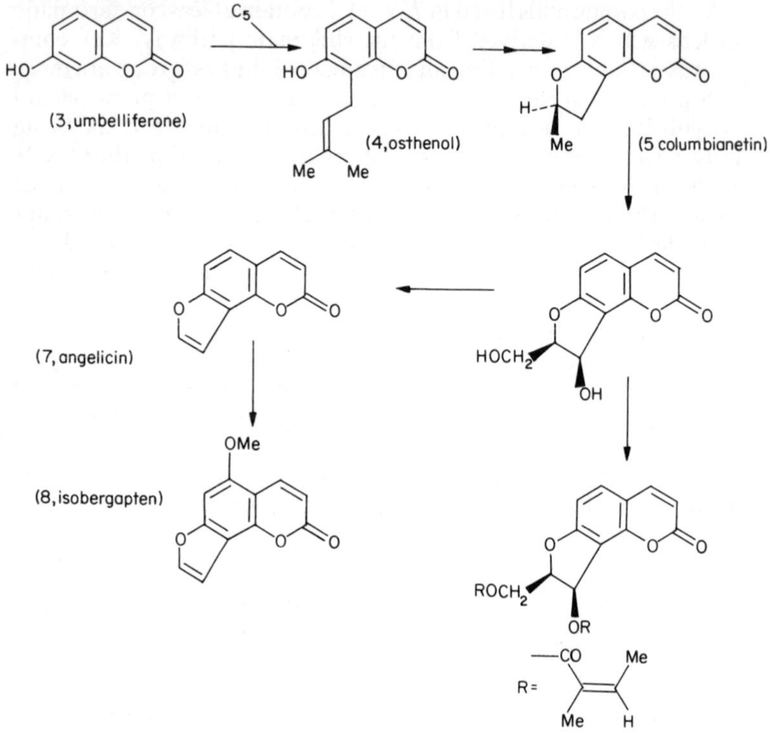

Figure 5.1. The biosynthesis of furanocoumarins in the Umbelliferae[18]

or by the addition of acetate units and to the origins of the phenolic hydroxyl group.

A characteristic feature of the chemistry of plant phenols derived from the shikimate pathway is the pattern of hydroxyl groups present in the aromatic rings. Almost invariably a single phenolic hydroxyl group, if present, is located ortho or para to a carbon substituent and thus *p*-coumaric acid (9) and *o*-coumaric acid (usually as coumarin, 10) are amongst the commoner plant phenols, whilst *m*-coumaric acid (11) has not been reported from plants[22]. Similarly polyphenols formed from shikimate generally have an ortho or para-dihydroxy or ortho-trihydroxy configuration and hence caffeic (12), ferulic (13), sinapic (14) acids, but not 3′,5′-dihydroxycinnamic acid (15), are common plant constituents. These patterns of hydroxylation are also reproduced wherever the $C_6 . C_3$ fragment is incorporated into more complex structures such as the

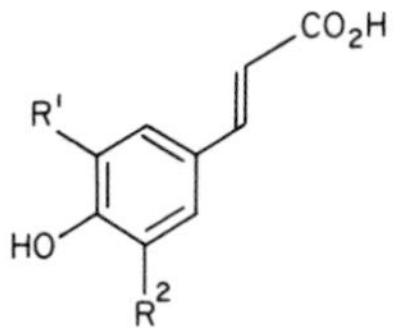

(9, R¹ = R²= H ; p -coumaric)

(12, R¹ = OH, R²=H; caffeic)

(13, R¹ = OMe, R²= H; ferulic)

(14, R¹= R² = OMe , sinapic)

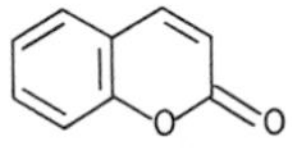

(10, coumarin)

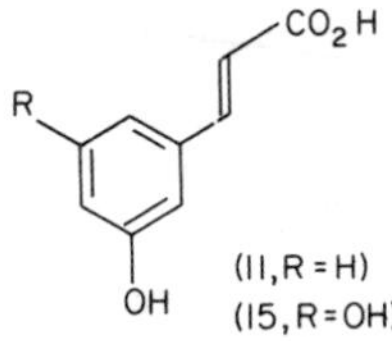

(11, R = H)
(15, R = OH)

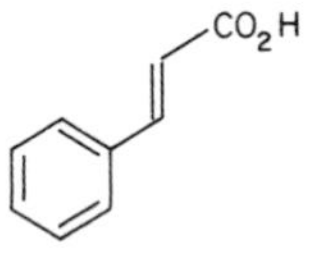

(17, cinnamic acid)

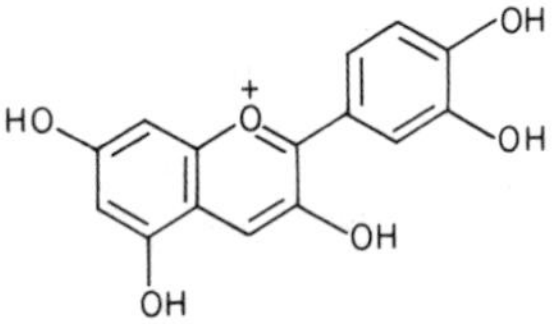

(16, cyanidin)

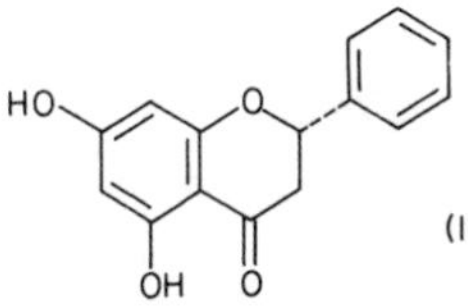

(18, pinocembrin)

ubiquitous floral pigment cyanidin (16). Paradoxically, the shikimate pathway is also the principal origin of aromatic nuclei which
are devoid of phenolic hydroxyl groups, for example cinnamic acid
itself (17) and in more elaborate structures such as the flavanone
pinocembrin (18).

5.2.2 PHENOLIC ESTERS AND GLYCOSIDES

The cell vacuole is generally assumed to be the main storage organ
for the phenols present in the vegetative tissues of a plant. Here the
majority are found in combination with sugars or related compounds either as glycosides or esters. It has often been suggested
that this chemical association is a detoxification mechanism
analogous in many ways to conjugation with glucuronic acid in
mammals. It may also bring about an enhanced solubility of the
metabolite and lead to a decrease in reactivity towards aerobic and
enzymic oxidation[23]. Examples which typically illustrate the
various modes of chemical combination are shown in the accompanying formulae (19–27).

The principal form of combination of phenols and sugars is that of
O-glycosides (e.g. 19, 25, 26, 27)[24–27] although several C-glycosides
have been identified and of these vitexin (22), from the wood of
Vitex lucens, is probably the best known[28]. Among the hexoses, the
three most stable ones glucose, mannose and galactose are very

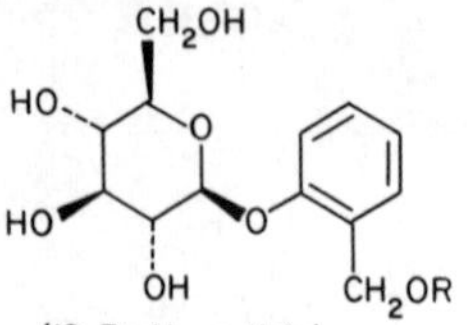

(19, R = H ; salicin)

(20, R = 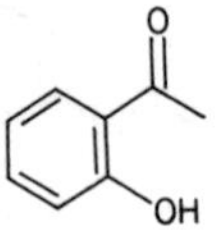 , ω - salicoylsalicin)

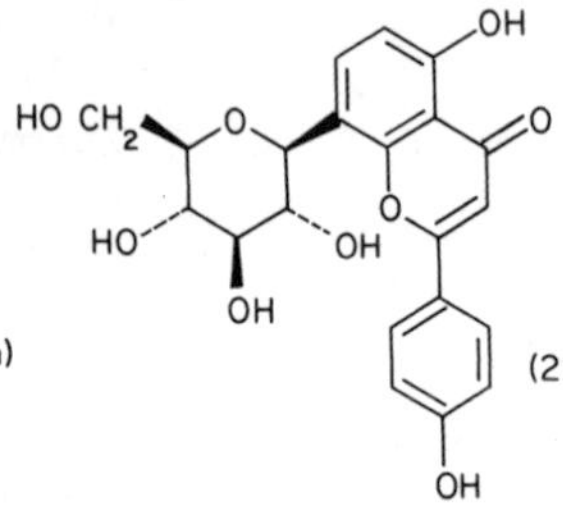

(21, R = 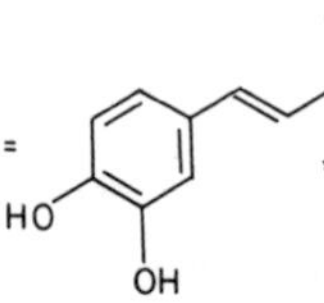 , ω - caffeoylsalicin)

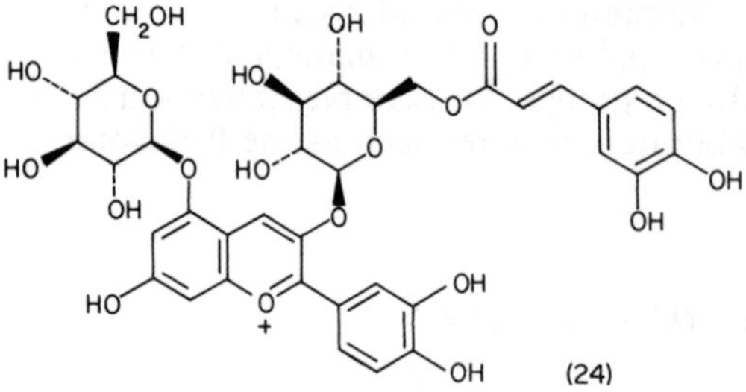

(22, vitexin)

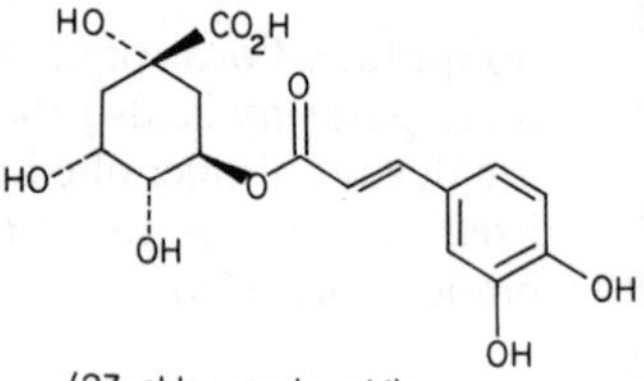

(23, chlorogenic acid)

(24)

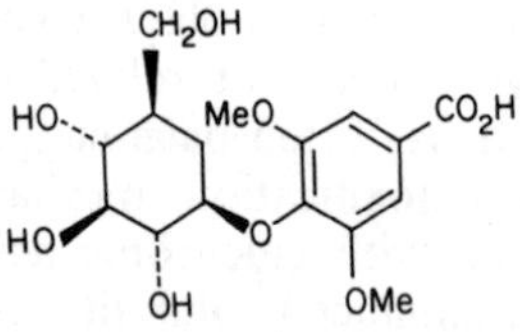

(25)

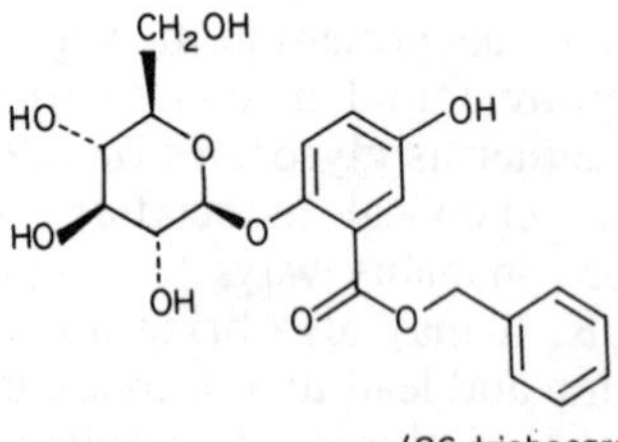

(26, trichocarpin)

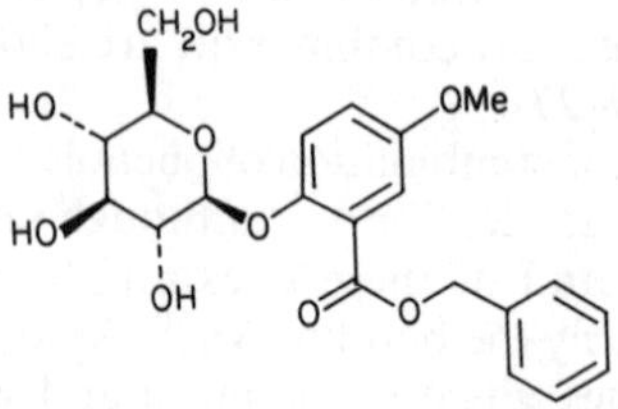

(27, trichoside)

widely distributed in nature and glucose and galactose along with rhamnose are frequently found conjugated with phenols in plants. It is curious, however, that until very recently allose, of only slightly higher free energy, was unknown in nature. Beylis, Howard and Perold[29], however, recently reported the isolation of the 6-O-benzoyl and 6-O-cinnamoyl esters of 2'-hydroxy-4'-hydroxymethyl-phenyl-β-D-alloside (28 and 29) from the South African shrub *Protea rubropilosa* and this remains the only proven occurrence of this aldohexose in a higher plant. Significantly in the related plant *Protea lacticolor*, the 6-O-benzoate of 2'-hydroxy-4'-hydroxymethyl-phenyl-β-D-glucoside (30), an isomer of salireposide (31) and nigracin (32)[30, 31], is a major leaf constituent.

Esters of phenolic acids, particularly the hydroxycinnamic acids (9, 12, 13 and 14), with (−)-quinic acid, (−)-shikimic acid and D-glucose are also widely distributed in the plant kingdon[32, 33]. Chlorogenic acid (23), 5-O-caffeoyl-quinic acid, is for example

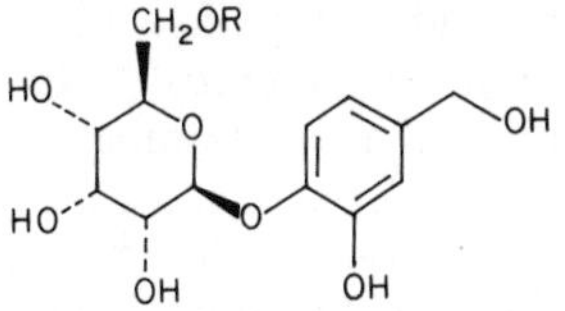

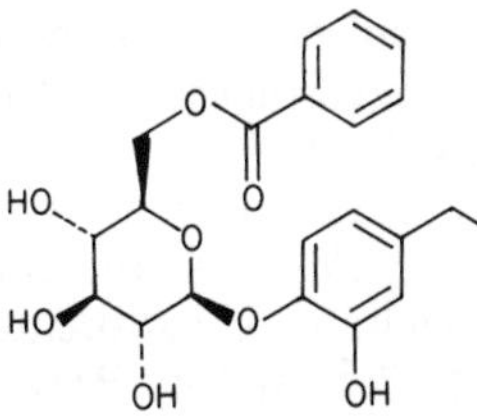

(28, R = 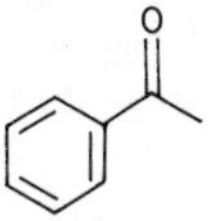 , pilorubrosin)

(29, R = 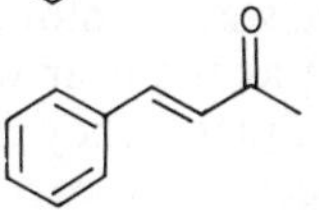 ,rubropilosin)

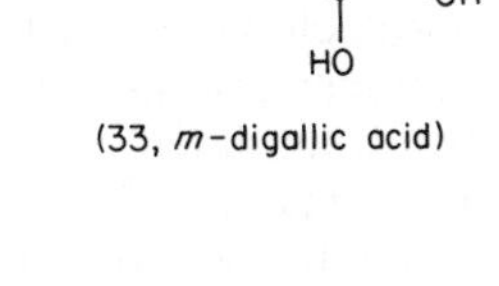

(30)

(31, R' = ,R² =H , salireposide)

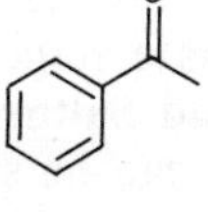

(32, R² = ,R¹ =H , nigracin)

(33, *m*-digallic acid)

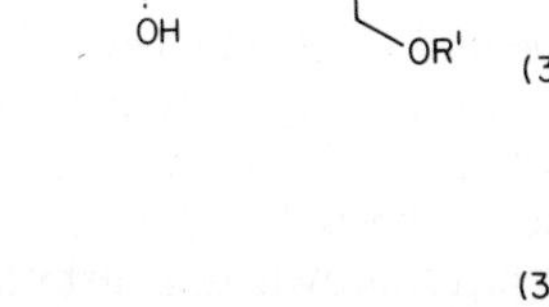

frequently isolated from plant tissues and in addition hydroxy-cinnamoyl esters of tartaric acid[34], L-malic acid[33] and 3′,4′-dihydroxyphenyl lactic acid[33], phenolic glycosides (e.g. 21 and 29)[25, 26, 35-38], glycerides[39], terpenes[40], flavonoids[41], anthocyanidins (e.g. 24)[41, 42], choline[43], and several alkaloids[44] have been described. Esters or glycosides of the hydroxybenzoic acids, e.g. salicylic acid, *p*-hydroxybenzoic acid, vanillic, syringic and gallic acid, are said to occur in all higher plants[45], but in many cases they appear, according to present data, to be particulate bound or associated with the insoluble lignin fraction of the plant cell[46, 47]. Unlike the hydroxycinnamic acids, there are only a limited number of reports of clearly defined molecular species involving hydroxybenzoic acids in higher plants. Pre-eminent amongst these are the esters of gallic acid and hexahydroxydiphenic acid with D-glucose which in earlier times had considerable commercial importance as natural tanning materials for animal hides and skins[48, 49]. Gallic acid is also unique amongst acidic metabolites of the shikimate class in its ability to form depside linkages and *m*-digallic acid (33) is thus a structural component of several gallotannins[48]. It is probable, nevertheless, that detailed studies of plant families will reveal a wider range of derivatives of other hydroxybenzoic acids. Thus glucosyringic acid (25) has recently been isolated from *Anodendron affine* and the systematic and detailed characterisation of the phenolic constituents of *Salix* and *Populus* species by Thieme[24, 30, 31] and by Pearl and Darling[25, 26, 35] has revealed the presence of several distinctive derivatives of benzoic acid (28, 31, 32), salicylic acid (20) and gentisic acid (26, 27).

Phenols are occasionally found in the free state in storage tissue and often in the dead or dying tissues, such as the heartwood and bark of trees[50], but in the vegetative tissues of plants they are almost invariably found associated with substances of a carbohydrate nature. The principal exceptions to this generalisation are the flavan-3-ols (+)-catechin (34), (+)-gallocatechin (35), (−)-epicatechin (36) and (−)-epigallocatechin (37) and the co-occurring oligomeric proanthocyanidins (condensed tannins). Apart from one unsubstantiated claim[51, 52], these phenols have always been found free in plants[53-55] but the metabolic significance of this observation, if any, has yet to be revealed.

The biochemical pathways leading to phenol glycoside formation have been reviewed by Pridham[23] and it is generally assumed that the glycosyl residue(s) are transferred to the phenol via a nucleoside diphosphate sugar at a late stage of biosynthesis[56]. Cardini and Leloir[57] and Pridham and Saltmarsh[58] have shown that arbutin (quinol β-D-glucoside, 1) can be synthesised by incubating quinol

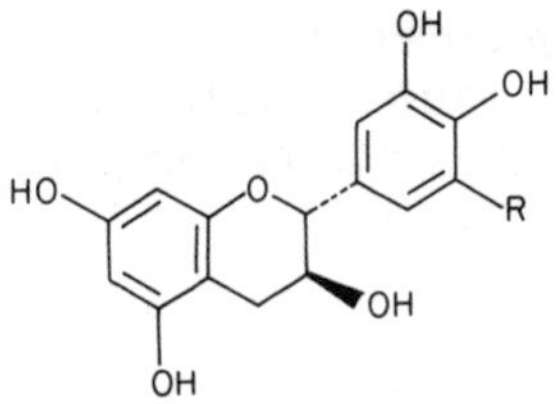

(34, R=H, (+) – catechin)

(35, R=OH, (+) – gallocatechin)

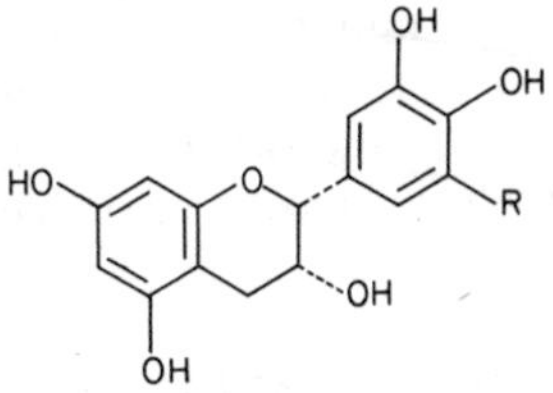

(36, R=H, (−) – epicatechin)

(37, R =OH, (−) – epigallocatechin)

with UDPG in the presence of enzyme preparations from wheat germ or *Vicia faba* seeds, and Barber[59] has demonstrated the ability of mung bean preparations to glucosylate the flavonol quercetin in the presence of UDPG and TDPG. In an elegant series of experiments utilising cell cultures of parsley (*Petroselinum hortense*), Grisebach and his collaborators[60, 61] were able to characterise the two enzymes UDP-glucose-apigenin and UDP-apiose-7-O-glucosyl apigenin (1 → 2) apiosyl transferase and they showed that these two enzymes catalysed the final two steps (38 → 39 → 40) in the biosynthesis of the flavone glycoside apiin (40). A UDP-apiose synthetase was also demonstrated in the cell suspension cultures of parsley which catalyses the NAD^+ dependent synthesis of UDP-apiose from UDP-D-glucuronic acid.

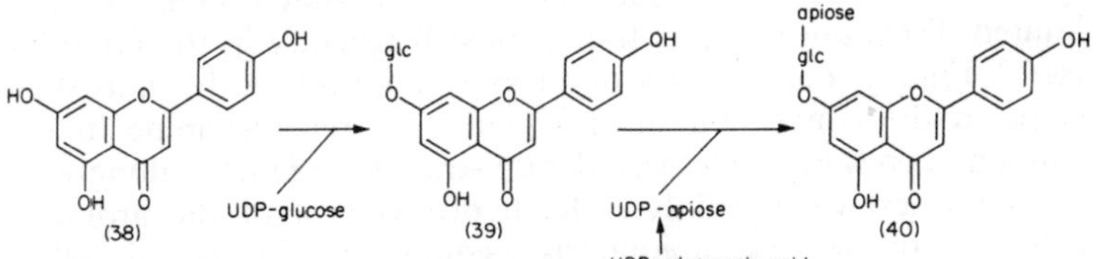

Little work at the biochemical level has been carried out on the synthesis of phenolic esters. The enzymic transfer of glucose from UDPG to carboxylic acids has been demonstrated and it is possible that the 1-O-hydroxycinnamoyl esters of β-D-glucose are formed by this means[23]. Although there is little experimental evidence yet to confirm this supposition, other natural phenolic esters are believed to be derived by transfer of the acyl group from co-enzyme A precursors. The activation of cinnamic acid in the presence of co-enzyme A has thus been reported to occur with extracts of *Beta*

vulgaris[62] and of cinnamic acid and *p*-coumaric acid by *Cicer arietinum* and *Apium petroselinum*[63] but the formation of co-enzyme A esters of the other naturally occurring hydroxycinnamic acids and hydroxybenzoic acids has yet to be observed.

5.3   PHENYLALANINE AMMONIA LYASE AND THE PHENYLPROPANOID POOL

5.3.1   L-PHENYLALANINE AMMONIA LYASE

It is generally agreed that L-phenylalanine (42), and in certain plants L-tyrosine (44), is an obligatory intermediate in the biosynthetic transformations which lead from the shikimate pathway to the phenylpropanoid pool and to almost all of the wide range of phenolic compounds characterised in plants as secondary metabolites[64]. Several well-documented facts form the basis for this opinion. Thus isotopically labelled L-phenylalanine is incorporated, usually with comparatively low isotopic dilution, into a variety of phenylpropanoid compounds and their derivatives in a variety of diverse taxa. Secondly, the enzyme L-phenylalanine ammonia lyase (EC 4.3.1.5) which catalyses the first step in phenylpropanoid biosynthesis—the loss of ammonia from the aromatic amino acid to give trans-cinnamate (17)—has been detected in a large number of vascular plants and in several genera of Basidiomycetes[65-68]. The corresponding enzyme L-tyrosine ammonia lyase which mediates the formation of *p*-coumarate (9) from L-tyrosine (44) has a more limited distribution and is found most frequently in the Gramineae[65]. Finally, the majority of enzymes involved in the individual steps in the conversion of shikimate to L-phenylalanine in the shikimate pathway have been demonstrated in plants[66] along with the enzymes which catalyse the hydroxylation of the aromatic ring[70-74] to yield the ubiquitous *p*-coumaric acid (9) and caffeic acid (12), widely believed to be the ultimate precursors of most plant phenolics.

L-Phenylalanine ammonia lyase activity[75] has been partially purified in protein fractions from many plants and some microorganisms. Its distribution in higher plants has been extensively catalogued[65] and evidence has been deduced to substantiate the view that its presence, and that of the corresponding L-tyrosine ammonia lyase, may be correlated with the ability of plants to synthesise phenylpropanoid compounds related to lignin. The enzyme has thus not been detected in bacteria, algae, lichens, mosses and fungi—with the exception of certain Basidiomycetes. Although the enzyme catalyses what is generally considered to be the initial step in the biosynthesis of lignin and a whole range of distinctive phenols in higher plants, in some micro-organisms, such as *Sporobolomyces roseus*[76], L-phenylalanine ammonia lyase appears to catalyse the first step in a series of reactions which degrade the amino acid via cinnamic, benzoic, *p*-hydroxybenzoic and protocatechuic acids and make it available as a general substrate for growth. This pathway of metabolism contrasts with the more frequently encountered pathway by which mammals and some microorganisms degrade L-phenylalanine and L-tyrosine via homogentisic acid.

The enzyme L-phenylalanine ammonia lyase has been isolated and extensively purified from potato tuber[77], corn[78], the yeast *Rhodotorula glutinis*[79, 80] and the procaryotic organism *Streptomyces verticillatus*[81]. Havir and Hanson[77] purified and characterised the enzyme from acetone powders of light exposed slices of potato tuber. The major species of the extract was isolated by chromatography (Sephadex G-200), sucrose density gradient centrifugation and acrylamide gel electrophoresis. Measurements indicated that the enzyme had a molecular weight of $\sim 330\,000$ and was appreciably aspherical. The enzyme was not inhibited by sulphydryl reagents and no clear requirement for metal ions was demonstrated. Treatment with carbonyl reagents such as sodium borohydride inhibited the enzyme. The pH optimum for the enzyme was shown to be 8.75 but at pH 6.8 and $30°$ the equilibrium coefficient at zero ionic strength was estimated to be 4.7, corresponding to a standard free energy change of -3.9 KJ mole^{-1}. L-Phenylalanine ammonia lyase isolated and purified from maize seedlings showed very similar properties to the potato enzyme and it was concluded[78] that this close similarity in properties of enzymes isolated from plants which belong to quite separate branches of the phylogenetic tree suggested a constancy in function. Yeast L-phenylalanine ammonia lyase isolated by Hodgins[79, 80] had a molecular weight of around 275 000 but in contrast to the potato enzyme it was deactivated by reagents which attacked sulphydryl groups.

It is reasonable to assume from the available evidence that the enzyme acts at a switching point in metabolism and diverts L-phenylalanine from the general pool of amino acids used for protein synthesis to the biosynthesis of phenylpropanoid compounds. Since initial steps are probable sites for overall pathway regulation, it is therefore not surprising that the factors which influence L-phenylalanine ammonia lyase activity have been subject to detailed scrutiny. Thus phytochrome control in dark grown seedlings, temperature changes and feedback repression of synthesis and inhibition of activity have all been examined. In the region of neutrality, cinnamate is an effective inhibitor of its own synthesis from L-phenylalanine. Thus at pH 6.6 and 30° using 1 mM substrate concentration, 0.167 mM cinnamate gave 80 per cent inhibition of L-phenylalanine ammonia lyase activity, and Havir and Hanson[77] postulated that in the whole organism feedback control takes place and operates to maintain a small but fairly constant pool of cinnamate available for further metabolism. Since cinnamate may also be channelled to several end-products, significant feedback control may also be exercised at later branch points in the pathways of phenylpropanoid biosynthesis.

Chemical inhibition of L-phenylalanine ammonia lyase activity may be achieved by the use of typical carbonyl reagents such as sodium borohydride and potassium cyanide. Treatment of the enzyme with tritiated sodium borohydride and subsequent hydrolysis gave alanine in which the majority of the radioactivity was confined to the β-methyl group[82]. Similarly reaction with ^{14}C potassium cyanide and hydrolysis gave aspartic acid labelled exclusively in the β-carboxyl group[80]. These observations led to the proposal[82] that the active site of the enzyme, like that of the related L-histidine ammonia lyase[83], contains a dehydro-alanine residue

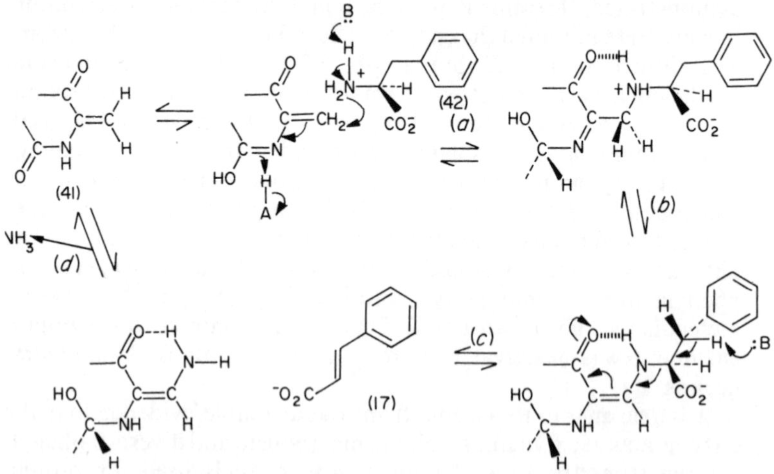

Figure 5.2. Mode of action of L-phenylalanine ammonia lyase[82]
Steps in mechanism:
(a) Addition of amino group of L-phenylalanine to dehydroalanine
(b) 1,3-Prototropic shift
(c) Elimination to generate *trans*-cinnamate and amino-enzyme
(d) Regeneration of enzyme and loss of ammonia

(41). Havir and Hanson[82] have proposed a mode of action for the enzyme in which the first step in the enzymic catalysis is the addition of the amino group of L-phenylalanine to the methylene group of the dehydro-alanine residue, *Figure 5.2*. Cinnamate (17) is first released from the enzyme in the elimination step and the 'amino enzyme' (44) can then either hydrolyse to release ammonia or react with cinnamate to regenerate L-phenylalanine.

Two groups of workers[84–86] have demonstrated that the enzyme operates in a clearly defined stereochemical manner and exclusively by loss of the *pro*-3s-hydrogen (42, H_A) of L-phenyl-

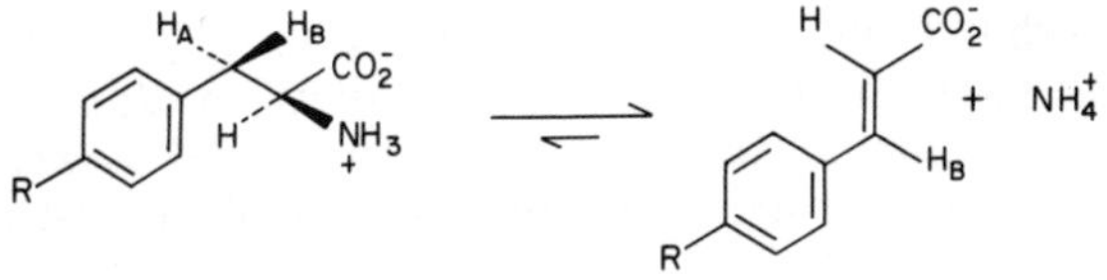

(42, R=H)
(44, R=OH)

(17, R=H)
(9, R = OH)

alanine and ammonia to give trans-cinnamic acid. The groups eliminated are therefore in the trans-antiperiplanar relationship to one another and this stereochemical feature of the enzyme's action is thus quite analogous to that observed in other ammonia lyase reactions[87–89].

5.3.2 L-TYROSINE AMMONIA LYASE

The presence in some plants of an enzyme which converts L-tyrosine (44) into *p*-coumarate (9) was first demonstrated by Neish[90] in 1961. The enzyme, which was originally named tyrase, is generally found in grasses[90] and some fungi[67]. Barley (*Hordeum vulgare*), maize (*Zea mays*)[78] and rice (*Oryza sativa*) were found to be particularly suitable sources of the enzyme as is expected from the ability of these plants to convert L-tyrosine to lignin with efficiency. In their plant survey, Young, Towers and Neish[65] observed that, unlike the analogous L-phenylalanine ammonia lyase, L-tyrosine ammonia lyase never occurred by itself but generally was found in conjunction with L-phenylalanine ammonia lyase and in much lower amount. Several lines of inquiry have shown[91, 92] that in maize a single enzyme is responsible for the elimination of ammonia from both L-phenylalanine and L-tyrosine and that the same active site acts for both substrates. This contrasts with the L-phenylalanine ammonia lyase from potatoes which possesses almost no L-tyrosine ammonia lyase activity. Battersby, Hanson and their associates[93] have confirmed these suggestions using the enzyme from maize. They showed, as predicted, that the 'amino-enzyme' formed from L-tyrosine could react with cinnamate to yield L-phenylalanine and similarly that the 'amino-enzyme' formed from L-phenylalanine reacted with p-coumarate to give L-tyrosine. This reversibility of the ammonia lyase enzymes and their ability to interconvert substrates have clear implications for more general biosynthetic work with higher plants in which the aromatic amino acids are used as substrates. In the yeast strains which have been studied purified L-phenylalanine ammonia lyase has been found to deaminate both L-phenylalanine and L-tyrosine[66]. However, whilst Parkhurst and Hodgins[94] have postulated the existence of one ammonia lyase with bisubstrate activity in *Sporobolomyces pararoseus*, in *Sporobolomyces roseus* Camm and Towers[95] suggested that there were separate ammonia lyases for L-phenylalanine and L-tyrosine.

L-Tyrosine ammonia lyase acts in a similarly highly stereospecific manner upon its typical substrate as does L-phenylalanine ammonia lyase, and Battersby and Hanson and their collaborators[93] showed that the reaction proceeded with removal of the *pro*-S hydrogen (44, H_A) from the β-methylene carbon atom of L-tyrosine to generate trans-p-coumarate (9).

5.3.3   HYDROXYCINNAMIC ACIDS AND THE PHENYLPROPANOID POOL

The first formed product following the enzyme catalysed loss of ammonia from L-phenylalanine is trans-cinnamic acid (17) and in the majority of higher plants there is good reason to believe that the remaining hydroxycinnamic acids—p-coumaric (9), caffeic (12), ferulic (13) and sinapic (14)—are formed sequentially from (17)[96, 97], *Figure 5.3*. In addition, all members of the Gramineae which have been tested can also convert L-tyrosine (44) to p-coumaric acid directly[65] and in this property they appear to be unique amongst higher plants. The phenolic acids (9, 12, 13 and 14) once formed then play a central role in the synthesis of lignin, flavonoids and the whole range of phenolic secondary metabolites (*Table 5.1*)[98–104]. They themselves (or as activated esters) are the principal constituents of the phenylpropanoid pool.

The biosynthetic sequence (17 → 9 → 12 → 13 → 14) was inferred from tracer studies in which ^{14}C labelled L-phenylalanine was fed to *Salvia splendens*[105]. All four hydroxycinnamic acids became labelled and measurements of the rate of incorporation of the

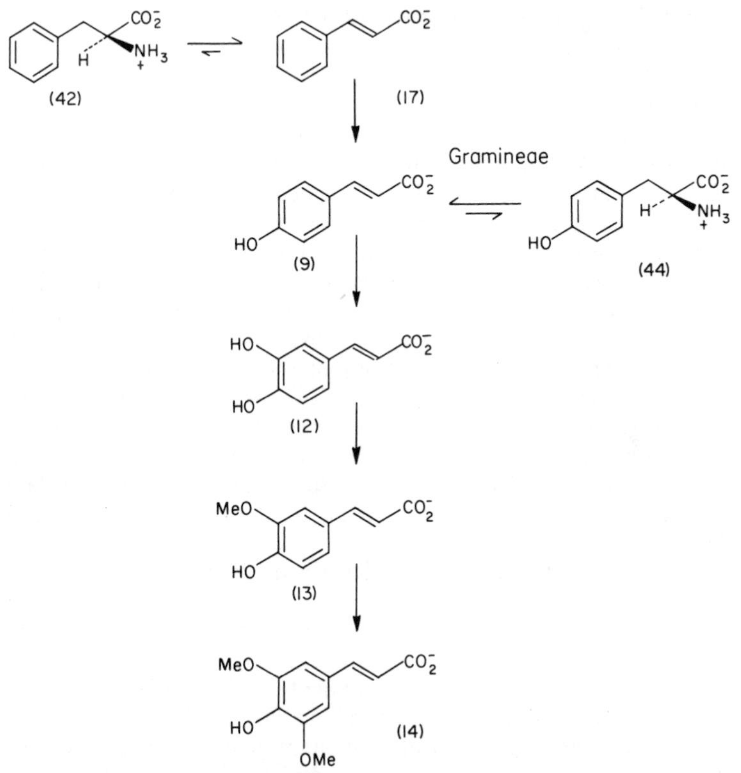

Figure 5.3. Biosynthesis of the hydroxycinnamic acids[96, 97, 105]

radioisotope into the individual acids suggested that they were formed successively in the manner shown, *Figure 5.3*. The general occurrence of the hydroxycinnamic acids as conjugates in higher plants led Levy and Zucker[106] and later Hanson and Zucker[107, 108] to question whether the steps subsequent to the formation of cinnamic acid (17) occur with the acids combined as esters. In an examination of the biosynthesis of chlorogenic acid (23) in potato tubers, they obtained evidence to substantiate this view and proposed a sequence: L-phenylalanine→cinnamic acid→5-O-cinnamoylquinate → 5-O-p-coumaroylquinate → 5-O-caffeoylquinate (chlorogenic acid). They were also able to show the feasibility of the final hydroxylation step using a cell free enzyme preparation. Neish[97] has, however, concluded that it is more probable that if conjugates of the cinnamic acids are involved in the hydroxylation sequence then these are most probably the activated co-enzyme A esters. These could then be envisaged as giving rise to the various esters, such as those of quinic acid, which are normally isolable as natural products, if they were not utilised further for other biosynthetic purposes.

Although these experiments do not exclude the possibility that ferulic (13) and sinapic (14) acid may be formed by direct methoxylation, the consensus of opinion supports the hypothesis that the acids (9, 12, 13, 14) are formed by a successive series of hydroxylation and methylation reactions. The oxygen methylation reactions are thought to occur by transmethylation from L-methionine, and a meta-O-methyltransferase, which utilises S-adenosylmethionine in the conversion of caffeic to ferulic acid (12 → 13), has been identified in several plants[103, 109, 110]. The position regarding the enzymology of the hydroxylation reactions postulated (*Figure 5.3*) is now rapidly being clarified. Enzymes catalysing two types of aromatic hydroxylation in higher plants are recognised. These are the reactions which involve the introduction of a phenolic hydroxyl group either into a non-hydroxylated aromatic ring (e.g. 17 → 9) or ortho to a hydroxyl group already present (e.g. 9 → 12).

Russell and Conn[111] have described an enzyme preparation from the apical buds of pea seedlings which catalyses the hydroxylation of cinnamic acid (17) to *p*-coumaric acid (9). Maximum activity of the enzyme, giving yields of 25 per cent of product, was observed at pH 7.5 and a requirement for NADPH[99, 111] as cofactor was also noted. The pea seedling enzyme differs markedly in its properties from the only other cinnamate hydroxylase which has been isolated, from spinach[112], and which showed a pH optimum of 4.6. Cinnamate hydroxylase has also been observed in *Fagopyrum esculentum*[113] and it showed an increase in activity in response to illumination. The mode of action of the enzyme from pea seedlings shows several similarities to the aromatic hydroxylases found in bacteria and some mammalian organs. In particular it has been possible to demonstrate, using the enzyme and its typical substrate, the distinctive NIH shift[114]. Thus an *in vitro* enzyme catalysed conversion of 4'-[³H]-cinnamate to *p*-coumarate gave[115] the product with over 88 per cent retention and migration of the tritium ortho to the hydroxyl function. Similar *in vivo* studies in *Catalpa hybrida* gave an 85 per cent retention of tritium when 4'-[³H]-cinnamate was converted to *p*-coumarate[116], and further demonstrations of the NIH shift with cinnamate as substrate have been made by Kirby and his colleagues[117] in relation to alkaloid biosynthesis in higher plants and by Zenk[118] and Grisebach[119] with various flavonoids.

The hydroxylation of L-phenylalanine to L-tyrosine which is

important in higher organisms occurs to only a very limited extent in plants and does not seem to provide a major route to L-tyrosine. Thus tracer experiments have shown that there is little if any conversion in *Narcissus incomparabilis*[120], *Colchicum* sp.[121, 122] and *Erythrina berterona*[123]. However, more recently a hydroxylase has been isolated from spinach which catalyses the direct conversion of L-phenylalanine to L-tyrosine[124]. Studies on the biosynthesis of selected alkaloids using isotopically labelled ^{14}C L-phenylalanine as a precursor also indicated that hydroxylation of L-phenylalanine to give L-tyrosine is possible in peyote[125] and barley[126]. In barley, for example, it was established that L-phenylalanine, L-tyrosine and tyramine (45) were precursors of the alkaloid hordenine (47) and the biosynthetic sequence (42 → 44 → 45 → 46 → 47) was inferred. Support for these observations, and hence the initial hydroxylation step (42 → 44), was obtained by Leete, Bowman and Manuel[127] who fed DL-3-[^{14}C]-4′-[^{3}H]-phenylalanine to germinating seedlings (*Hordeum distichum*) and isolated N-methyltyramine (46) which had retained 88 per cent of the tritium ortho to the hydroxyl group—a further example of the NIH shift accompanying aromatic hydroxylation in higher plants.

The introduction of a hydroxyl group ortho to one already present in an aromatic ring, as in the conversion of *p*-coumaric acid (9) to caffeic acid (12), can be catalysed by the monophenolase (cresolase)

(42) (44) (45)

(47, hordenine) (46, N-methyltyramine)

(*i*)

(*ii*)

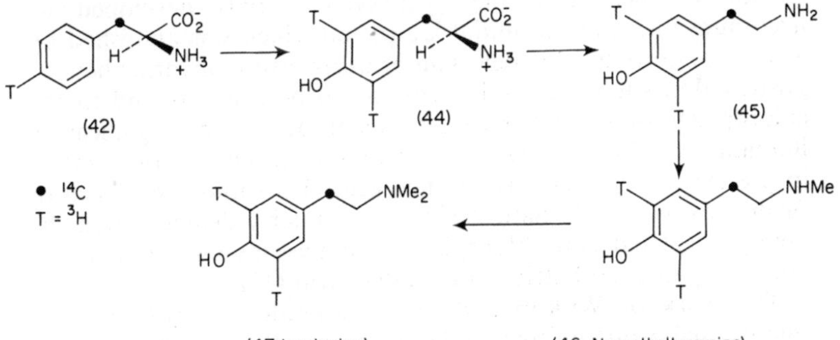

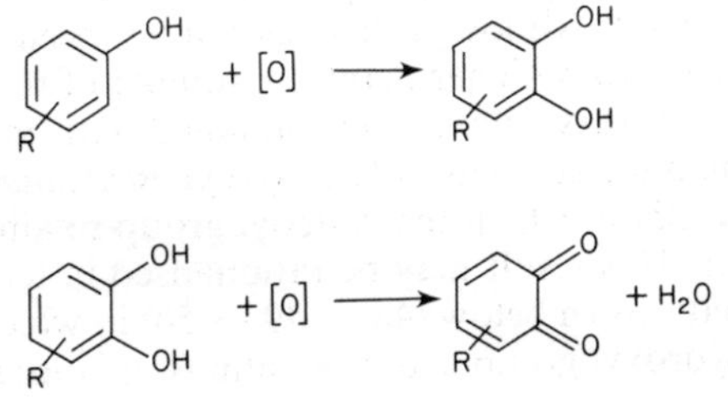

activity of the phenolase complex (tyrosinase, polyphenoloxidase) which is widely distributed throughout the phylogenetic scale[128]. The phenolase complex almost invariably consists of two enzymic activities, (*i*) cresolase and (*ii*) catechol oxidase. The two activities appear to be functions of the same enzyme complex and mechanistic proposals to account for these observations have been discussed earlier. The hydroxylation of phenols to ortho-dihydroxyphenols has been reported with phenolase preparations from many sources but much of this work has been difficult to interpret because of this dual function. As a consequence, there have been no more than very limited studies to date with enzymes isolated from plants on specific reactions, such as *p*-coumaric acid (9) → caffeic acid (12), involved in the biosynthesis of plant phenols. Potato phenolase isolated by Patil and Zucker[129] showed a ratio of catechol oxidase (*ii*) to cresolase (*i*) activity of 8:1 and was able to catalyse the conversion of 5-O-*p*-coumaroylquinic acid to the caffeoyl analogue, chlorogenic acid (23). An enzyme from the leaves of spinach beet (*Beta vulgaris*)[73, 74] which catalyses the hydroxylation of *p*-coumaric acid (9) to caffeic acid (12) has been purified by about a thousandfold. The preparations contained copper (0.17–0.33 per cent), showed a cresolase (*i*) to catechol oxidase (*ii*) activity ratio of 0.37, and required for activation an electron donor such as ascorbate, NADH or dimethyltetrahydropteridine. In a manner characteristic of phenolases, the hydroxylation only developed its maximum rate after an initial lag period which was abolished by low concentrations of caffeic acid or other ortho-dihydric phenols. Distribution studies show the enzyme to be firmly bound to the chloroplasts of spinach beet and sugar beet[74] and significantly illumination of chloroplast suspensions with *p*-coumaric acid (9) in presence of oxygen gave caffeic acid (12) without evidence of further oxidation[130]. Sato[132, 133] has similarly demonstrated the ability of isolated plant chloroplasts to catalyse the enzymic hydroxylation of *p*-coumaric acid (9) to caffeic acid (12).

The work of Witkop and his collaborators[114] with various microsomal aromatic hydroxylases predicts that the introduction of the 3′-hydroxyl group into *p*-coumaric acid (9) will not give rise to an NIH shift. Although this has not yet been tested specifically with enzymes from higher plants, a number of results in agreement with this hypothesis have been obtained with intact plants. Thus 4′-[³H]-cinnamic acid fed to *Fagopyrum esculentum*[118] gave chlorogenic acid (23) in which the caffeoyl group retained 50 per cent of the tritium. This result may be rationalised in terms of a sequence such as that shown below (48 → 49 → 50) in which introduction of the first hydroxyl group causes an almost quantitative migration of the tritium from the 4' position to the ortho positions. Substitution of the second phenolic group results, however, in loss of the tritium at the position of substitution in an analogous manner to the conversion of L-tyrosine to L-dopa.

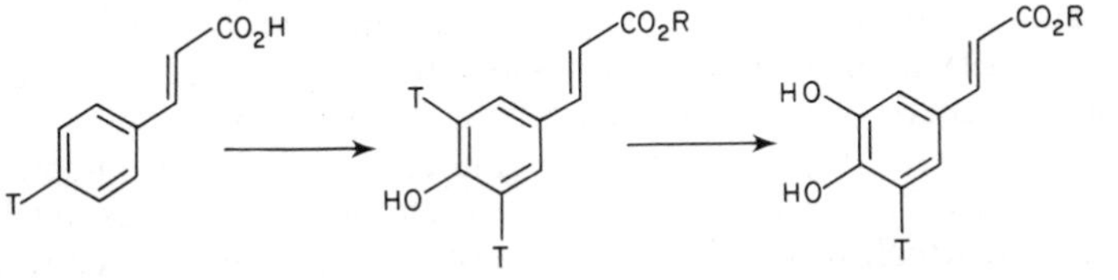

(48)   (49)   (50)

T = ^{3}H

Similar results have been obtained with several secondary plant metabolites such as alkaloids and flavonoids which contain an aromatic structural fragment which is derived from caffeic acid or its equivalent. Thus 3-[^{14}C]-4'-[^{3}H]-cinnamic acid (^{3}H : ^{14}C = 10.4) was administered to pea seedlings *Pisum sativum* in one typical experiment[119] and quercetin (52) and kaempferol (51) were isolated. From earlier related studies, it was deduced that the introduction of the 4'-hydroxyl group in the flavonoids occurs at the cinnamic acid stage, i.e. *p*-coumaric acid is the direct precursor of flavonoids and of the C_6 . C_3 fragment in kaempferol (51). The results of Grisebach and Sutter were in agreement with this proposal since the retention of tritium in kaempferol was approximately the same as

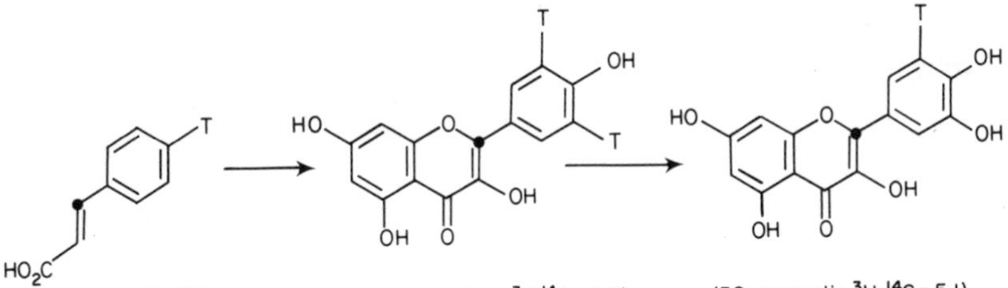

(48, ^{3}H : ^{14}C = 10·4)   (51, kaempferol ^{3}H:^{14}C = 8·8)   (52, quercetin ^{3}H:^{14}C = 5·1)

T = ^{3}H , ● = ^{14}C

in the hydroxylation of cinnamic acid to *p*-coumaric acid. Introduction of the second hydroxyl group into the B-ring of flavonoids has been assumed to occur at the intermediate dihydroflavonol stage and this should lead to loss of approximately half the original tritium as was observed by Grisebach and Sutter[119].

5.3.4   3,4,5-TRIHYDROXYCINNAMIC ACID

From time to time views have been expressed[64, 134] which in essence question the assumption that the route from the shikimate pathway to *all* the $C_6.C_3$, $C_6.C_1$ and $C_6.C_3.C_6$ compounds in higher plants invariably involves L-phenylalanine and cinnamic acid as obligatory intermediates. Particular attention is usually focused on the biogenetic origins of the 3,4,5-trihydroxyaryl system as it occurs, for example, in myricetin, (+)-gallocatechin (35), (−)-epigallo-catechin (37) and delphinidin (52) since no enzyme has yet been isolated which will hydroxylate a mono- or dihydroxylated phenol to give this vicinal-trihydroxy system. According to the 'zimtsaure hypothese'[135] the $C_6.C_3$ structural fragment, present for example in delphinidin (52), is derived from 3,4,5-trihydroxycinnamic acid (53) and experimental evidence based on tracer studies[136] and the genetic control of floral pigmentation[137, 138] supports this view. However, comparative studies on the incorporation of 3-[^{14}C]-cinnamic acid and 1,6-[^{14}C$_2$]-(±)-shikimic acid into delphinidin (52) in *Viola cornuta*[139] have shown that the shikimic acid substrate is much the more superior precursor of the trihydroxylated B-ring of the anthocyanidin. It has been suggested, as an explanation of this observation, that the 3,4,5-trihydroxycinnamic acid unit is in this case derived directly from (−)-shikimic acid (or a closely related intermediate) *without* the intervention of L-phenylal-anine[99, 139].

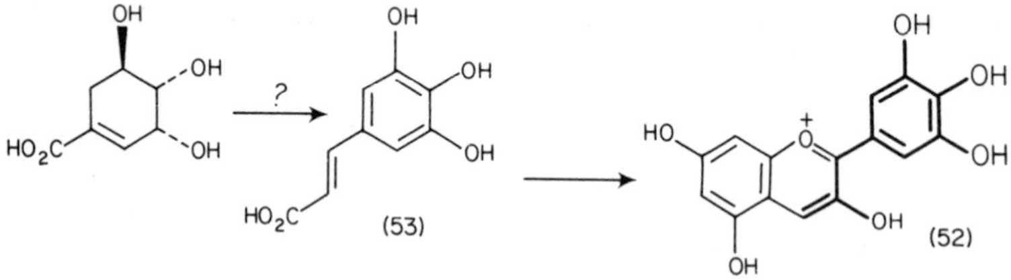

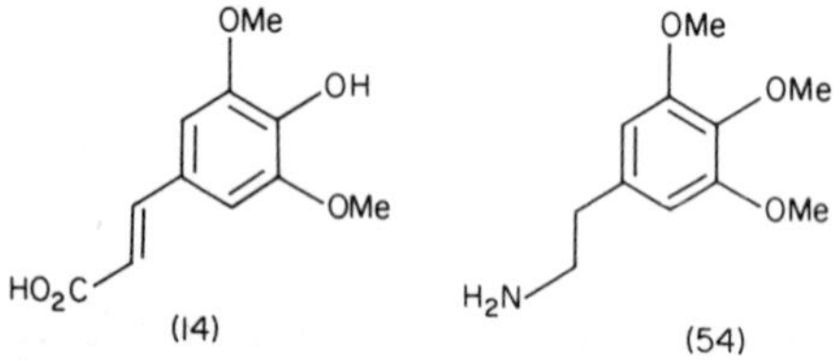

Moreover, it has also been noted that the apparently closely related syringyl (as in sinapic acid, 14) and trimethoxyaryl (as in mescalin, 54) patterns of aryl substitution are formed by a distinct series of hydroxylation and methylation steps of an aromatic ring and that the 3,4,5-trihydroxyaryl grouping is not formed as an intermediate[96, 97, 127]. Significantly perhaps, the parent 3,4,5-trihydroxycinnamic acid is itself something of an enigma, for, although it is frequently postulated as a biosynthetic intermediate, it has never been found free or combined in higher plants[3, 134].

One further example in the literature, which suggests that alternative pathways to other $C_6 . C_3$ intermediates may exist in higher plants, concerns the coniferyl units of lignin[102, 103]. These structural fragments (55) of the polymer are believed to be ultimately derived from ferulic acid (13)[102, 103]. Oxidation of the lignin releases them as vanillin (56)[140, 141]. In a study of lignin formation in Norway spruce[142] radioactive vanillin was isolated after feeding both 1-[^{14}C] and 6-[^{14}C]-D-glucose. Subsequent degradation of the vanillin revealed a pattern of distribution of radioactivity in the aromatic ring which was asymmetric. Thus the ratio of activities in the C-6 and C-2 positions in the vanillin obtained after feeding 1-[^{14}C]-D-glucose was, for example, 0.63 and although the authors did not comment on this aspect of their results, two interpretations

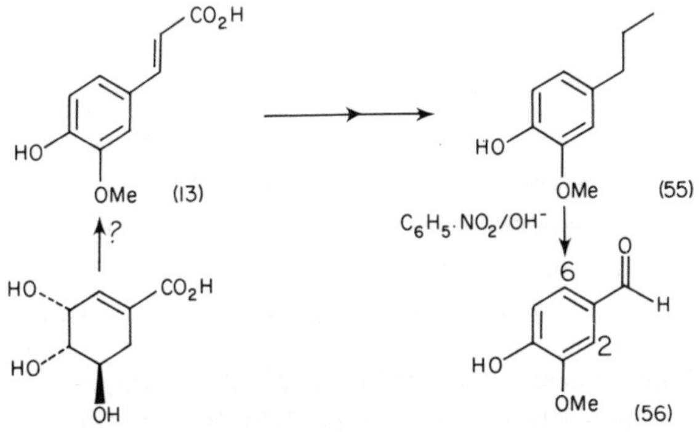

are possible. It may be inferred, on the one hand, that a direct transformation of (−)-shikimic acid to ferulic acid (13) could occur without the intervention of L-phenylalanine and cinnamic acid. Alternatively, it may be implied that all the transformations from (−)-shikimic acid through cinnamic acid to ferulic acid (13) occur in the plant on a multi-enzyme complex and from which at no stage does the substrate become detached from the enzyme surface.

5.3.5   COUMARINS

Coumarins form a chemically distinct group of compounds which are closely related to the hydroxycinnamic acids but are rather less widely distributed in higher plants. The four coumarins most commonly found in nature are: coumarin (10), umbelliferone (3), aesculetin (57) and scopoletin (58)[143] and there is good evidence to sustain the proposal that they are formed from cinnamic acid and the corresponding hydroxycinnamic acids (9, 12 and 13)[144]. Possible exceptions to this general rule are the coumarin daphnetin (59) which occurs as the glucoside daphnin (60) in *Daphne odora* and aesculetin (57)[145] which occurs as the glucosides aesculin (61) and cichorin (62) in *Cichorium intybus*[146]. Studies with labelled precursors showed *p*-coumaric acid to be a much more efficient precursor of these coumarins than caffeic acid[147]. The elaborate variations of the coumarin structure which are found in some higher plant families (e.g. the Umbelliferae) most commonly represent further biosynthetic variations (hydroxylation, isoprenylation, cyclisation and oxidation of the isoprenyl group)[18-21] upon one of the four basic structures (10, 3, 57 and 58)[143].

(10, all R's = H, coumarin)

(3, R¹ = R³ = H, R² = OH, umbelliferone)

(57, R¹ = R² = OH, R³ = H, aesculetin)

(58, R¹ = OH, R² = OMe, R³ = H, scopoletin)

(59, R¹ = H, R² = R³ = OH, daphnetin)

(60, R¹ = H, R² = Oglc, R³ = OH, daphnin)

(61, R¹ = Oglc, R² = OH, R³ = H, aesculin)

(62, R¹ = OH, R² = Oglc, R³ = H, cichorin)

Most simple coumarins occur in plants predominantly or exclusively as glucosides and a unique feature of the biosynthesis of coumarins in intact higher plants is the involvement of such glucosides in the biosynthetic sequence[144]. The β-D-glucosides of *o*-coumaric acid (66) and *o*-coumarinic acid (67) are found for example in plants such as sweet clover (*Melilotus alba*)[148, 149] and lavender (*Lavandula officinalis*)[150] and coumarin itself is only formed by hydrolysis of the glucoside (67) and subsequent lactonisation of the cis-*o*-coumaric acid[148]. Haskins and Gorz[148] have concluded that free coumarin does not occur in *Melilotus alba* and is only observed as an artefact of extraction when appropriate action is not taken to rapidly inactivate the β-glucosidase also present in the plant tissue. Similar observations have been made in relation to the occurrence of umbelliferone in *Hydrangea macrophylla*[151] and scopoletin in *Nicotiana tabacum*[152, 153]. In contrast, Brown[150] using kinetic measurements, suggested that coumarin (10) and *o*-coumarinic-β-D-glucoside (67) probably have a common precursor in sweet grass, *Hierochloe odorata*, and that the lactone is not formed from the glucoside. These observations are in a sense similar to those of Ourisson and his collaborators[154] who utilised tobacco tissue cultures to study the biosynthesis of scopoletin (58), *Figure 5.4*. The free coumarin (58) occurs in significant quantities in tissue cultures and studies were made by Ourisson of the biosynthetic relationship between 'free' and 'bound' coumarins in the cultures. Kinetic studies demonstrated that the main pathway to scopoletin (58) does not involve scopolin (64) or the glucoside (63) as intermediates. Scopolin (64), it was proposed, might be a storage form of the coumarin scopoletin (58) or might equally well arise by an independent pathway from glucisidoferulic acid (63).

Studies[144, 155, 156] on the biosynthesis of coumarin itself in clover nevertheless favour the pathway shown in *Figure 5.5*, and analogous

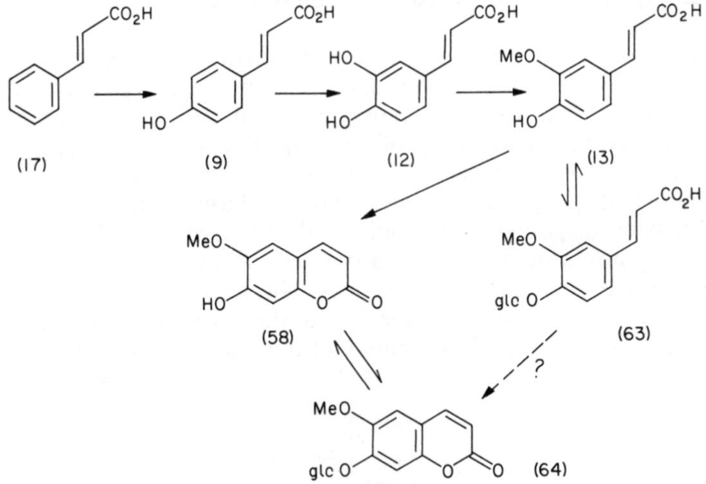

Figure 5.4. Biosynthesis of scopoletin (58) in Tobacco tissue culture[154]

routes have been proposed for the biosynthesis of herniarin (68) in lavender[157] and for umbelliferone (3) in *Hydrangea macrophylla*[151]. A variation on this scheme of coumarin biosynthesis was proposed by Stoker and Bellis[158] in which *cis*-cinnamic acid is hydroxylated to give coumarinic acid which in turn lactonises to give coumarin.

Key steps common to all pathways, which remain only partially

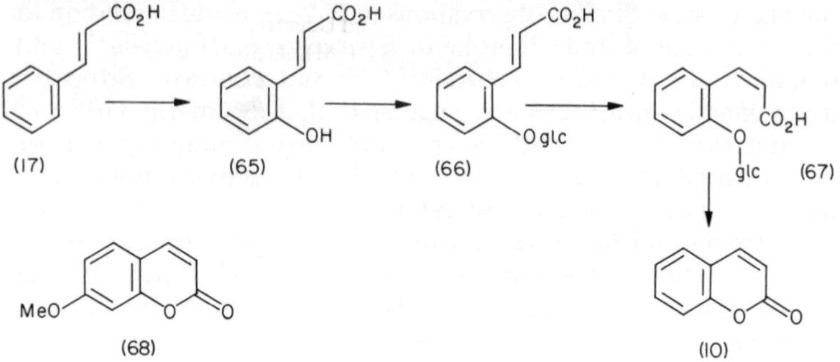

(17)   (65)   (66)   glc   (67)

MeO

(68)   (10)

Figure 5.5. Biosynthesis of coumarin (10) in sweet clover[144, 155, 156]

resolved, are the mode of introduction of the hydroxyl group *ortho* to the C_3 side chain in the cinnamic acid and the *trans* → *cis* isomerisation of the side chain double bond. The first oxidative step (17 → 65, *Figure 5.5*) has been the subject of much speculation[159]. The reaction has yet to be demonstrated with an isolated enzyme system from a higher plant but in a model study, Meyers[151, 160] has accomplished the *ortho*-hydroxylation of *cis-p*-coumaric acid using a copper-ascorbate complex in the presence of oxygen at pH 4.6. The *trans* → *cis* isomerisation has been attributed to the occurrence of a purely photochemical process[161, 162] or the action of a specific isomerase[163]. Ourisson[154], however, found no significant changes in the extent of labelling of phenylpropanoid intermediates in scopoletin biosynthesis when tissue culture experiments were conducted in light or darkness. On the basis of his own work, Ourisson suggested that ferulic acid (13) itself may be cyclised directly into scopoletin (58) via free radical forms generated in the presence of peroxidase like enzymes. This suggestion recalls earlier speculative views on this topic[159] and the work of Kenner and his collaborators on the biosynthesis of the antibiotic novobiocin[164].

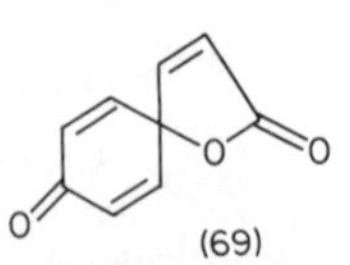

(69)

Others have also raised the question whether an oxidative cyclisation may not also operate in higher plants. Scott and his collaborators[165] on the basis of model reactions thus postulated the involvement of an intermediate spirolactone (69) in umbelliferone biosynthesis but subsequent tracer studies[151] did not support this theory.

5.3.6 LIGNANS AND LIGNIN

Records show that the French chemist Payen[166] early in the nineteenth century was probably the first scientist to undertake the study of lignin, the most abundant and widely distributed phenolic natural product in higher plants. Now almost a century and a half later, scientists are still unable to assign a definitive structure to this amorphous polymeric substance[140, 141]. This inability stems in part from the fact that lignin has resisted all efforts at its isolation in the natural state and also from the very limited success which has attended standard methods of chemical degradation[140, 141, 167]. Somewhat paradoxically there is nevertheless general, although not absolute, agreement concerning the principal structural features of lignin and the manner in which lignin is formed in the plant cell[102, 103, 140, 141, 167–169].

Plant lignins can be divided into three broad classes which are usually referred to as softwood (gymnosperm), hardwood (dicotyledonous angiosperm) and grass or annual plant (monocotyledonous angiosperm) lignins. According to the Freudenberg hypothesis[102, 103, 140, 141], all are polymers formed largely, if not entirely, from monomers with a $C_6 . C_3$ phenylpropane skeleton. Conifer or softwood lignins are a very homogeneous group and are derived almost exclusively (> 80 per cent) from coniferyl alcohol (71). Hardwood lignins are formed correspondingly from coniferyl and sinapyl alcohols (71 and 72)—for example an estimate of the proportion of these two alcohols utilised in the generation of beech wood lignin is $1:1$[170]. Finally, lignins from grasses and plants such as bamboo and palm exhibit considerable variability between species and they are thought to be formed from p-coumaryl alcohol (70) in addition to the coniferyl and sinapyl monomers (71 and 72). The aromatic residues associated with particular lignins are most conveniently detected and identified by alkaline nitrobenzene oxidation when they yield respectively p-hydroxybenzaldehyde, vanillin and syringaldehyde.

Investigations of the biogenesis of lignin have concentrated upon conifer and softwood lignin and only these are referred to in the following discussion. Klason first argued[171] in 1897 that lignin was a polymeric product derived from a $C_6 . C_3$ building unit in which the C_6 fragment had the 4-hydroxy-3-methoxyphenyl, or guaiacyl, substitution pattern. In 1933, following model studies of the oxidative transformation of compounds such as isoeugenol (73) to the crystalline dimer (74), Erdtman[172–174] suggested that lignin is formed by a similar oxidative polymerisation of phenolic $C_6 . C_3$ precursors like coniferyl alcohol. He proposed that polymerisation would occur via coupling reactions, not only in the *ortho* position to the phenolic group but also at the β-carbon atom in the side chain. Ten years later, Freudenberg and Richtzenhain[175] achieved the formation of crude lignin-like polymers by the action of an extract of mushroom on coniferyl alcohol and this led to a brilliant and elegant series of experimental studies on the structure and biogenesis of conifer lignin. Freudenberg and his collaborators[102, 169, 170, 176-180] showed that an insoluble lignin (molecular weight 6000–10 000) is formed from coniferyl alcohol by the enzyme laccase[181] in water at 20°, pH 6.0 in the presence of oxygen. The polymer resembles very closely the lignin from conifer wood and the process by which it is formed is regarded by Freudenberg, and others, as bearing a very close resemblance to the natural process of lignification in plants.

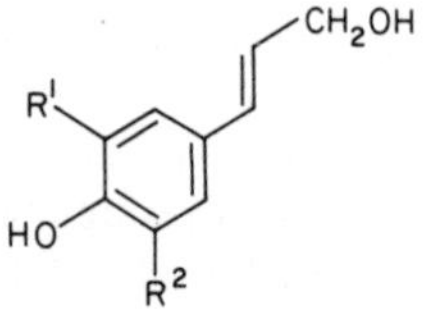

(70, all R² = H, p-coumaryl alcohol)
(71, R¹ =OMe, R²= H, coniferyl alcohol)
(72, R¹ = R²= OMe, sinapyl alcohol)

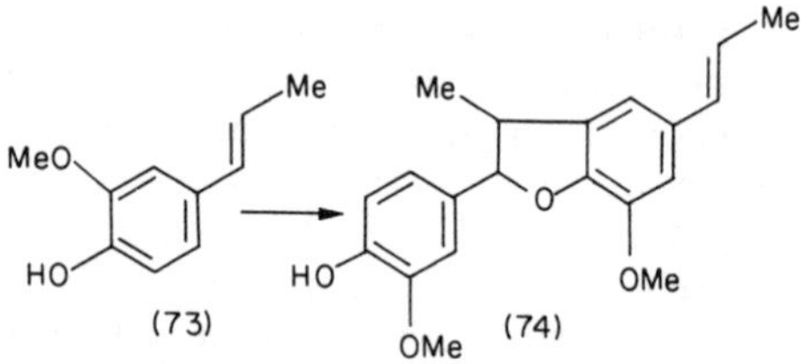

(73)   (74)

In addition to the polymer, intermediate products of lower molecular weight ('lignols') were also isolated by Freudenberg and his group when the oxidation of coniferyl alcohol was not allowed to proceed to completion. Further oxidation of these lignols also yielded amorphous lignin-like products and confirmed their role as intermediates in the polymerisation process. Quantitatively predominant among these intermediates were the dimeric products or dilignols such as (±)-pinoresinol (80), dehydrodiconiferyl alcohol (79) and guaiacyl glycerol-β-coniferyl ether (81), but trilignols, tetralignols, and a pentalignol and a hexalignol were also obtained. These lignols, Freudenberg proposed, were also the natural intermediates formed during lignification and, from a consideration of the ways in which they are apparently formed, a hypothesis was evolved to show how the polymer grows and the major structural features which it possesses.

The formation of all the lignols was rationalised in terms of the dehydrogenation of coniferyl alcohol by a free radical mechanism and plausible reaction pathways to all the intermediates such as (±) pinoresinol (80) were formulated[180] (*Figure 5.6*). The evidence

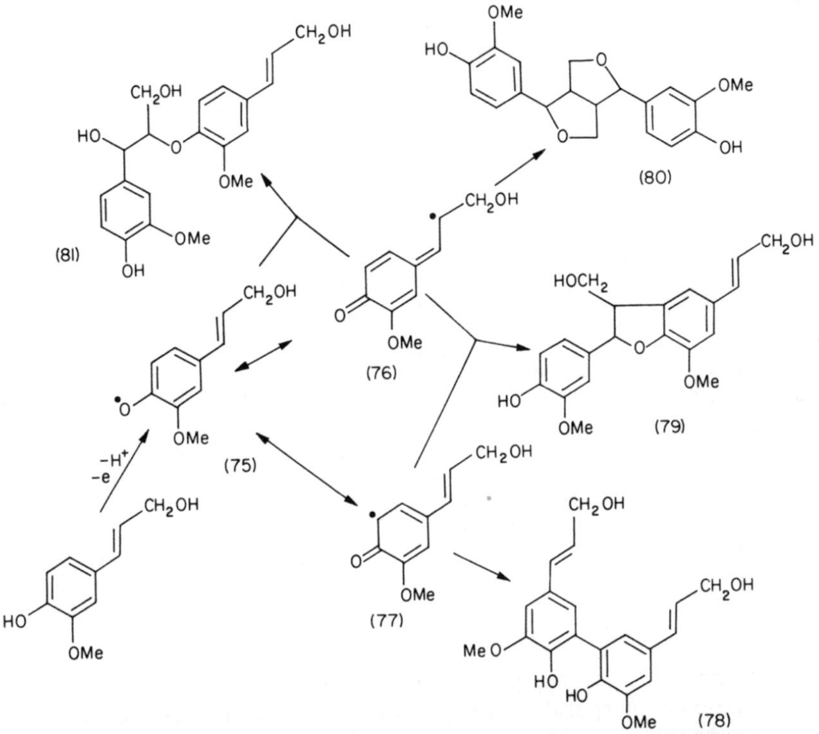

Figure 5.6. Formation of dilignols by oxidation of coniferyl alcohol[180]

from these model reactions and the analytical data relating to the number and type of functional groups in native conifer lignin were combined and used to deduce the frequency of occurrence of particular structural elements and patterns in the natural polymer and hence to formulate structural proposals for conifer lignin[102, 103, 140, 141, 167, 170, 180, 182, 183]. The molecular architecture is complex but not completely random. According to these various formulations, there is a methodical distribution of certain types of linkage and certain building units throughout the structure. Although it is not the most recent and up-to-date, the structural formulation of Freudenberg in 1961 (*Figure 5.7*) is typical of the style of these proposals.

There is little doubt that lignin is ultimately a product of the shikimate pathway of metabolism and that the lignin building units, the hydroxycinnamyl alcohols (70–72), are derived from the corresponding cinnamic acids by reductive means. Several workers have for example observed that $[^{14}C]$-shikimic acid is readily metabolised

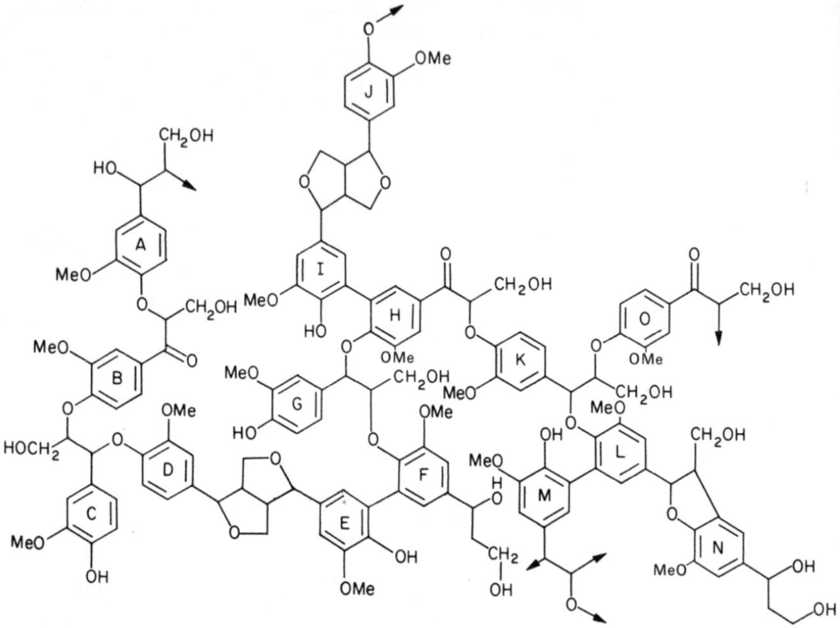

Figure 5.7. Freudenberg's 1961 formulation for lignin

into lignin by growing plants and that specifically labelled [^{14}C]-D-glucose is incorporated into the aryl units in a manner such as to suggest that (—)-shikimic acid is an intermediate in their formation[142, 186–188]. Evidence for the intervention of the various cinnamic acids (9, 12–14) along the pathway has also been obtained from various tracer studies. Thus, for example, ^{14}C labelled ferulic acid (13), which has the ring oxygenation pattern of coniferyl alcohol (71), is efficiently incorporated into the conifer lignin of young wheat without degradation[189–192]. The necessary reduction of the cinnamic acid side chain may be envisaged (*Figure 5.8*) as a two step process analogous to the reduction of β-methyl-β-hydroxy-glutaric acid to mevalonic acid and proceeding via the co-enzyme A ester (82).

Coniferyl alcohol (71) occurs in the cambial sap of conifers as the β-D-glucoside coniferin (83) and a great deal of effort has been expended to unravel the latter's relationship to coniferyl alcohol in the process of lignification. Although many experiments have been reported to show that radioactive coniferin (83) can be incorporated into lignin[193–195], there is no evidence to suggest that the formation of the glucoside is an obligatory step in lignification. Freudenberg has proposed that coniferin and other glycosides form a reservoir

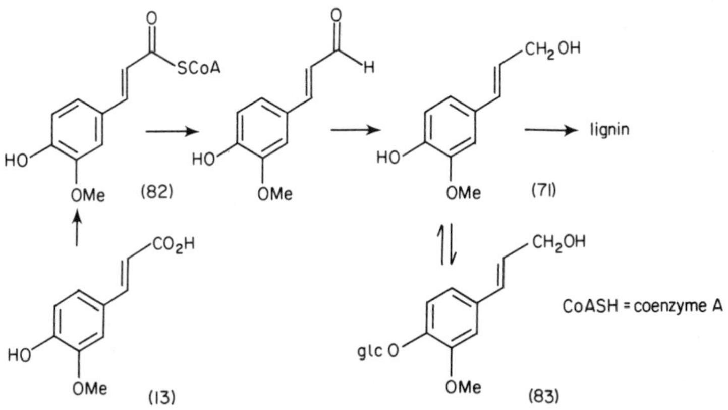

Figure 5.8. Probable mode of formation of coniferyl alcohol and coniferin

of the monomer coniferyl alcohol (71) and that glycosylation increases the solubility of the phenolic alcohol in the storage vessel and protects it from attack by the phenol dehydrogenases which initiate lignification. A β-glycosylase (EC 3.2.1.21) has been located[194] in the xylem of some plants and it is possible that this enzyme regulates the level of phenol in the glycoside pool and thus controls the rate and the location of lignification.

Lignin is initially deposited at the outer layers of the cell wall and the enzymes which control the oxidative process must therefore be situated near these sites since the half-life of the typical free radical intermediates (75–77) which have been postulated is approximately 45 seconds. It has been suggested that a carbohydrate matrix is required for the process of lignification, and that the lignin may be developed by polymerisation from a coniferyl alcohol unit covalently bound to the carbohydrate structure. Freudenberg and Grion[196] have produced evidence in support of this hypothesis. Thus they found that when coniferyl alcohol is dehydrogenated in a concentrated sucrose or sorbitol solution, carbohydrate linked p-hydroxybenzyl ethers (84) are formed, probably by addition to the quinone methide system of intermediates such as (76). Although they have suggested that this reaction is probably the main way in which lignin is attached to the cell wall polysaccharide, Harkin[197] has also discussed the possibility of the formation of sugar-lignol complexes via attachment at the β-carbon atom of the phenylpropane side chain (85). He proposed that coniferyl radicals (75) abstract a hydrogen atom from a sugar molecule by a radical exchange and that the required linkage is then formed by radical recombination.

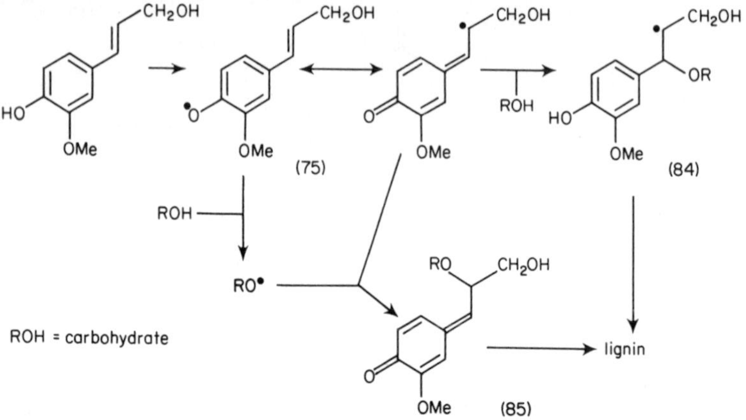

(75)

(84)

ROH = carbohydrate

RO•

(85)

lignin

    There has been substantial discussion of the nature of the enzyme systems which catalyse the polymerisation process of lignification. Mason[197, 198] believes the enzyme to have a phenoldehydrogenase nature and Higuchi[103] has supported this conclusion, and among the enzymes of cambial sap in spruce wood, Freudenberg[176] has demonstrated the presence of a plentiful supply of phenoldehydrogenase and some peroxidase activity.

    Additional support for the theory of lignin formation by dehydrogenation of $C_6 . C_3$ cinnamyl alcohol precursors is often derived from the widespread occurrence of dimers of the $C_6 . C_3$ carbon skeleton, joined by the β-carbon atoms in the side chains, in woods and wood exudates. This group of natural products— lignans—has long been recognised and has stimulated a considerable research interest[199, 200]. Typical members of this class are for example (−)-guaiaretic acid (86) and (+)-pinoresinol (87), the

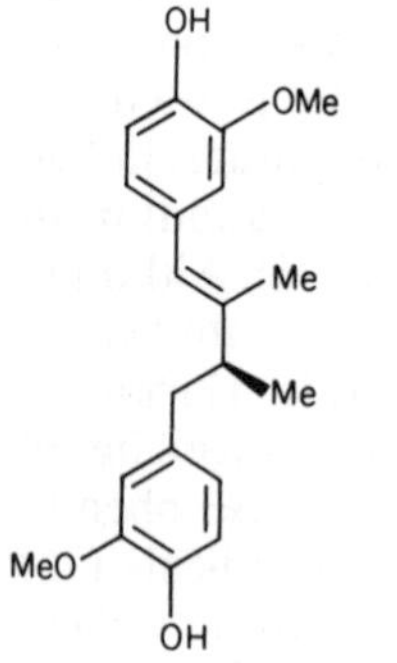

(86, (−)-guaiaretic acid)

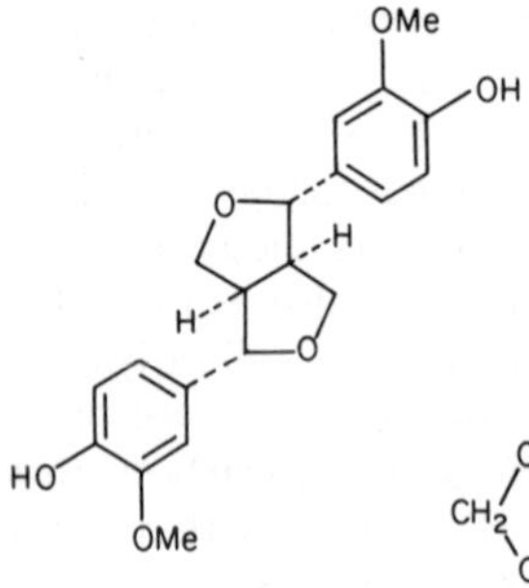

(87 , (+)-pinoresinol)

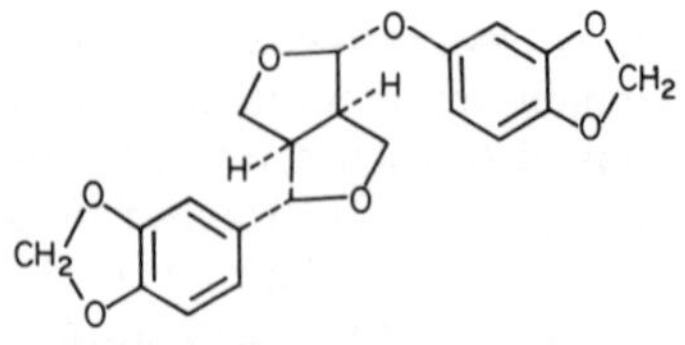

(88, sesamolin)

racemic form of which was isolated by Freudenberg during the oxidative polymerisation of coniferyl alcohol. Two unique variations on the lignan structure are represented by sesamolin (88)[201] and phrymarolin I (89)[202].

(89, phrymarolin I)

Lignans appear to be closely related to probable intermediates in the biosynthesis of lignin and it is believed, without as yet there being any experimental proof, that they are formed by analogous oxidative pathways. However, a point of some significance which still requires explanation is that the majority of lignans occur in distinctive chiral forms but there is presently no evidence available to show that chiral centres of only one type are created in the formation of lignin. Thus no optically active degradation product of lignin has yet been isolated.

5.3.7   ALLYL AND PROPENYL PHENOLS

The allyl and propenyl phenols form a small group of compounds of the $C_6 . C_3$ type which may be regarded as further reduction products of the cinnamyl alcohols (70–72). It is generally assumed that they are formed in this way in plants and some evidence for this view has been obtained by Kaneko[203, 204] in the case of anethole (90) in *Foeniculum vulgare*. However, more recent work by Canonica and his collaborators[205] has removed support for this particular

(90, anethole)   (91, eugenol)

pathway. Thus evidence was obtained in *Ocymum basilicum* to support the biosynthetic pathway L-phenylalanine → cinnamic acid → ferulic acid → eugenol (91) but also to show that during this process C-1 is lost and is replaced by a carbon atom derived from the C-1 pool (such as from L-methionine). It is clear that further work is required to clarify this position.

5.4   DEGRADATION PRODUCTS OF THE PHENYLPROPANOID POOL

5.4.1   HYDROXYBENZOIC ACIDS

Esters or glycosides of a number of hydroxy and methoxy benzoic acids are thought to occur in all higher plants[45, 206–211]. Some plant species are particularly rich in these compounds and thus extracts of *Gaultheria procumbens* when subjected to alkaline hydrolysis yield[211] *p*-hydroxybenzoic (93), vanillic (95), syringic (96), protocatechuic (94), salicylic (97), gentisic (98) and 2,3-dihydroxybenzoic acid (99). These acids, along with benzoic acid (92) itself, are the most commonly occurring in plants. Although work with micro-organisms show conclusively that *p*-hydroxybenzoic (93), 2,3-dihydroxybenzoic (99) and protocatechuic acid (94) are formed in such organisms directly from non-aromatic precursors in the shikimate pathway (chorismic acid[212], *iso*chorismic acid[213] and 3-dehydroshikimic acid[214] respectively) there has been little direct evidence, except in the case of gallic acid[215, 216] to suggest that similar paths exist in higher plants to the hydroxybenzoic acids.

The possibility that β-oxidation of cinnamic acid derivatives and removal of the terminal C_2 fragment of the side chain would form a biosynthetic route to the acids such as (92–96) was postulated by Geissman and Hinreiner[217] in 1952 and experimental evidence to support this hypothesis was obtained in the 1960s by several groups[45, 100, 136, 206, 218–226] independently. A detailed study by Neish and his associates[45] in which *p*-coumaric (9), caffeic (12), ferulic (13) and sinapic (14) acids, all ^{14}C labelled in the β-position of the side chain, were administered to shoots of both monocotyledenous and dicotyledenous plants led in each case to the formation of the corresponding hydroxy- or hydroxymethoxy-substituted benzoic acids in which the major proportion of the radioactivity was located in the carboxyl function. The validity of this biosynthetic pathway was demonstrated[100, 136, 218–226] by analogous experiments on the various substituted benzoic acids (93–99). On the basis of this work Zenk[136, 223] and others[226] formulated a general biosynthetic scheme (*Figure 5.9*) for the formation of the substituted benzoic acids by degradation of the corresponding cinnamic acids.

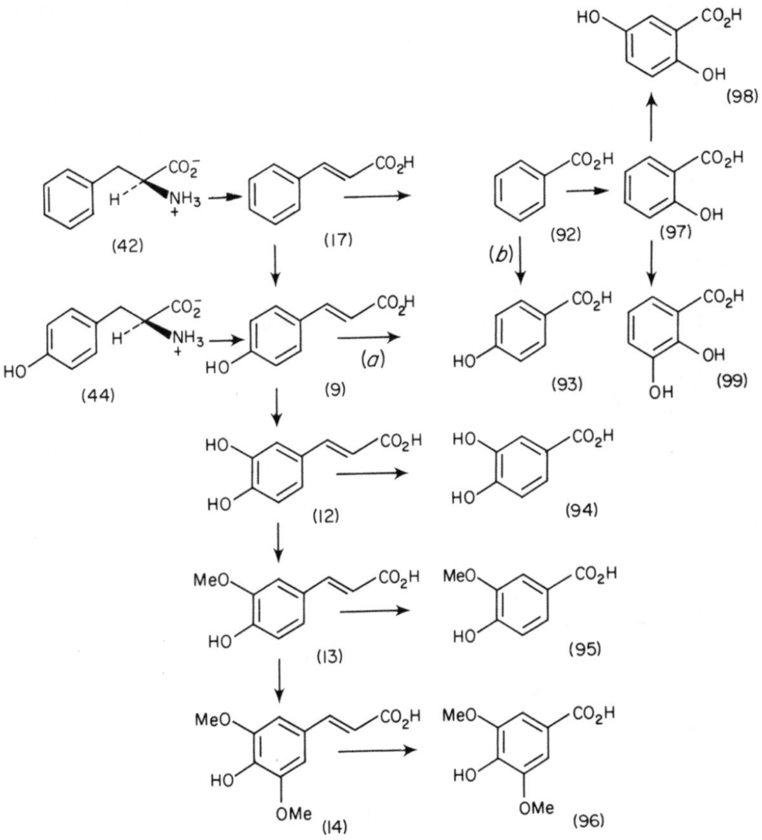

Figure 5.9. Biosynthesis of the hydroxybenzoic acids in plants[136, 223, 226]

Evidence has also been obtained in higher plants[45, 218, 226] for the operation of alternative pathways of biosynthesis for the formation of some of the substituted benzoic acids (93, 95, 97, 98 and 99) in which the required hydroxylation or methylation occurs at the stage of a $C_6 . C_1$ substrate (*Figure 5.9*). Some attempts to establish the relative importance of the two pathways have been made in some cases, but whether the distinctive hydroxylation patterns of the benzoic acids have been established at the cinnamic acid level before β-oxidation or is determined at the stage of a benzoic acid derivative is still not clear. Zenk and Muller[223] thus used kinetic methods to study the formation of p-hydroxybenzoic acid (93) in *Catalpa bignoniodes* and they concluded that the acid (93) is mainly produced by degradation of p-coumaric acid (*Figure 5.9*, path a) and is formed, only in a minor way, by hydroxylation of benzoic acid (*Figure 5.9*, path b). In a similar study of the NIH shift in plant phenol biosynthesis, Zenk[227] observed that 4'-[³H]-cinnamic acid fed to *Catalpa hybrida* gave p-coumaric acid and p-hydroxybenzoic acid in which very similar extents of retention and migration of the isotope to the ortho position to the hydroxyl group (respectively 85 per cent and 83 per cent) were observed. Zenk[136] has thus expressed the opinion that in the living plant under normal conditions the interconversion of p-hydroxylated benzoic acids does *not* occur. In contrast, Ellis and Amrhein[228] in an analogous study of salicylic acid (97) biosynthesis in *Gaultheria procumbens* suggested that the principal route of biogenesis occurred via cinnamic acid (17) and benzoic acid (92) and that o-coumaric acid was not an intermediate.

A further pathway of biosynthesis of salicylic acid (97) may involve its formation from the unusual alicyclic acid (102) which is found in the glucosides tremulacin (101) and salicortin (100) which have been isolated from *Populus tremuloides* and *Salix purpurea*

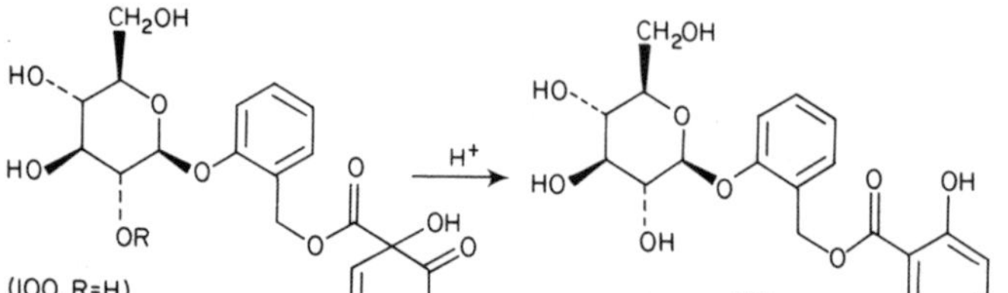

(100, R=H)
(101, R=CO·C₆H₅)

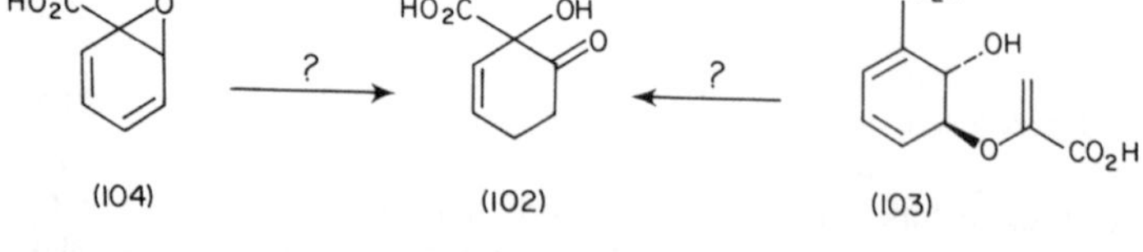

(19)

(104)

(102)

(103)

respectively[229, 230]. The alicyclic addendum in these glycosides is readily aromatised to salicylic acid by acid, and salicortin (100) itself is readily transformed with dilute acid to ω-salicoylsalicin (20) which is an abundant glucoside in both *Salix* and *Populus* species. There have been no suggestions as to the biosynthetic origins of this distinctive alicyclic structure (102) but it may be envisaged as being derived from isochorismic acid (103) by a series of 1,5-sigmatropic shifts, or the arene oxide (104) derived from benzoic acid.

There has as yet been little work at the enzyme level concerning the biosynthesis of the hydroxybenzoic acids in higher plants and its regulation. Boudet, Alibert and Ranjeva have, however, presented[232] data obtained with the leaves and roots of *Quercus* species to show that the biogenesis of the hydroxycinnamic acids and the hydroxybenzoic acids is compartmentalised in different organelles. The phenylpropanoid compounds are formed mainly in microsomes whereas the $C_6 . C_1$ acids are synthesised in microbodies. Associated with this physical separation of these two types of enzymic synthesis, two L-phenylalanine ammonia lyase isoenzymes were shown to exist in *Quercus pedunculata*. One enzyme was found in the microsomal fraction and was sensitive to trans-cinnamic acid; the second enzyme, which was sensitive to benzoic acid, was located in the microbody and mitochondrial fraction. Benzoate synthase, an enzyme believed to be involved in the formation of benzoic acid (92) from cinnamic acid (17), was also found to be confined to the same mitochondrial fraction. On the basis of these observations, the French workers were able to suggest some plausible ways in which regulation of the two synthetic pathways could be achieved in higher plants.

5.4.2 HYDROXYBENZALDEHYDES AND HYDROXYBENZYL ALCOHOLS

Aromatic $C_6 . C_1$ aldehydes and alcohols are relatively infrequently encountered in higher plants and their occurrence has something of a novelty value. The most prominent of the $C_6 . C_1$ aldehydes is probably vanillin (105) and of the alcohols saligenin, as the β-D-glucoside salicin (20), and the two glucosides of gentisyl alcohol (31 and 32) are certainly the most noteworthy.

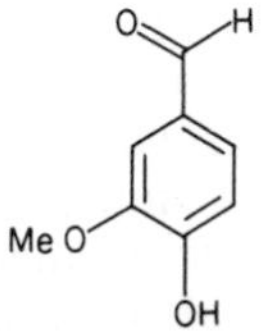

(l05, vanillin)

Salicin (20) was the first glucoside to be found in nature (1828) and its discovery is associated with Buchner, one of the great names of chemistry. It is widely distributed in the Salicaceae and numerous derivatives have been described by Thieme[24, 30, 31, 229] and by Pearl and Darling[25, 26, 35, 230, 231]. In an investigation of the biogenetic origins of saligenin (106) in *Salix purpurea*, ^{14}C tracer studies showed L-phenylalanine, cinnamic (17) and o-coumaric acid to be good precursors of the phenol[233]. Saltmarsh and Pridham[234] made a further important observation when they showed that saligenin (106) fed to plant tissues is converted to o-hydroxybenzyl-β-D-glucoside (107) but not salicin (20). This prompted the suggestion that saligenin (106) was not the direct precursor of the natural glucoside but rather salicylic acid-β-D-glucoside. However, salicylic acid itself proved an inferior intermediate for the biosynthesis of (20) and Zenk was therefore led to propose[233] that salicin was formed in the biosynthetic sequence (*Figure 5.10*) via the aldehyde (108) and helicin (109).

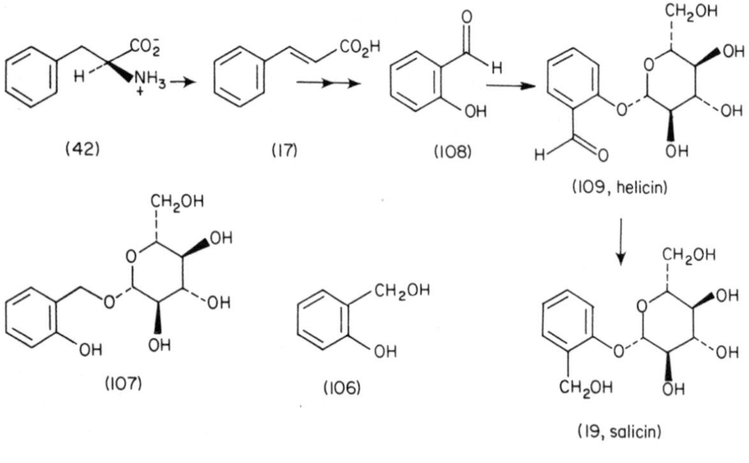

Figure 5.10. Biosynthesis of salicin[223, 234]

Similar results have been obtained in a study of the biogenesis of vanillin (105) in *Vanilla planifolia* and a similar direct breakdown of the phenylpropanoid precursor ferulic acid (13) was proposed. The mechanism of breakdown of the cinnamic acids to $C_6.C_1$ compounds has most reasonably been formulated[136] by Zenk as a β-oxidation involving the removal of acetate (or more probably acetyl co-enzyme A) from the side chain of the phenylpropanoid substrate. A scheme based on these ideas is shown in *Figure 5.11.*

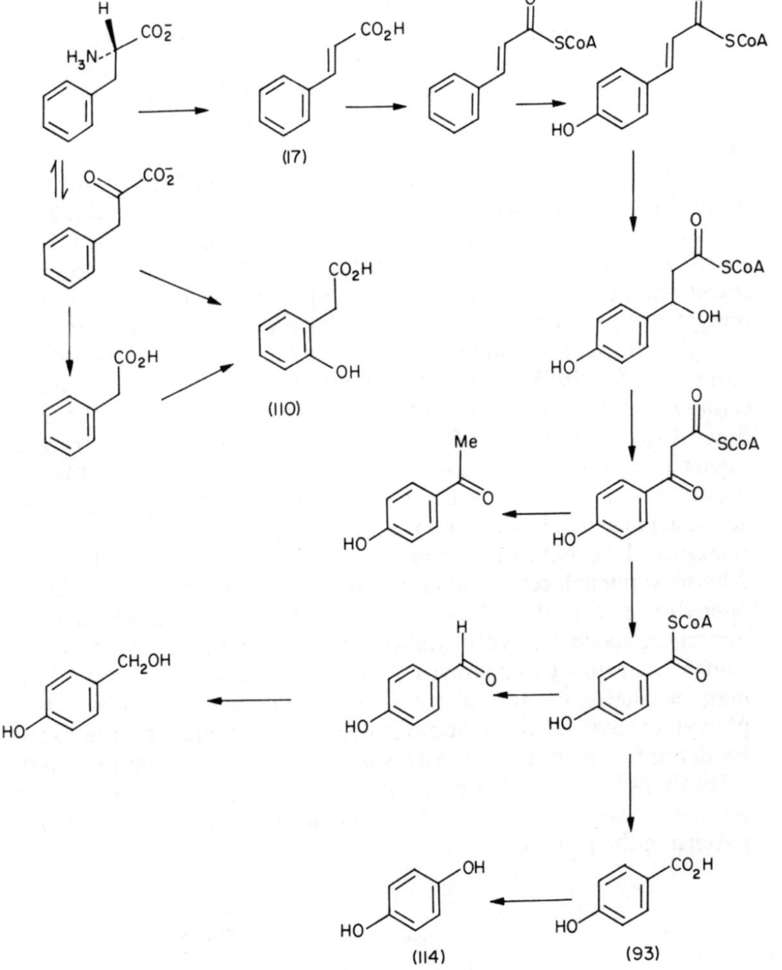

Figure 5.11. Suggested routes for the biosynthesis of some $C_6 . C_2$, $C_6 . C_1$ and C_6 phenols[136, 217]

5.4.3   SIMPLE PHENOLS AND $C_6.C_2$ COMPOUNDS

Embodied in the hypothetical scheme of biogenesis (*Figure 5.11*) which has evolved to describe the modes of biosynthesis of the various $C_6.C_1$ plant phenols, are related proposals which have been discussed to explain the origins of the various $C_6.C_2$ and C_6 compounds which occur in nature. Neither of these two groups of phenols is particularly common and definitive biosynthetic studies have therefore been few.

The typical pattern of aromatic hydroxylation found with the naturally occurring acetophenones suggests a derivation allied to the shikimate pathway, and in the case of pungenoside (111) in *Picea pungens*, this has been substantiated by standard isotopic tracer methods[235, 236]. The results support the idea that acetophenones are formed by decarboxylation of the corresponding benzylacetic acid as shown in *Figure 5.11* and as was proposed by Robinson[11] in 1955. In connection with this hypothesis, it is interesting to note the recent isolation and characterisation for the first time of the methyl ester of *p*-hydroxybenzoyl acetate (112) from a higher plant *Bellendena montana*[237]. A further group of naturally occurring phenols with a $C_6.C_2$ carbon skeleton are the phenylacetic acids which have been investigated by Kindl[238–240] in *Astilbe chinensis*. Two pathways were established for the biosynthesis of 2-hydroxyphenylacetic acid (110, *Figure 5.11*). In the first of these, phenylpyruvic acid is degraded to phenylacetic acid which then undergoes specific *o*-hydroxylation. In the second, a direct oxidative transformation of phenylpyruvic acid is postulated to occur in a manner analogous to the well-known conversion of 4-hydroxyphenyl pyruvic acid to homogentisic acid in higher organisms. Evidence for the latter pathway was obtained by the administration of both 2-[^{14}C]-D-L-phenylalanine and 4'-[^{3}H]-L-phenylalanine to *Astilbe chinensis* and determination of the radioactive labelling pattern in the product.

(111)   (112)

Derivatives of β-phenylethylamine are perhaps the most significant compounds of the $C_6.C_2$ group found in higher plants. Pridham[23] has reviewed the major compounds of this class which occur naturally. Many are familiar alkaloids and others are important intermediates in the formation of other more complex alkaloidal structures. Most of these phenolic amines appear to be derived from L-phenylalanine and/or L-tyrosine, by decarboxylation, hydroxylation of the aromatic ring and methylation (N and O) with L-methionine. However, the order in which the individual steps are performed has been the subject of some debate. The case of mescalin (113), the hallucinogenic alkaloid from the peyote cactus, *Lophophora williamsii*, is typical and only very recently have the steps in the pathway from L-phenylalanine (42) and L-tyrosine (44) been fully elucidated (*Figure 5.12*)[241-246]. The importance of postulated intermediates in the pathway was demonstrated by trapping experiments and by an evaluation of their relative efficiencies as precursors of the alkaloid. Thus normescalin and 3',4'-dimethoxyphenylethylamine, both of which figured in earlier schemes, were

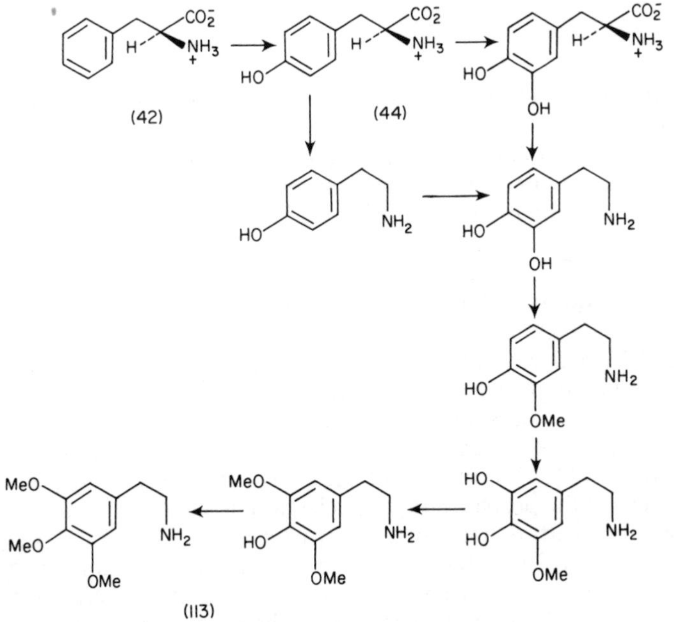

Figure 5.12. Biosynthesis of mescalin[241-246]

excluded on these grounds and of the remaining possibilities the biogenetic scheme in *Figure 5.12* was formulated. A defined sequence of hydroxylation and methylation is followed which parallels that formulated for the biogenesis of the various hydroxycinnamic acids.

Relatively few simple phenols with a six carbon skeleton are known in nature and amongst these hydroquinone-β-D-glucoside (arbutin, 1) has been the most frequently encountered and studied. The biosynthetic origin of arbutin (1) as a metabolite of the shikimate pathway was established by Grisdale and Towers[247] and Zenk[248, 249] later showed that p-hydroxybenzoic acid (93) was an excellent precursor of the aromatic ring of hydroquinone (114), *Figure 5.11.* Zenk[248, 249] proposed that hydroquinone was formed by an oxidative decarboxylation of p-hydroxybenzoic acid (93) and

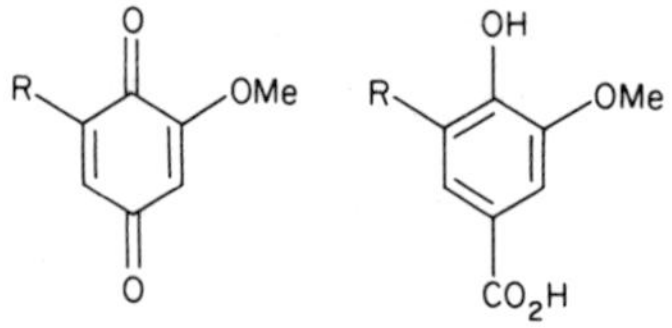

(II5, R=H)   (II7, R=H)
(II6, R=OMe)   (II8, R=OMe)

that arbutin (1) itself was formed by subsequent glucosylation. Similar observations and proposals have been made by Ellis and Towers[250] in connection with the biosynthesis of catechol from salicylic acid (97) in *Gaultheria* sp. and of the plant quinones (115 and 116) from the related acids (117 and 118)[251].

5.4.4   GALLIC ACID AND ITS DERIVATIVES

3,4,5-Trihydroxycinnamic acid (119) has never been found in the plant kingdom except as its trimethyl ether which occurs occasionally in association with some alkaloids and as a constituent of the root of *Polygala senega*[252]. Bate-Smith[3] has suggested that ellagic acid (129) is the taxonomic equivalent of the missing acid but in view of the probable origin of ellagic acid (129), (or hexahydroxydiphenic acid its unlactonised form) it seems preferable to regard gallic acid (120) as the systematic equivalent. From this viewpoint it is of possible significance that gallic acid—as its numerous esters—is only metabolised in plants where the steady state concentration of the hydroxycinnamic acids is small[215, 216]. Studies of the biosynthesis of gallic acid have been numerous and at least three pathways have been proposed for its biosynthesis in higher plants (*Figure 5.13*).

Zenk[222] formulated a conventional pathway (*Figure 5.13, b*) from L-phenylalanine (42) to 3,4,5-trihydroxycinnamic acid (119) followed by β-oxidation to give gallic acid (120). This conclusion was based on comparative feeding studies using ^{14}C-labelled L-phenylalanine, L-tyrosine, D-glucose and benzoic acid in *Rhus typhina*. Neish, Towers and their associates[45] favoured a variation on this scheme (*Figure 5.13, c*) in which β-oxidation at the caffeic acid stage (12) gives protocatechuic acid (94) which undergoes further hydroxy-

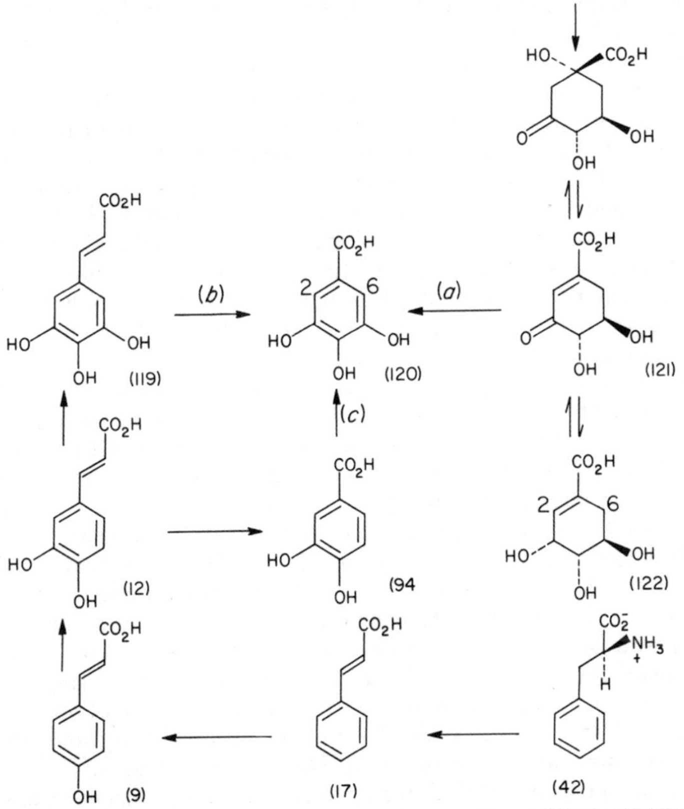

Figure 5.13. Postulated pathways for the biosynthesis of gallic acid[215, 216, 222, 253, 255]

lation to yield gallic acid. These workers were thus able to demonstrate that [^{14}C]-protocatechuic acid is transformed to gallic acid in *Geranium hortorum* in support of the hypothesis. However, work first with the mould *Phycomyces blakesleeanus*[253] and later higher plants[215, 216, 254] (*Rhus typhina, Acer saccharinum* and *Acer pseudoplatanus*) showed that a third route (*Figure 5.13, a*) to gallic acid—the direct dehydrogenation of 3-dehydroshikimic acid (121)— existed in plants and Conn and Swain[255] supported this conclusion in a study of the biosynthesis of gallic acid in *Geranium pyrenaicum*.

Although it is possible that a multiplicity of pathways may exist for the biosynthesis of any metabolite[1], an extensive study of the biosynthesis of gallic acid in plants[215, 216] has led to the conclusion that pathways *b* and *c* (*Figure 5.13*) are minor if not aberrant and that the major route to the production of gallic acid (120) is that in which an intermediate in the shikimate pathway, probably 3-dehydroshikimic acid (121) is dehydrogenated (*Figure 5.13, a*). Several lines of evidence form the basis for this view. As in the mould *Phycomyces blakesleeanus* direct conversion of [^{14}C]-protocatechuic acid to gallic acid could not be detected in *Rhus typhina* and in one important series of experiments designed to establish the relative merits of the two other pathways (*a* and *b*), 1-[^{14}C]-D-glucose and 6-[^{14}C]-D-glucose were administered to several plants which metabolised gallic acid (120). The gallic acid was isolated and degraded to determine the distribution of the radioactive tracer in the molecule. In all the experiments utilising these two substrates, generally less than 10 per cent of the radioactivity was located in the carboxyl group but up to 75 per cent was found in the C_2 and C_6 positions (*Figure 5.14*). This labelling pattern is very similar to that established by Sprinson and his collaborators[256] in the corresponding carbon atoms of (−)-shikimic acid (122) formed in *Escherichia coli* from 1-[^{14}C]-D-glucose or 6-[^{14}C]-D-glucose. As further proof of this apparent relationship, (−)-shikimic acid (122) was also obtained from experiments in *Rhus typhina* and degraded to show an almost identical pattern of labelling to that of the gallic acid. On the other hand, comparison of the distribution of radioactivity in veratric acid, obtained by chemical conversion from caffeic acid (12), and gallic acid (120) isolated concurrently from *Acer pseudoplatanus* revealed no obvious relationships such as might be expected if gallic acid were derived by degradation of this acid (*Figure 5.13*, pathway *b* or *c*). The consensus of these and other observations was interpreted by Dewick and Haslam[215, 216] in terms of the proposal that during normal modes of metabolism the carbon skeleton and hydroxyl groups of gallic acid are derived directly from (−)-shikimic acid probably via its 3-dehydro derivative (121).

Gallic acid occurs in higher plants almost invariably in the form of esters with D-glucose (or related polyols) which range in complexity from the simple β-D-glucogallin (123) to the characteristic gallotannin from *Rhus semialata* and *Rhus typhina* (124)[48, 257]. However, little is yet known concerning the way in which these substances are formed and stored in the plant cell where they may constitute over half its dry weight.

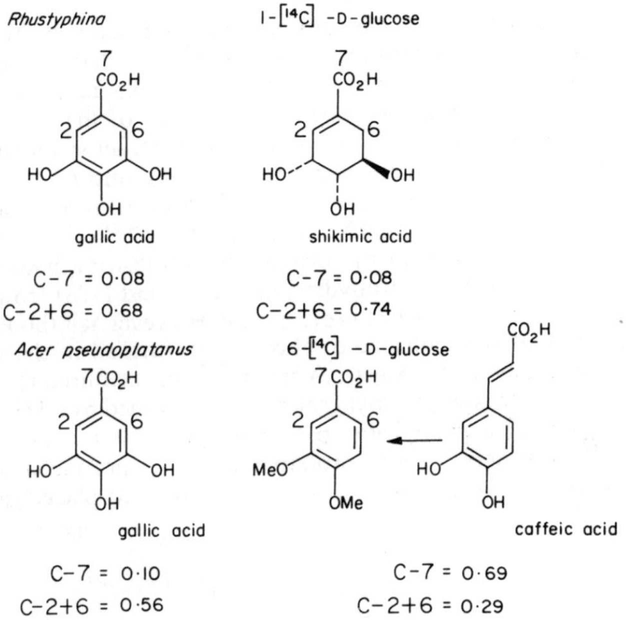

Figure 5.14. Distribution of radioactivity in gallic acid, (−)-shikimic acid and veratric acid after feeding 6-[^{14}C]-D-glucose and 1-[^{14}C]-D-glucose[215, 216]

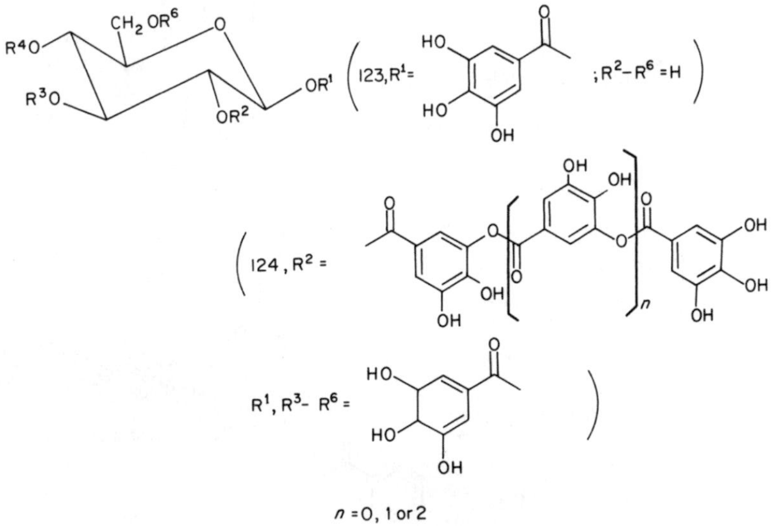

$n = 0$, 1 or 2

Closely related both structurally and biogenetically to the gallotannins are ellagitannins[49]. Schmidt[49, 263] has advanced elegant schemes for the biosynthesis of the ellagitannins, and although there has been no experimental proof of their validity, these hypotheses have been used to support the structural formulations put forward for many members of this group such as the brevilagins[258, 259], terchebin[260, 261], brevifolin carboxylic acid[262], chebulagic acid[263, 264] and chebulinic acid (127)[263, 264]. Ellagic acid (129) is derived from the ellagitannins following hydrolytic cleavage of (+) or (−)-hexahydroxydiphenic acid (128) from the appropriate ellagitannin structure (e.g. 126). However, it is thought[49] improbable that the molecule of (128) is pre-formed in the plant and then esterified to the sugar. Wenkert[265] has favoured the idea that the hexahydroxydiphenoyl group may be formed by a Michael-type addition of a galloyl group to 3-dehydroshikimic acid, but Schmidt[49] considers it more probable that it is produced by an intramolecular dehydrogenation of two appropriately placed galloyl

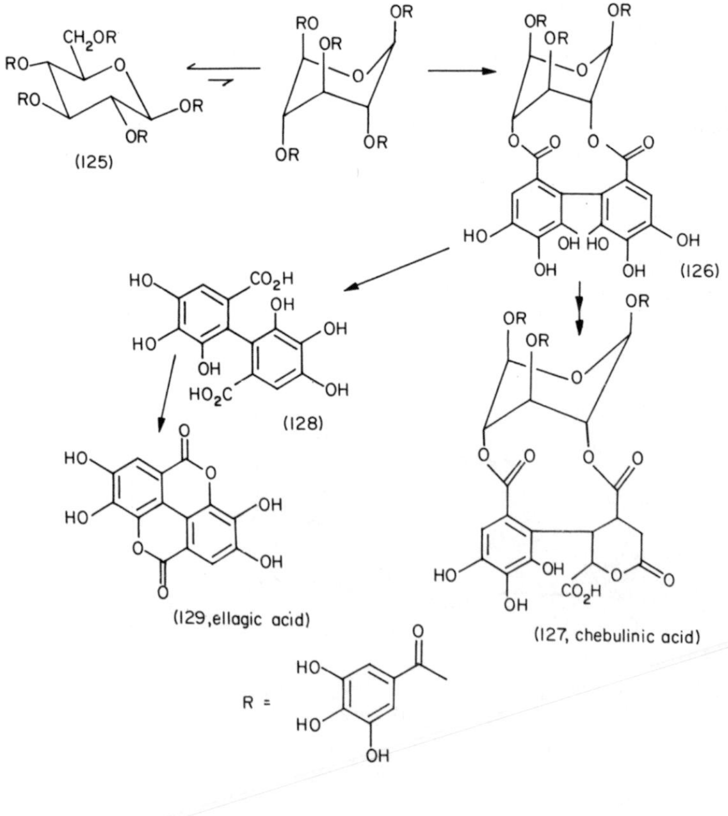

(125)

(126)

(128)

(129, ellagic acid)

(127, chebulinic acid)

R =

groups in a polygalloyl ester. Subsequent steps in the biosynthetic scheme advanced by Schmidt and his collaborators[49, 263, 266] (oxidation, reduction and hydrolytic cleavage of one aryl group in the hexahydroxydiphenoyl unit) permit the formulation of each of the distinctive and diverse structural types found within the ellagi-tannin class. Chebulinic acid (127) may for example be envisaged as derived from β-penta-O-galloyl-D-glucose (125) via the hexa-hydroxydiphenoyl intermediate (126)[266].

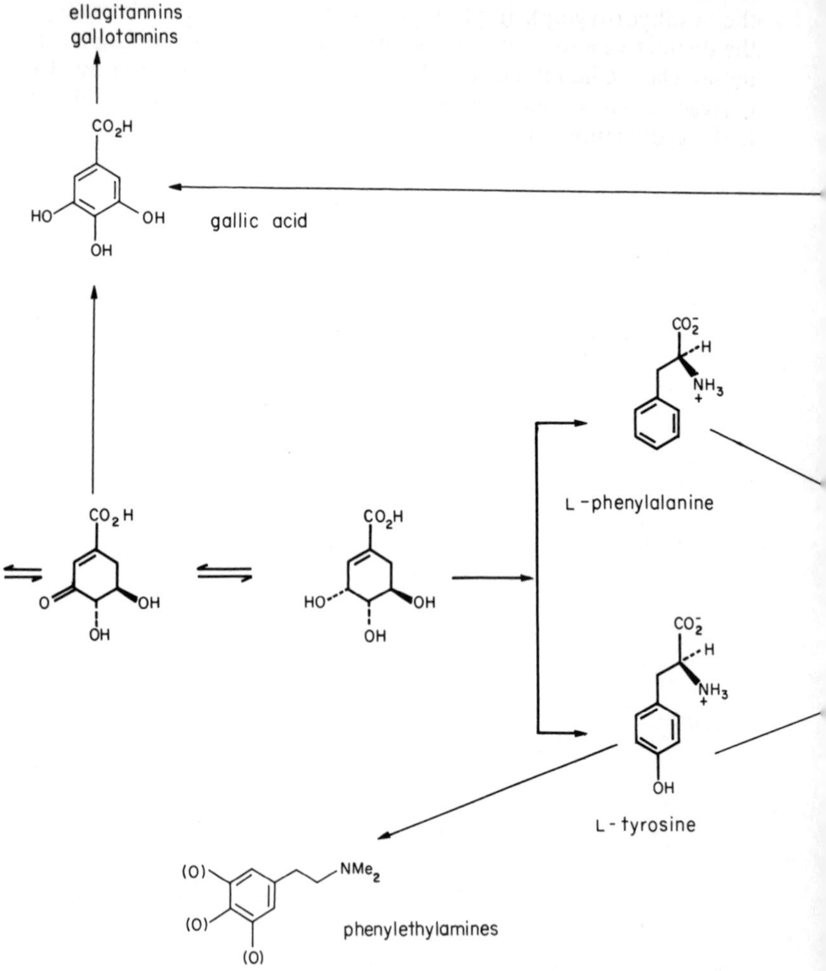

Figure 5.15. The Shikimate pathway in higher plants: phenylpropanoid compounds and their derivatives: summary

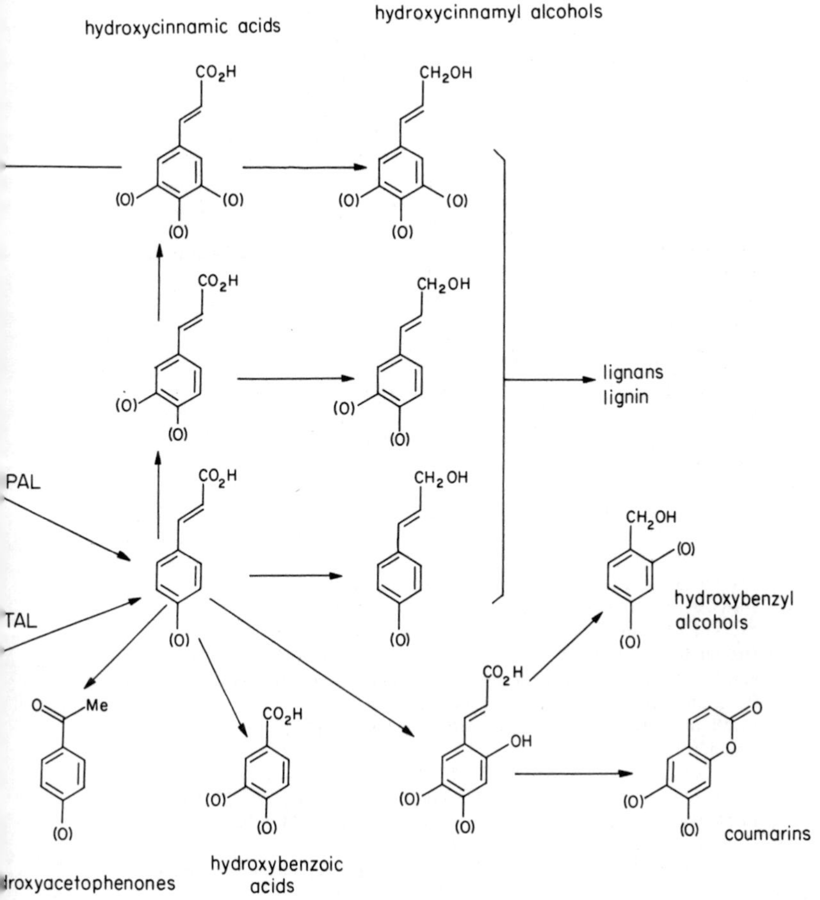

hydroxycinnamic acids

hydroxycinnamyl alcohols

CO$_2$H

CH$_2$OH

(O) (O)
(O)

(O) (O)
(O)

CO$_2$H

CH$_2$OH

(O)
(O)

(O)
(O)

lignans
lignin

PAL

TAL

CO$_2$H

CH$_2$OH

(O)

(O)

CH$_2$OH
(O)

hydroxybenzyl
alcohols

(O)

Me

CO$_2$H

CO$_2$H

(O)

(O) (O)

OH

(O)
(O)

(O)

(O)

O

O

(O) coumarins

roxyacetophenones

hydroxybenzoic
acids

234   Higher Plants: Phenylpropanoid Compounds

REFERENCES

1. Bu'Lock, J. D. (1965). *The Biosynthesis of Natural Products*, p. 2. London; McGraw-Hill
2. Geissman, T. A. and Crout, H. D. G. (1969). *Organic Chemistry of Secondary Plant Metabolism*, p. 5. San Francisco; Freeman, Cooper and Co.
3. Bate-Smith, E. C. (1962). *J. Linn. Soc.* (Bot), **58**, 95
4. Fowden, L. (1965). *Science Progress*, **53**, 583
5. Riberau-Gayon, P. (1972). *Plant Phenolics*. Edinburgh; Oliver and Boyd
6. Geissman, T. A. and Crout, H. D. G. (1969). *Organic Chemistry of Secondary Plant Metabolism*, p. 429. San Francisco; Freeman, Cooper and Co.
7. Cadman, C. H. (1960). *Phenolics in Plants in Health and Disease*, p. 101. Edited by J. B. Pridham. Oxford; Pergamon
8. Rothschild, M. (1972). *Phytochemical Ecology*, p. 1. Edited by J. B. Harborne. London and New York; Academic Press
9. Deverall, B. J. (1972). *Phytochemical Ecology*, p. 217. Edited by J. B. Harborne. London and New York; Academic Press
10. Salisbury, F. B. and Ross, C. (1969). *Plant Physiology*, p. 397. Belmont, California; Wadsworth
11. Robinson, R. (1955). *Structural Relations of Natural Products*. Oxford; Clarendon Press
12. Ruzicka, L. and Stoll, M. (1922). *Helv. Chim. Acta*, **5**, 929
13. Schopf, C. and Bayerle, H. (1934). *Annalen*, **513**, 190
14. Collie, J. N. (1907). *J. Chem. Soc.*, **91**, 1806
15. Geissman, T. A. and Crout, H. D. G. (1969). *Organic Chemistry of Secondary Plant Metabolism*, p. 150, 169. San Francisco; Freeman, Cooper and Co.
16. Aneja, R., Mukerjee, S. K. and Seshadri, T. R. (1958). *Tetrahedron*, **4**, 256
17. Ollis, W. D. and Sutherland. I. O. (1961). *Chemistry of Natural Phenolic Compounds*, p. 74. Oxford; Pergamon
18. Brown, S. A. and Steck, W. (1970). *Can. J. Biochem.*, **48**, 872
19. Floss, H. G. and Mothes, U. (1966). *Phytochemistry*, **5**, 161
20. Floss, H. G. and Paikert, H. (1969). *Phytochemistry*. **8**, 589
21. Brown, S. A., El-Dakhakny, M. and Steck, W. (1970). *Can. J. Biochem.*, **48**, 863
22. Karrer, W. (1958). *Konstitution und Vorkommen der Organischen Pflanzenstoffe*. Basel and Stuttgart; Birkhauser Verlag
23. Pridham, J. B. (1965). *Ann. Rev. Plant Physiol.*, **16**, 13
24. Thieme, H. (1964). *Pharmazie.*, **19**, 725
25. Pearl, I. A. and Darling, S. F. (1968). *Phytochemistry*, **7**, 1845, 1851, 1855
26. Pearl, I. A. and Darling, S. F. (1968). *Phytochemistry*, **7**, 825, 831
27. Loeschke, V. and Francksen, H. (1964). *Naturwiss*, **51**, 140
28. Koeppen, B. H. (1962). *Chem. and Ind.*, 2145
29. Beylis, P., Howard, A. S. and Perold, G. W. (1971). *Chem. Commun.*, 597
30. Thieme, H. (1966). *Pharmazie*, **21**, 769
31. Thieme, H. (1967). *Pharmazie*, **22**, 59
32. Harborne, J. B. and Corner, J. J. (1961). *Biochem. J.*, **81**, 242
33. Harborne, J. B. (1964). *Biochemistry of Phenolic Compounds*, p. 129. Edited by J. B. Harborne. London and New York; Academic Press
34. Scarpati, M. L. and Oriente, G. (1958). *Tetrahedron*, **4**, 43
35. Pearl, I. A. and Darling, S. F. (1962). *J. Org. Chem.*, **27**, 1806
36. Britton, G., Haslam, E. and Naumann, M. O. (1964). *J. Chem. Soc.*, 5649
37. Challice, J. S. and Westwood, M. N. (1972). *Phytochemistry*, **11**, 37
38. Challice, J. S. and Williams, A. H. (1968). *Phytochemistry*, **7**, 119
39. Daniels, D. G. H., King, H. G. C. and Martin, H. F. (1963). *J. Sci. Food Agric.*, **14**, 385

40. Gardner, P. D., Park, G. J. and Albers, C. C. (1961). *J. Amer. Chem. Soc.*, **83**, 1511
41. Harborne, J. B. (1967). *Comparative Biochemistry of the Flavonoids*. London; Academic Press
42. Birkhofer, L., Kaiser, C. and Kosmol, H. (1965). *Z. Naturforsch.*, **20B**, 605, 923
43. Kjaer, A. (1960). *Fortschr. Chem. org. Naturstoffe*, **18**, 122
44. Klohs, M. W., Draper, M. D. and Keller, F. (1954). *J. Amer. Chem. Soc.*, **76**, 2843
45. El-Basyouni, S. Z., Chen, D., Ibrahim, R. K., Neish, A. C. and Towers, G. H. N. (1964). *Phytochemistry*, **3**, 485
46. Ibrahim, R. K., Towers, G. H. N. and Gibbs, R. D. (1962). *J. Linn. Soc.* (Bot.), **58**, 223
47. Smith, D. C. C. (1955). *J. Chem. Soc.*, 2347
48. Haworth, R. D. and Haslam, E. (1964). *Progress in Organic Chemistry*, Vol. 6, p. 1. Edited by J. W. Cook and W. Carruthers. London; Butterworths
49. Schmidt, O. Th. (1956). *Fortschr. Chem. org. Naturstoffe*, **14**, 71
50. Hillis, W. E. (1962). *Wood Extractives*, p. 60. Edited by W. E. Hillis. London; Academic Press
51. Shima, K., Hisada, S. and Inagaki, I. (1971). *Phytochemistry*, **10**, 894
52. Uvarova, N. I., Tizba, J. and Herout, V. (1963). *Coll. Czech. Chem. Commun.*, **32**, 325
53. Weinges, K. and Wild, R. (1970). *Annalen*, **734**, 46
54. Haslam, E. (1969). *J. Chem. Soc.* (C), 1824
55. Thompson, R. S., Jacques, D., Haslam, E. and Tanner, R. J. N. (1972). *J. Chem. Soc.* (Perkin I), 1387
56. Wallace, J. W., Mabry, T. J. and Alston, R. E. (1969). *Phytochemistry*, **8**, 93
57. Cardini, C. E. and Leloir, L. F. (1957). *Ciencia Invest.* (Buenos Aires), **13**, 514
58. Pridham, J. B. and Saltmarsh, M. J. (1963). *Biochem. J.*, **87**, 218
59. Barber, G. A. (1962). *Biochemistry*, **1**, 463
60. Hahlbrock, K., Ebel, J., Ortmann, R., Sutter, A., Wellmann, E. and Grisebach, H. (1971). *Biochim. Biophys. Acta*, **244**, 7
61. Hahlbrock, K., Sutter, A., Wellmann, E., Ortmann, R. and Grisebach, H. (1971). *Phytochemistry*, **10**, 109
62. Walton, E. and Butt, V. S. (1971). *Phytochemistry*, **10**, 295
63. Grisebach, H., Barz, W., Hahlbrock, K., Kellner, S. and Patschke, L. (1966). *Proc. 2nd Meeting Fed. Eur. Biochem. Soc.* (Vienna), **3**, 25
64. Swain, T. and Williams, C. A. (1970). *Phytochemistry*, **9**, 2115
65. Young, M. R., Towers, G. H. N. and Neish, A. C. (1966). *Can. J. Bot.*, **44**, 341
66. Ogata, K., Uchiyama, K. and Yamada, H. (1967). *Agr. Biol. Chem.*, **31**, 200
67. Bandoni, R., Moore, R. J. K., Subba-Rao, P. V. and Towers, G. H. N. (1968). *Phytochemistry*, **7**, 205
68. Power, D. M., Towers, G. H. N. and Neish, A. C. (1965). *Can. J. Biochem.*, **43**, 1397
69. Gamborg, O. L. (1966). *Can. J. Biochem.*, **44**, 791
70. Russell, D. W. and Conn, E. E. (1967). *Arch. Biochem. Biophys.*, **122**, 256
71. Amrhein, N. and Zenk, M. H. (1968). *Naturwiss.* **55**, 394
72. Vaughan, P. F. T., Butt, V. S., Grisebach, H. and Schill, L. (1969). *Phytochemistry*, **8**, 1373
73. Vaughan, P. F. T. and Butt, V. S. (1969). *Biochem. J.*, **113**, 109
74. Vaughan, P. F. T. and Butt, V. S. (1970). *Biochem. J.*, **119**, 89
75. Kokoul, J. and Conn, E. E. (1961). *J. Biol. Chem.*, **236**, 2692
76. Moore, K., Subba-Rao, P. V. and Towers, G. H. N. (1968). *Biochem. J.*, **106**, 507
77. Havir, E. A. and Hanson, K. R. (1968). *Biochemistry*, **7**, 1896, 1904

78. Marsh, H. V., Havir, E. A. and Hanson, K. R. (1968). *Biochemistry*, **7**, 1915
79. Hodgins, D. S. (1971). *J. Biol. Chem.*, **246**, 2977
80. Hodgins, D. S. (1972). *Arch. Biochem. Biophys.*, **149**, 91
81. Emes, A. V. and Vining, L. C. (1970). *Can. J. Biochem.*, **48**, 613
82. Havir, E. A. and Hanson, K. R. (1970). *Arch. Biochem. Biophys.*, **141**, 1
83. Winkler, R. B. (1969). *J. Biol. Chem.*, **244**, 6550
84. Hanson, K. R., Wightman, R. H., Staunton, J. and Battersby, A. R. (1971). *Chem. Commun.*, 185
85. Hanson, K. R., Wightman, R. H., Staunton, J. and Battersby, A. R. (1972). *J. Chem. Soc.* (Perkin I), 2355
86. Ife, R. and Haslam, E. (1971). *J. Chem. Soc.* (C), 2818
87. Krasna, A. I. (1958). *J. Biol. Chem.*, **233**, 1010
88. Sprecher, M. and Sprinson, D. B. (1966). *J. Biol. Chem.*, **241**, 864
89. Givot, I. L., Smith, T. A. and Abeles, R. H. (1969). *J. Biol. Chem.*, **244**, 6341
90. Neish, A. C. (1961). *Phytochemistry*, **1**, 1
91. Havir, E. A., Reid, P. D. and Marsh, H. V. (1971). *Plant Physiol.*, **48**, 130
92. Havir, E. A., Reid, P. D. and Marsh, H. V. (1972). *Plant Physiol.*, **49**, 480
93. Strange, P. G., Staunton, J., Wiltshire, H. R., Battersby, A. R., Hanson, K. R. and Havir, E. A. (1972). *J. Chem. Soc.*, 2364
94. Parkhurst, J. R. and Hodgins, D. S. (1971). *Phytochemistry*, **10**, 2997
95. Camm, E. L. and Towers, G. H. N. (1969). *Phytochemistry*, **8**, 1407
96. Neish, A. C. (1960). *Ann. Rev. Plant Physiol.*, **11**, 55
97. Neish, A. C. (1964). *Biochemistry of Phenolic Compounds*, p. 295. Edited by J. B. Harborne. London; Academic Press
98. Zenk, M. H. (1967). *Ber. Dtsch. Bot. Ges.*, **80**, 573
99. Grisebach, H. and Barz, W. (1969). *Naturwiss*, **56**, 539
100. Kindl, H. (1971). *Naturwiss*, **58**, 554
101. Grisebach, H. (1968). *Recent Advances in Phytochemistry*, p. 379, Vol. 1. Edited by T. J. Mabry, R. E. Alston and V. C. Runeckles. New York; Appleton-Century-Crofts
102. Freudenberg. K. (1962). *J. Pure Appl. Chem.*, **5**, 9
103. Higuchi, T. (1971). *Adv. Enzymol.*, **34**, 207
104. Brown, S. A. (1965). *Lloydia*, **28**, 332
105. McCalla, D. R. and Neish, A. C. (1959). *Can. J. Biochem. Physiol.*, **37**, 531, 537
106. Levy, C. C. and Zucker, M. (1960). *J. Biol. Chem.*, **235**, 2418
107. Hanson, K. R. and Zucker, M. (1963). *J. Biol. Chem.*, **238**, 1105
108. Hanson, K. R. (1966). *Phytochemistry*, **5**, 491
109. Finkle, B. J. and Nelson, R. F. (1963). *Biochim. Biophys. Acta*, **78**, 747
110. Finkle, B. J. and Masri, M. S. (1964). *Biochim. Biophys. Acta*, **92**, 424
111. Russell, D. W. and Conn, E. E. (1967). *Arch. Biochem. Biophys.*, **122**, 256
112. Nair, P. M. and Vining, L. C. (1965). *Phytochemistry*, **4**, 161
113. Amrhein, N. and Zenk, M. H. (1968). *Naturwiss*, **55**, 394
114. Guroff. G., Daly, J. W., Jerina, D. M., Renson, J., Witkop, B. and Udenfriend, S. (1967). *Science*, **158**, 1524
115. Russell, D. W., Conn, E. E., Sutter, A. and Grisebach, H. (1968). *Biochim. Biophys. Acta*, **170**, 210
116. Zenk, M. H. (1967), *Z. Pflanzenphysiol.*, **57**, 477
117. Bowman, W. R., Bruce, I. T. and Kirby, G. W. (1969). *Chem. Commun.*, 1075
118. Zenk, M. H. and Amrhein, N. (1969). *Phytochemistry*, **8**, 107
119. Sutter, A. and Grisebach, H. (1969). *Phytochemistry*, **8**, 101
120. Suhadolnik, R. J., Fisher, A. G. and Zulalian, J. (1962). *J. Amer. Chem. Soc.*, **84**, 4348
121. Leete, E. and Nemeth, P. E. (1960). *J. Amer. Chem. Soc.*, **82**, 6055
122. Battersby, A. R. and Herbert, R. B. (1960). *Proc. Chem. Soc.*, 346

123. Leete, E. and Ahmad, A. (1966). *J. Amer. Chem. Soc.*, **88**, 4722
124. Nair, P. M. and Vining, L. C. (1965). *Phytochemistry*, **4**, 401
125. Rosenberg, H., McLaughlin, J. C. and Paul, A. G. (1967). *Lloydia*, **30**, 100
126. Massicot, J. and Marion, L. (1957). *Can. J. Chem.*, **35**, 1
127. Leete, E., Bowman, R. M. and Manuel, M. F. (1971). *Phytochemistry*, **10**, 3029
128. Dawson, C. R. and Tapley, W. B. (1951). *The Enzymes*, Vol. 2, p. 454. Edited by J. B. Sumner and K. Myrback. New York; Academic Press
129. Patil, S. S. and Zucker, M. (1965). *J. Biol. Chem.*, **240**, 3938
130. Mayer, A. M. and Friend, J. (1960). *Nature*, **185**, 464
131. Butt, V. S. (1972). *Phytochemistry*, **11**, 861 (Abstract)
132. Sato, M. (1966). *Phytochemistry*, **5**, 385
133. Sato, M. (1969). *Phytochemistry*, **8**, 353
134. Swain, T. and Bate-Smith, E. C. (1965). *Lloydia*, **28**, 313
135. Hess, D. (1968). *Biochemische Genetik*. Berlin, Heidelberg and New York; Springer Verlag
136. Zenk, M. H. (1965). *Biosynthesis of Aromatic Compounds*, p. 45. Edited by G. Billek. Oxford; Pergamon
137. Hess, D. (1965). *Z. Pflanzenphysiol.*, **53**, 1
138. Meyer, C. (1964). *Z. Vererbungslehre*, **95**, 171
139. Pla, J. (1967). *Bull. Soc. Chim. Biol.*, **49**, 395
140. Schubert, W. J. (1965). *Lignin Biochemistry*. New York; Academic Press
141. Pearl, I. A. (1967). *The Chemistry of Lignin*, London and New York; Edward Arnold and Marcel Dekker
142. Acerbo, S. N., Schubert, W. J. and Nord, F. F. (1960). *J. Amer. Chem. Soc.*, **82**, 735
143. Dean, F. W. (1963). *Naturally Occurring Oxygen Ring Compounds*, p. 176. London; Butterworths
144. Conn, E. E. (1964). *Biochemistry of Phenolic Compounds*, p. 399. Edited by J. B. Harborne. London; Academic Press
145. Sato, M. and Hasegawa, M. (1971). *Phytochemistry*, **10**, 2367
146. Harborne, J. B. (1960). *Biochem. J.*, **74**, 262
147. Sato, M. and Hasegawa, M. (1972). *Phytochemistry*, **11**, 657
148. Haskins, F. A. and Gorz, J. H. (1961). *Crop Sci.*, **1**, 320
149. Kosuge, T. (1961). *Arch. Biochem. Biophys.*, **95**, 211
150. Brown, S. A. (1962). *Science*, **137**, 977
151. Austin, D. J. and Meyers, M. B. (1965). *Phytochemistry*, **4**, 245, 255
152. Runeckles, V. C. (1963). *Can. J. Biochem. Physiol.*, **41**, 2259
153. Steck, W. (1968). *Phytochemistry*, **7**, 1711
154. Fritig, B., Hirth, L. and Ourisson, G. (1970). *Phytochemistry*, **9**, 1963
155. Kosuge, T. and Conn, E. E. (1959). *J. Biol. Chem.*, **234**, 2133
156. Kosuge, T. and Conn, E. E. (1961). *J. Biol. Chem.*, **236**, 1617
157. Brown, S. A. (1963). *Phytochemistry*, **2**, 137
158. Stoker, J. R. and Bellis, D. M. (1962). *J. Biol. Chem.*, **237**, 2303
159. Grisebach, H. and Ollis, W. D. (1961). *Experentia*, **15**, 4
160. Meyers, M. B. (1963). *Proc. Chem. Soc.*, 243
161. Haskins, F. A. and Gorz, H. J. (1961). *Biochem. Biophys. Res. Commun.*, **6**, 298
162. Edwards, K. G. and Stoker, J. R. (1967). *Phytochemistry*, **6**, 655
163. Stoker, J. R. (1964). *Biochem. Biophys. Res. Commun.*, **14**, 17
164. Bunton, C. A., Kenner, G. W., Robinson, M. J. T. and Webster, B. R. (1963). *Tetrahedron*, **19**, 1001
165. Scott, A. I., Dodson, P. A., McCapra, F. and Meyers, M. B. (1963). *J. Amer. Chem. Soc.*, **85**, 3702
166. Payen, A. (1838). *Compt. Rend.*, **7**, 1052

167. Brauns, F. E. (1952). *The Chemistry of Lignin*, Supplement (1960). New York; Academic Press
168. Brown, S. A. (1964). *Biochemistry of Phenolic Compounds*, p. 361. Edited by J. B. Harborne. London; Academic Press
169. Freudenberg, K. (1954). *Fortschritt. chem. org. Naturstoffe*, **11**, 43
170. Freudenberg, K. and Sidhu, G. S. (1961). *Holzforschung.*, **15**, 33
171. Klason, P. (1897). *Svensk. Kem. Tidskr.*, **9**, 133
172. Erdtman, H. (1933). *Annalen*, **503**, 283
173. Erdtman, H. (1933). *Biochem. Z.*, **258**, 172
174. Erdtman, H. (1935). *Svensk. Kem. Tidskr.*, **47**, 223
175. Freudenberg, K. and Richtzenhain, H. (1943). *Ber.*, **76**, 997
176. Freudenberg, K., Harkin, J. M., Reichert, M. and Fukuzumi, T. (1958). *Chem. Ber.*, **91**, 581
177. Freudenberg, K., Kraft, R. and Heimberger, W. (1951). *Chem. Ber.*, **84**, 472
178. Freudenberg, K., Reznik, H., Boesenberg, H. and Rasenack, D. (1952). *Chem. Ber.*, **85**, 641
179. Freudenberg, K. and Tausend, H. (1964). *Chem. Ber.*, **97**, 3418
180. Freudenberg, K. (1965). *Science*, **148**, 595
181. Higuchi, T. and Ito, Y. (1958). *J. Biochem.* (Japan). **45**, 575
182. Adler, E. (1957). *Ind. Eng. Chem.*, **49**, 1377
183. Adler, E. (1961). *Papier*, **15**, 604
184. Freudenberg, K. (1964). *Holzforschung*, **18**, 3
185. Freudenberg, K. and Harkin, J. M. (1964). *Holzforschung*, **18**, 166
186. Eberhardt, G. and Schubert, W. J. (1956). *J. Amer. Chem. Soc.*, **78**, 235
187. Acerbo, S., Schubert, W. J. and Nord, F. F. (1958). *J. Amer. Chem. Soc.*, **80**, 1990
188. Nord, F. F. and Schubert, W. J. (1959). *Experentia*, **15**, 245
189. Brown, S. A. and Neish, A. C. (1959). *J. Amer. Chem. Soc.*, **81**, 2419
190. Brown, S. A., Wright, D. and Neish, A. C. (1959). *Can. J. Biochem. Physiol.*, **37**, 25
191. Freudenberg, K. (1956). *Angew. Chem.*, **68**, 84, 508
192. Higuchi, T. (1962). *Can. J. Biochem. Physiol.*, **40**, 31
193. Freudenberg, K. and Bittner, F. (1953). *Chem. Ber.*, **86**, 155
194. Freudenberg, K., Reznik, H., Fuchs, W. and Reichert, M. (1955). *Naturwiss.*, **42**, 29
195. Kratzl, K. and Puschmann, G. (1960). *Holzforschung.* **14**, 1
196. Freudenberg, K. and Grion, G. (1959). *Chem. Ber.*, **92**, 1355
197. Mason, H. S. and Cronyn, M. (1955). *J. Amer. Chem. Soc.*, **77**, 491
198. Mason, H. S. (1965). *Adv. Enzymol.*, **16**, 105
199. Haworth, R. D. (1942). *J. Chem. Soc.*, 448
200. Birch, A. J., Macdonald, P. L. and Pelter, A. (1967). *J. Chem. Soc.* (C), 1968
201. Haworth, R. D. and Haslam, E. (1955). *J. Chem. Soc.*, 827
202. Tamiguchi, E. and Oshima, Y. (1972). *Agr. Biol. Chem.*, **36**, 1013
203. Kaneko, K. (1960). *Chem. and Pharm. Bull.* (Japan), **8**, 611
204. Kaneko, K. (1961). *Chem. and Pharm. Bull.* (Japan), **9**, 108
205. Canonica, L., Manitto, P., Monti, D. and M. Sanchez, A. (1971). *Chem. Commun.*, 1108
206. Ibrahim, R. K. (1964). *Flora*, **154**, 481
207. Pearl, I. A., Beyer, D. L. and Laskowski, D. (1959). *Tappi*, **42**, 319
208. Pearl, I. A., Beyer, D. L., Laskowski, D. and Whitney, D. (1960). *Tappi*, **43**, 756
209. Griffiths, L. A. (1959). *J. Exp. Bot.*, **10**, 437
210. Ibrahim, R. K., Towers, G. H. N. and Gibbs, R. D. (1962). *J. Linn. Soc.*, **52**, 223
211. Ibrahim, R. K. and Towers, G. H. N. (1960). *Arch. Biochem. Biophys.*, **87**, 125

212. Gibson, M. I. and Gibson, F. (1964). *Biochem. J.*, **90**, 248
213. Young, I. G., Batterham, T. J. and Gibson, F. (1969). *Biochem. Biophys. Acta*, **177**, 389
214. Gross, S. R. (1958). *J. Biol. Chem.*, **233**, 1146
215. Dewick, P. M. and Haslam, E. (1968). *Chem. Commun.*, 673
216. Dewick, P. M. and Haslam, E. (1969). *Biochem. J.*, **113**, 537
217. Geissmann, T. A. and Hinreiner, E. (1952). *Bot. Rev.*, **18**, 165
218. Kindl, H. and Billek, G. (1962). *Osterr. Chem. Ztg.*, **63**, 290
219. Grisebach, H. and Vollmer, K. O. (1963). *Z. Naturforsch.*, **18B**, 753
220. Grisebach, H. and Vollmer, K. O. (1964). *Z. Naturforsch.*, **19B**, 781
221. Grisebach, H. and Vollmer, K. O. (1966). *Z. Naturforsch.*, **21B**, 435
222. Zenk, M. H. (1964). *Z. Naturforsch.*, **19B**, 83
223. Zenk, M. H. and Muller, G. (1964). *Z. Naturforsch.*, **19B**, 398
224. Kindl. H. (1964). *Monats. Chem.*, **95**, 439
225. Gross, D. and Schütte, H. (1963). *Arch. Pharm.*, **396**, 1
226. Billek, G. and Schmook, F. P. (1966). *Osterr. Chem. Ztg.*, **67**, 401
227. Zenk, M. H. (1967). *Z. Pflanzenphysiol.*, **57**, 477
228. Ellis, B. E. and Amrhein, N. (1971). *Phytochemistry*, **10**, 3069
229. Thieme, H. (1964). *Pharmazie.*, **19**, 725
230. Pearl, I. A. and Darling, S. F. (1971). *Phytochemistry*, **10**, 483
231. Pearl, I. A. and Darling, S. F. (1971). *Phytochemistry*, **10**, 3161
232. Alibert, G., Ranjeva, R. and Boudet, A. (1972). *Biochim. Biophys. Acta*, **279**, 282
233. Zenk, M. H. (1967). *Phytochemistry*, **6**, 245
234. Pridham, J. B. and Saltmarsh, M. J. (1963). *Biochem. J.*, **87**, 218
235. Neish, A. C. (1959). *Can. J. Bot.*, **37**, 1085
236. Takahashi, M., Ito, T. and Mitzutani, A. (1960). *J. Pharm. Soc.* (Japan), **80**, 782
237. Bick, I. R. C., Bremmer, J. B. and Gillard, J. W. (1971). *Phytochemistry*, **10**, 475
238. Kindl, H. and Billek, G. (1962). *Nature*, **194**, 579
239. Kindl, H. and Billek, G. (1962). *Monats. Chem.*, **93**, 85
240. Kindl, H. (1969). *Eur. J. Biochem.*, **7**, 340
241. Leete, E. (1966). *J. Amer. Chem. Soc.*, **88**, 4218
242. Rosenberg, H., McLaughlin, J. L. and Paul, A. G. (1967). *Lloydia*, **30**, 91, 100
243. Rosenberg, H., Paul, A. G. and Hanna, K. L. (1969). *Lloydia*, **32**, 36
244. Agurell, S., Lundstrom, J. and Sandberg, F. (1967). *Tetrahedron Lett.*, 2433
245. Agurell, S. and Lundstrom, J. (1968). *Tetrahedron Lett.*, 4437
246. Agurell, S. (1969). *Lloydia*, **32**, 40
247. Grisdale, S. K. and Towers, G. H. N. (1960). *Nature*, **188**, 1130
248. Zenk, M. H. (1964). *Z. Naturforsch.*, **19B**, 856
249. Zenk, M. H. and Bolkart, K. H. (1968). *Z. Pflanzenphysiol.*, **59**, 439
250. Ellis, B. E. and Towers, G. H. N. (1969). *Phytochemistry*, **8**, 1415
251. Zenk, M. H. and Leistner, E. (1968). *Lloydia*, **31**, 275
252. Corner, J. J., Harborne, J. B., Humphries, S. G. and Ollis, W. D. (1962). *Phytochemistry*, **1**, 73
253. Knowles, P. F., Haworth, R. D. and Haslam, E. (1961). *J. Chem. Soc.*, 1854
254. Cornthwaite, D. and Haslam, E. (1965). *J. Chem. Soc.*, 3008
255. Conn, E. E. and Swain, T. (1961). *Chem. and Ind.*, 592
256. Sprinson, D. B. (1961). *Adv. Carbohydrate Chem.*, **15**, 235
257. Haslam, E. (1967). *J. Chem. Soc.*, 1734
258. Schmidt, O. Th., Schanz, R., Eckert, R. and Wurmb, R. (1967). *Annalen*, **706**, 131
259. Schmidt, O. Th., Schanz, R., Groebke, W. and Wurmb, R. (1967). *Annalen*, **706**, 154
260. Schmidt, O. Th., Schulz, J. and Wurmb, R. (1967). *Annalen*, **706**, 169; 180

261. Schmidt, O. Th., Schulz, J. and Fiesser, H. (1967). *Annalen*, **706**, 187
262. Schmidt, O. Th., Eckert, R., Gunther, E. and Fiesser, H. (1967). *Annalen*, **706**, 205
263. Jochims, J., Taigel, G. and Schmidt, O. Th. (1968). *Annalen*, **717**, 169
264. Uddin, M. and Haslam, E. (1967). *J. Chem. Soc.* (C), 2381
265. Wenkert, E. (1959). *Chem. and Ind.*, 609
266. Schmidt, O. Th. and Mayer, W. (1956). *Angew. Chem.*, **68**, 103

6

The Shikimate Pathway in Higher Plants: Extension of the Phenylpropanoid Unit: Miscellaneous Metabolites

6.1 EXTENSION OF THE PHENYLPROPANOID UNIT

6.1.1 THE BIRCH–DONOVAN HYPOTHESIS

The two carbon unit of acetic acid, as acetyl co-enzyme A or its activated form malonyl co-enzyme A, is of central importance in plant biochemistry for it not only provides the starting-point for the biosynthesis of fatty acids, polyacetylenes and the whole range of isoprenoid compounds but it is also one of the fundamental building blocks for many characteristic groups of plant phenols. This possibility was recognised theoretically by Birch and Donovan in 1953 in their enunciation of the 'acetate hypothesis'[1]. These authors considered the various ways in which cyclisation of a poly-β-keto compound, formed by head to tail linkage of acetate units, could lead to different phenolic substances. One speculation which they made was that alternative modes of cyclisation of a common precursor, such as (1) formed by the condensation of three acetate units with cinnamic acid, could give rise to the stilbene (2) on the one hand and the flavanone (4), via the chalcone (3) on the other (*Figure 6.1*). This proposal accords well with the earlier suggestions of Robinson[2] and with the detailed structural work on flavonoids and stilbenes where it was noted that there were certain regularities in the patterns of hydroxyl and methoxyl substitution. Thus in the

241

flavonoid group of natural products the oxygen containing substituents (-OH, -OMe, -Oglc) are, with few exceptions, in the meta
positions in ring A but in the ortho relationship in ring C.

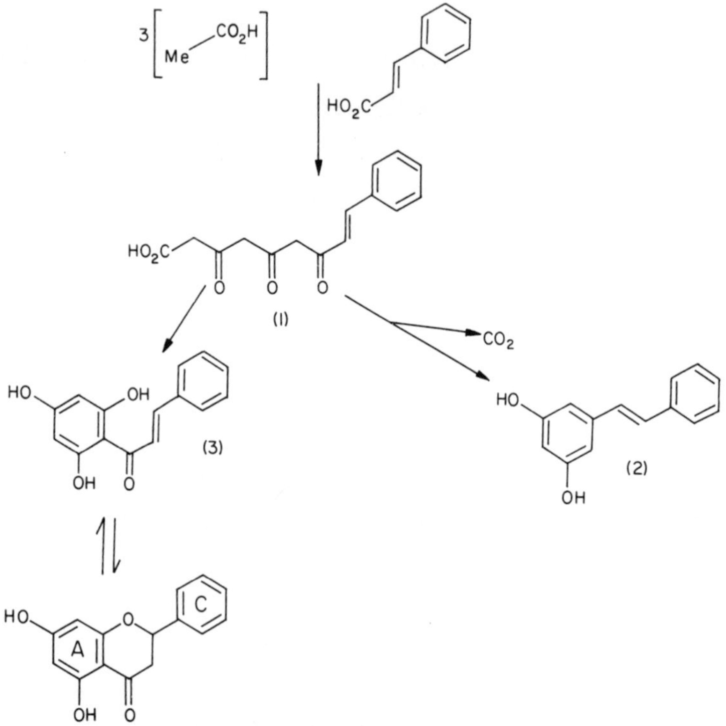

Figure 6.1. Biosynthesis of flavonoids and stilbenes from acetate and cinnamate—
Birch and Donovan[1]

Yangonin (5), an α-pyrone from *Piper methysticum* (Kawa),
could similarly be envisaged as derived by a variation of the same
process in which two acetate units are condensed with one of
p-hydroxycinnamate and with the subsequent cyclisation occurring
on oxygen rather than carbon. A further biogenetic relationship,
which Birch and Donovan emphasised, was that between phenyl-
coumalin (6) and cotoin (7), which co-occur in the Coto bark. They
suggested that these might be derived by the addition of one and two
acetate units respectively to benzoylacetic acid (8) or alternatively of
two or three acetate units to benzoic acid (9).

This type of biogenetic analysis, based essentially on the acetate
hypothesis, combined with an analysis of structural relationships

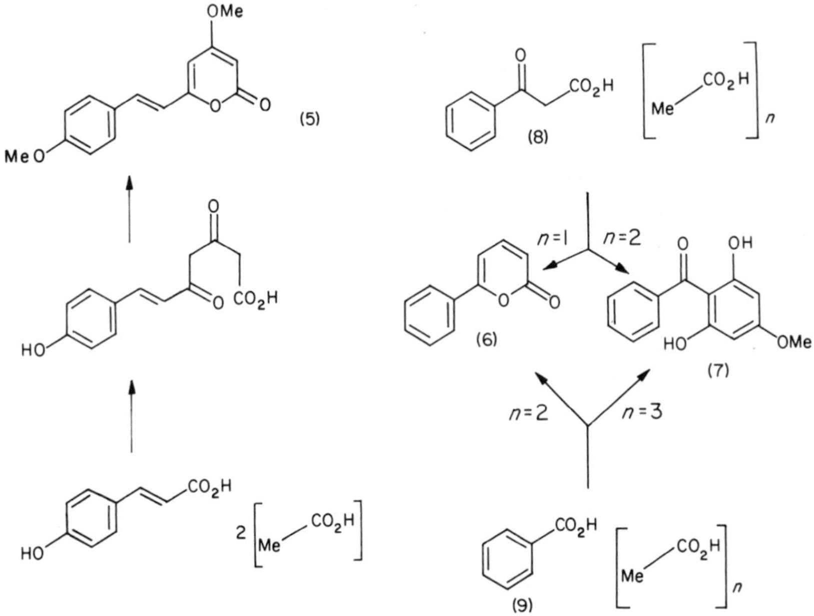

has since been extended and considerably amplified by others[3-10]. The biogenesis of flavonoids[3-7], neoflavonoids[8-10], pyrones[4,5], benzophenones[4,5,9] and xanthones[4,11] may thus be satisfactorily explained within the framework of these ideas. More importantly experimental work has, in many cases, fully verified the original theoretical proposals and it is now generally recognised that the formation of a large number of phenolic compounds in higher plants proceeds initially by an extension of a phenylpropanoid ($C_6 . C_3$) or phenylpropanoid derived (e.g. $C_6 . C_1$) unit by the addition of acetate units. Discussion is restricted here principally to those compounds for which firm experimental support of the proposed pathways of biosynthesis has been obtained.

6.1.2 FLAVONOIDS

The flavonoids comprise the largest single family of oxygen ring compounds occurring in nature and they are responsible as plant pigments for the majority of the red, violet and blue and for several of the yellow and orange colours of plant tissues. All flavonoids contain a 1,3-diarylpropane carbon skeleton ($C_6 . C_3 . C_6$) which is usually accommodated within a 2-phenylchroman structure (10).

The variety of oxidation levels which are possible within the heterocyclic ring B of (10) give rise to the numerous classes of natural product classified as flavonoids. The major groups found in plants are the anthocyanins, flavones and flavonols but less widely distributed are minor groups such as the chalcones, aurones, flavan-3-ols, flavanones and the oligomeric biflavonoids and proanthocyanidins. The number of known compounds of the flavonoid class is numerically impressive but this is due less to a high degree of structural diversity than to the manner in which the parent structural unit (10) is hydroxylated, alkoxylated or glycosylated.

(10)   (11)   (12)

Closely related both structurally and biogenetically to the flavonoids are two groups of natural products—the isoflavonoids and neoflavonoids. In contrast to the flavonoids however, these groups are of a more limited taxonomic distribution in higher plants and are found primarily in the Leguminosae. Isoflavonoids have in common the 1,2-diphenyl propane skeleton (11) and include many classes of natural product such as the isoflavones, isoflavanones, rotenoids, pterocarpans and coumestans. The term neoflavonoid was introduced by Ollis and his associates[8–10] to cover the third group of compounds which possess the 1,1-diarylpropane skeleton (12) and includes such compounds as the 4-phenylcoumarins and the various dalbergiones.

The structures of many of the naturally occurring flavonoids, isoflavonoids and neoflavonoids contain C_5 units which are apparently derived from mevalonate the precursor of the isoprenoid units of terpenes and steroids. The biosynthetic processes in which the additional units are incorporated into the structures of these compounds presumably involved the intervention of $\gamma\gamma$-dimethylallylpyrophosphate or isopentenylpyrophosphate as intermediates[12, 13]. An additional, and rather unusual structural variation, which has recently been encountered is that of the flavonolignan class in which a flavonoid (taxifolin) is apparently condensed with a phenylpropanoid intermediate (coniferyl alcohol)[14, 15]. Typical examples of the various classes and the considerable structural diversity which is to be found within them are shown in the accompanying formulae (13–33).

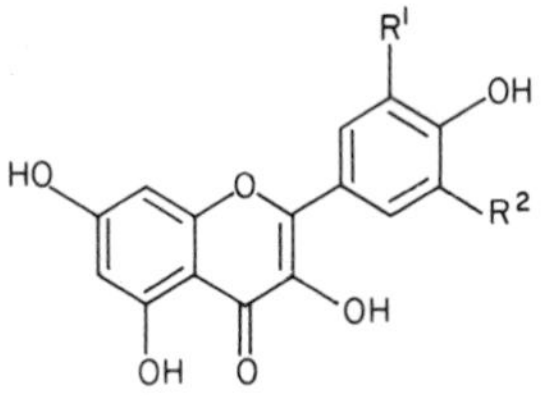

flavonol

(13, R^1=R^2=H ; kaempferol)
(14, R^1=OH, R^2=H; quercetin)
(15, R^1=R^2=OH; myricetin)

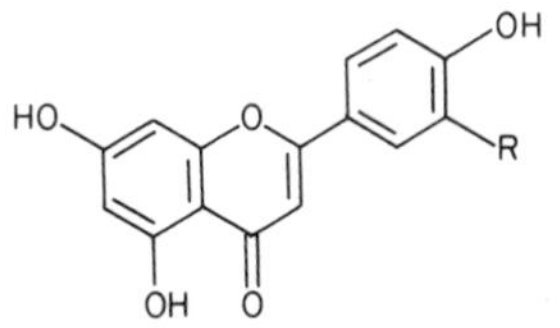

flavone

(16, R=H ; apigenin)
(17, R=OH; luteolin)

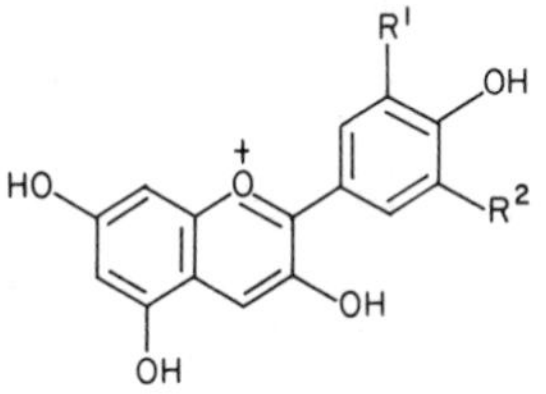

anthocyanidin

(18, R^1=R^2=H ; pelargonidin)
(19, R^1=OH, R^2=H; cyanidin)
(20, R^1=R^2=OH; delphinidin)

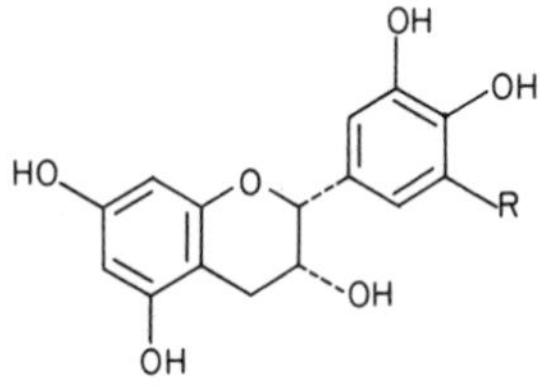

flavan-3-ol

(21, R=H ; (−)−epicatechin)
(22, R=OH ; (−)-epigallocatechin)

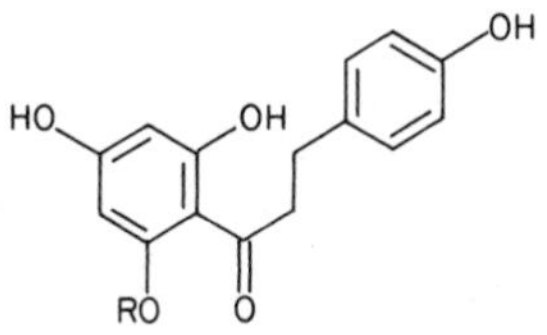

dihydrochalcone

(23, R=H ; phloretin)
(24, R=Glc ; phloridzin)

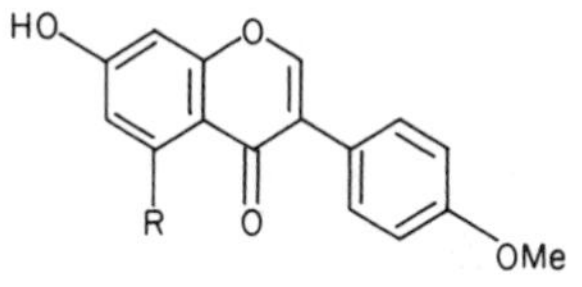

isoflavone

(25, R=H; formononetin)
(26, R=OH; biochanin A)

Early work on the formation of cyanidin (19)[16,17], quercetin (14)[18-20], (−)-epicatechin (21)[21, 22] and phloretin (23)[23] demonstrated quite clearly the biosynthetic pattern postulated by the Birch–Donovan hypothesis[1], i.e. ring A of the flavonoids is derived by a head to tail condensation of three acetate (or malonate) units, while ring C and the three carbon chain originate from a phenylpropanoid precursor. All subsequent experimental observations show that chalcones, with their respective isomeric flavanone, occupy a central position as intermediates in flavonoid biosynthesis[24-26]. Thus Grisebach and his collaborators have demonstrated the incorporation of appropriately labelled ^{14}C labelled chalcones (⇌ flavanones)

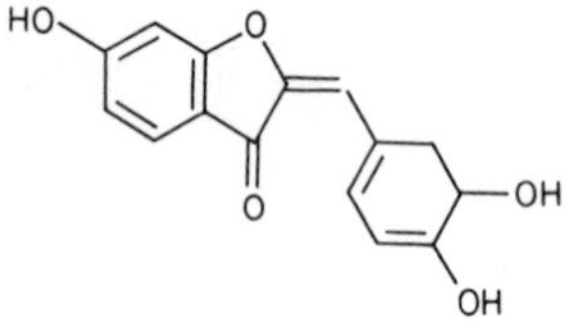

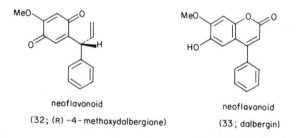

aurone

(27; sulphuretin)

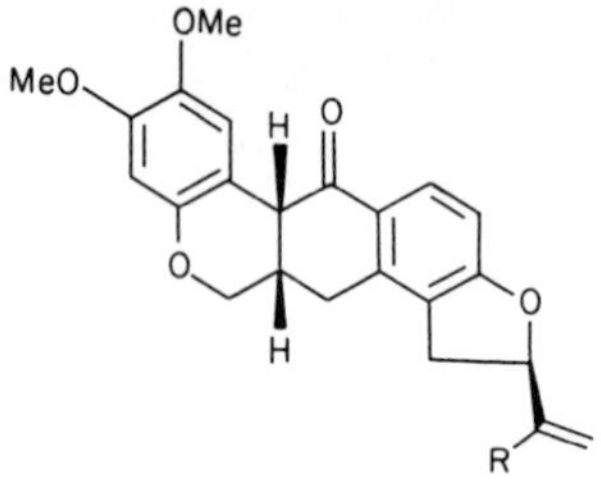

isoflavone – rotenoid

(28, R = Me, rotenone)

(29, R = CH$_2$OH, amorphigenin)

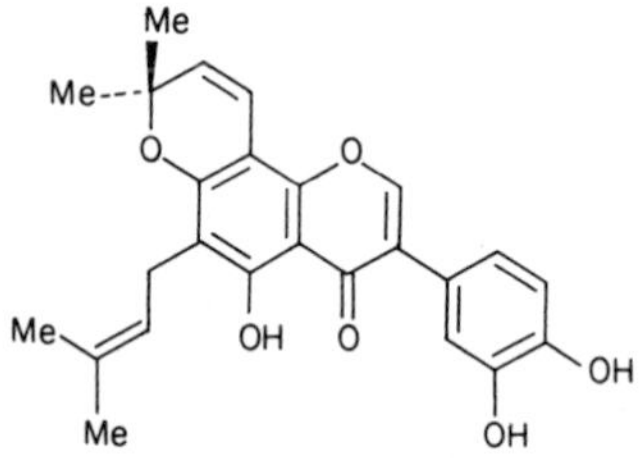

isoflavone

(30, pomiferin)

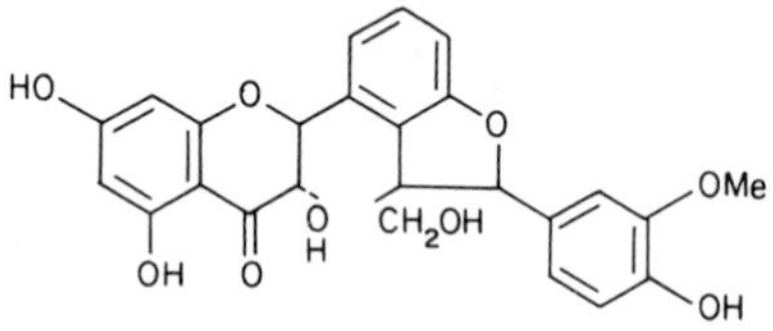

flavono –lignan

(31, silychristin)

neoflavonoid

(32; (R) –4– methoxydalbergione)

neoflavonoid

(33; dalbergin)

into cyanidin (19)[27] and quercetin (14)[27], into the isoflavones biochanin A (26)[28-30] and formononetin (25)[28-30] and of a chalcone glucoside into (−)-epicatechin (21)[31], apigenin (16)[32] and the dihydrochalcone phloridzin (24)[33]. In some cases an unexpected degradation to carbon dioxide (from ring A of the chalcone) and a phenylpropanoid derivative (usually the cinnamic acid) was observed during the feeding experiments and in later work the intact incorporation of chalcones into the flavonoids was demonstrated by double labelling experiments. Thus, for example, a sample of the chalcone (34) labelled both in ring A and the 2-position was fed to red cabbage seedlings to give radioactive cyanidin (19)[34]. The ratio of activity in the product at C-2 and in the phloroglucinol ring was shown to be identical to that in the precursor by degradation.

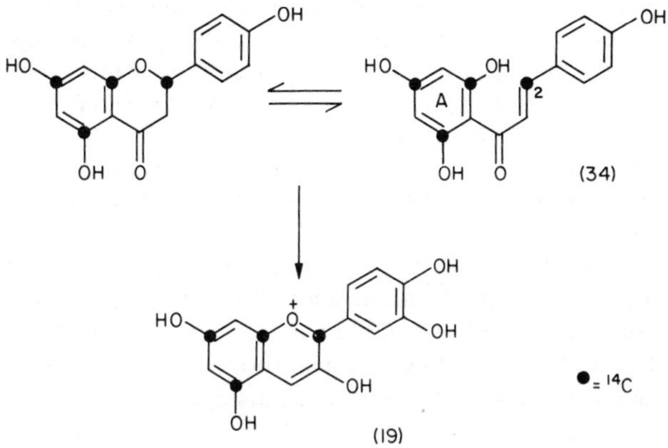

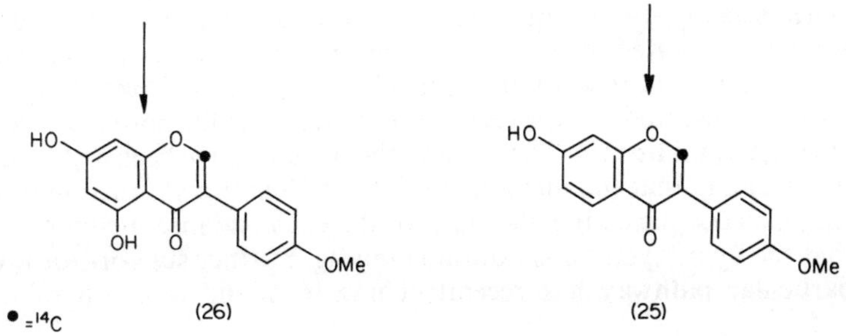

The substitution pattern in ring A of the chalcone ($\rightleftharpoons$ flavanone), i.e. the resorcinol or phloroglucinol structure, determines the specificity of the transformation. Thus in *Cicer arietinum*, the chalcone (35), appropriately labelled at C-2, is incorporated much more efficiently into the isoflavone with the phloroglucinol type of substitution in ring A (biochanin A, 26) and conversely the chalcone (36) served only as a precursor for formononetin (25), with the resorcinol orientation in ring A, in the same plant[29]. Analogously when the multiply labelled chalcone ($\rightleftharpoons$ flavanone, 34) was fed to *Cicer arietinum* shoots, its incorporation into biochanin A (26) was much higher than into formononetin (25) and occurred without degradation[34].

An important chemogenetic point which has been examined in studies of flavonoid biogenesis is the stage at which the substitution pattern in ring C of the final product is established. According to the 'zimtsaurestart hypothese' of Hess[35], this is thought to be controlled at the start of flavonoid synthesis by selection of the appropriate cinnamic acid from the phenylpropanoid pool for the formation of the chalcone. Thus when (methyl-^{14}C)-ferulic acid (40) or (methyl-[14]C$_2$)-sinapic acid (41) is fed to *Petunia hybrida*, the anthocyanidins with a substitution pattern in ring C (e.g. 37 and 38) corresponding to that of the precursor are most readily labelled. Furthermore, a methylating enzyme has been isolated from the flower petals of *Petunia hybrida* for which caffeic acid (39), but not cyanidin (19), is a good substrate[36]. Hess concluded from this evidence that the substitution pattern in ring C of the anthocyanidins (37 and 38) was already determined at the cinnamic acid stage.

(39)   (40, R=H)   (37, R=H)
       (41, R=OMe)   (38, R=OMe)

On the other hand the later substitution of the C ring at the stage of a C_{15} ($C_6 . C_3 . C_6$) intermediate has been postulated by many workers. Thus Neish and his colleagues[18] observed that although caffeic acid (39) has the same hydroxylation pattern as ring C of quercetin (14), it is a much poorer precursor of the flavonoid, based on isotopic dilution data, than either cinnamic or *p*-coumaric acid. Similarly, when the relative incorporations of the two chalcone glucosides (42 and 43) into quercetin (14) and cyanidin (19) in buckwheat seedlings was compared, it was found[37] that both chalcones were incorporated to the same extent. Finally, in experiments with excised buckwheat seedlings generally tritiated dihydrokaempferol (44) was observed to be a very efficient precursor for both quercetin (14) and cyanidin (19). These, and other results, have been explained by assuming that the introduction of the 3'-hydroxyl function into ring C takes place after the stage of the chalcone intermediate and that the hydroxylation is not rate limiting. Further support for this particular pathway has recently come from the isolation of an enzyme complex of the phenolase type from sugar beet (*Beta vulgaris*) which is able to hydroxylate some flavanonols with a 4'-hydroxyl group at the 3'-position[38]. Quantitative measurements[39] with the enzyme showed, in agreement with *in vivo* experiments[40], that dihydrokaempferol (44) is readily hydroxylated at the 3'-position and is a superior substrate for hydroxylation than kaempferol (13).

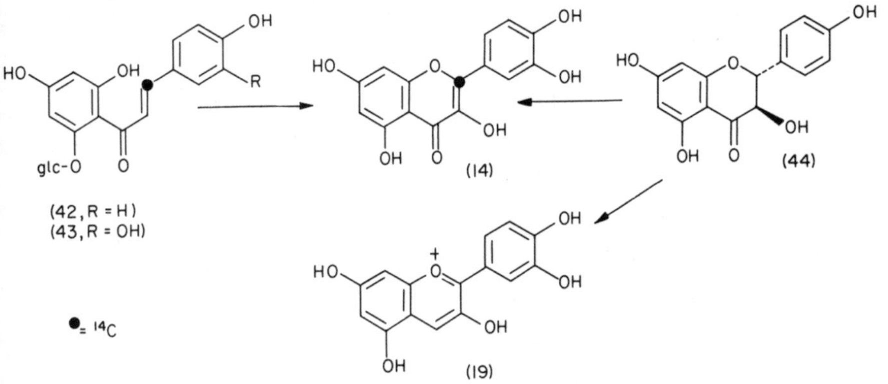

According to these, and earlier investigations, a specific role for dihydroflavones in flavonoid biosynthesis has been defined. Thus the biosynthetic pathway to quercetin (14) and cyanidin (19) in buckwheat has been formulated[40] as shown in *Figure 6.2*. Nevertheless, the possibility that the oxygenation pattern of ring C may be determined by other kinds of control mechanism still exists, and it has been tentatively proposed[24] that these may represent some of the differences between the routes for anthocyanidin biosynthesis and that for the other flavonoids and isoflavonoids.

The important question of the biosynthetic pathway to the chalcone is still an open one[24]. Two pathways have been discussed. That of Grisebach[25, 26] is a logical extension of the original idea of Birch and Donovan[1] and formulates the reaction as a condensation between a cinnamyl co-enzyme A ester and 3 molecules of a malonyl co-enzyme A to give a cinnamoyl poly-β-keto intermediate which then cyclises to the chalcone (*Figure 6.2*). An alternative pathway[24, 26], which has some chemical analogies[41], involves the acylation of phloroglucinol or resorcinol (45 or 46) and then rearrangement of the intermediate acylphenol (47) to yield the chalcone. More definitive proof in favour of either pathway is still desirable but the observation[18] that quercetin (14) biosynthesised from ^{14}C labelled carbon dioxide in the presence of phloroglucinol contained very similar ratios of activity in both rings A and C whilst

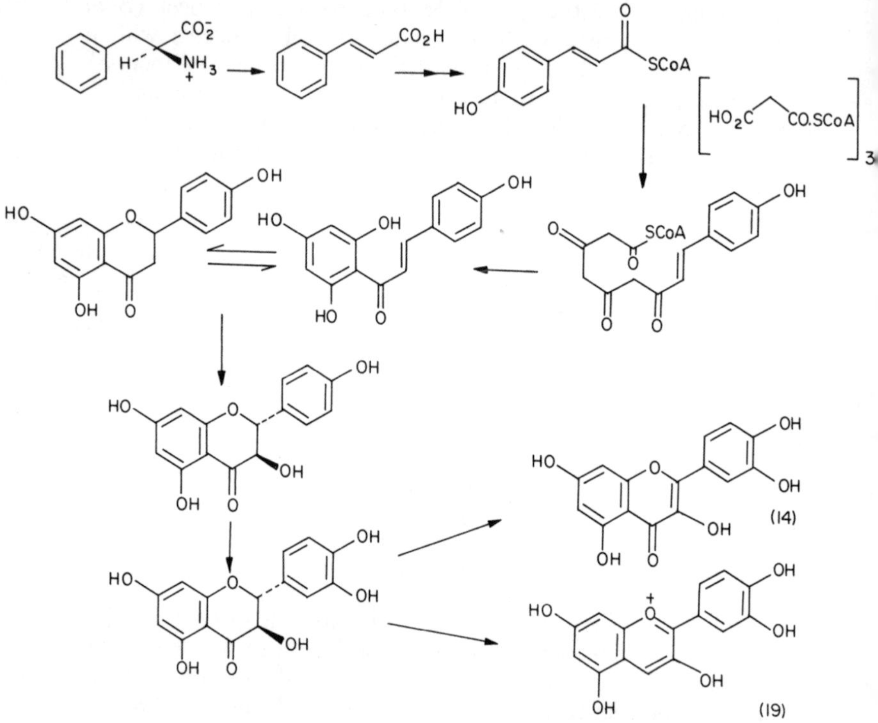

Figure 6.2. Biosynthesis of cyanidin and quercetin in buckwheat[40]

the addition of L-phenylalanine (a precursor of ring C via the phenylpropanoid pool) caused a tenfold reduction in the radioactivity incorporated into ring C has generally been taken as proof that phloroglucinol (or resorcinol) cannot be a precursor of ring A of flavonoids. Further evidence in support of the Grisebach pathway is the reported activation of cinnamic and p-coumaric acid in presence of co-enzyme A by extracts of chick pea (*Cicer arietinum*) and parsley (*Petroselinum hortense*)[42] and of cinnamic acid by preparations from *Beta vulgaris*[43].

In vitro an equilibrium exists in aqueous solution between the chalcone and flavanone, and in the case of chalcones, with hydroxyl groups in the 2′ and 6′ positions, this equilibrium lies entirely on the side of the flavanone. *In vivo* this reaction is under enzymic control since all flavanones which occur in nature have the 2(s) configuration[44,45]. Moustafa and Wong[46] were the first workers to isolate a flavanone-chalcone isomerase from soya bean (*Soja hispida*) and they showed that it catalysed the cyclisation of 2′,4,4′-trihydroxy-chalcone (48) to (—)-(2s)-4′,7-dihydroxyflavanone (49). Flavanone-chalcone isomerase activity has since been detected in a number of plants and purification of the enzymes from mung bean (*Phaseolus aureus*), chick pea (*Cicer arietinum*) and parsley (*Petroselinum hortense*) shows that they occur as iso-enzymes[47]. The enzymes

(45, R = OH)
(46, R = H)
(47)

from soya and mung bean have a low substrate specificity but none of the isomerases is able to catalyse the conversion of a chalcone glucoside to a flavanone. Because of this ready enzymatically controlled equilibrium between chalcones and flavanones, it has been in many cases very difficult to determine whether the further reactions to the individual flavonoids occurs from the chalcone or flavanone as substrate. Thus Grisebach, Patschke and Barz[48] demonstrated that (—)-(2s)-(2-^{14}C)-5,7,4′-trihydroxyflavanone-5-β-D-glucoside (naringenin, 5-β-D-glucoside, 50), in comparison to the 2R isomer, was preferentially incorporated into quercetin (14) and cyanidin (19) in buckwheat and into biochanin-A (26) in *Cicer arietenum*, and this evidence, they suggested, supported the assumption that the 2s-flavanones are intermediates in flavonoid biosynthesis. However, it has been noted[24] by Grisebach and Barz that this observation may merely be a measure of the relative ease of enzyme catalysed ring opening of the 2R and 2s flavanones to give the chalcone which is the true biosynthetic precursor. This latter conclusion was also reached by Wong[49] and by Wong and Grisebach[50] on the basis of kinetic and competitive double labelling experiments which strongly favoured the chalcone as the direct precursor of the various other flavonoids.

The biosynthetic correlations between different flavonoid classes (flavones, flavonols, flavanonols, anthocyanidins, flavan-3-ols, etc.) have been made mainly on the basis of genetic data. In many respects, however, the problems related to the formation of these differing categories is only partially solved and the pathways to some flavonoid compounds is based solely upon good chemical analogies. Plausible schemes have been elaborated (*Figure 6.3*) and in these Grisebach and his associates[24-26] have accorded 3-hydroxy-flavanones (dihydroflavonols) a key role. Thus for example this

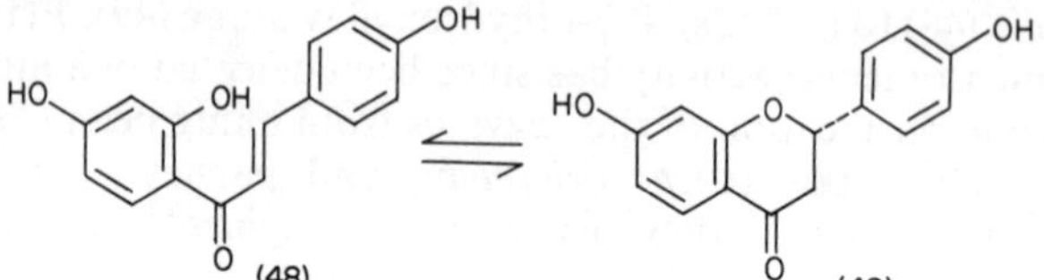

(48) ⇌ (49)

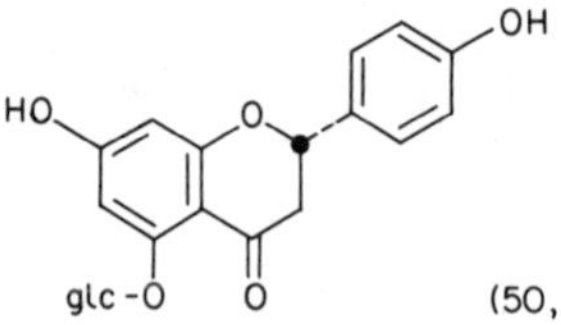

(50, ● = ^{14}C)

proposition was supported by experimental work with tritium labelled dihydrokaempferol (44) in buckwheat seedlings[40, 51] which showed the dihydroflavonol to be an excellent precursor of both quercetin (14) and cyanidin (19) and the role of (44) in the biosynthesis of cyanidin has been confirmed in a later study carried out with cell cultures of *Haplopappus gracilis*[52]. Some problems are still associated with the mode of formation of the dihydroflavonol intermediate from the chalcone (⇌ flavanone) but investigations have eliminated a number of chemically plausible routes which involve the corresponding flavone as an intermediate. In experiments with *Chamaecyparis obtusa*, Grisebach and Kellner[53] showed that the hydrogen in the 2-position of the chalcone (⇌ flavanone, 51, 52) was retained in the biosynthesis of taxifolin (54); evidence which these workers interpreted as favouring a direct hydroxylation of the flavanone or the formation of a chalcone epoxide (53) as an intermediate in the production of (54).

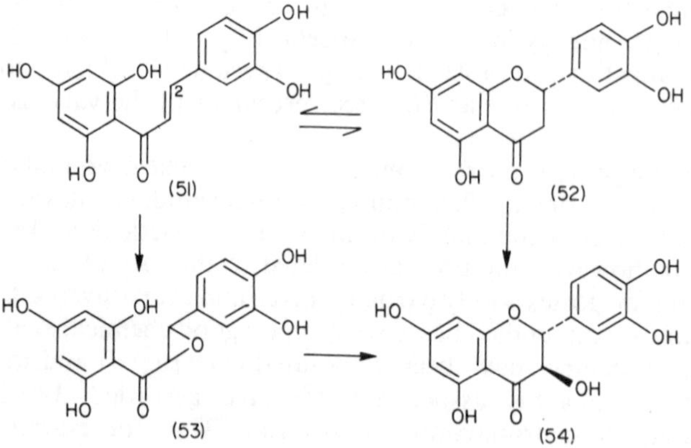

(51) ⇌ (52)

(53) → (54)

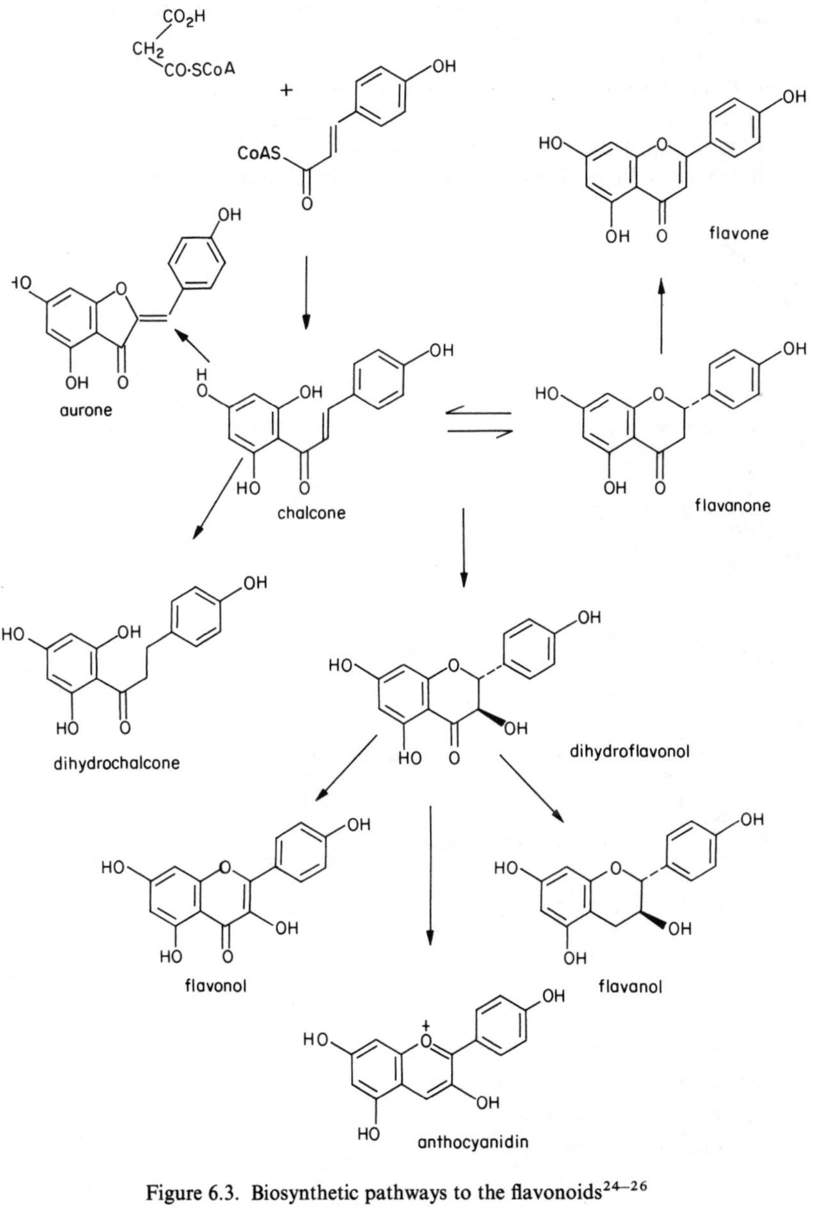

Figure 6.3. Biosynthetic pathways to the flavonoids[24-26]

Some preliminary observations on the control of flavonoid biosynthesis have been made with cell suspension cultures of parsley (*Petroselinum hortense*) by Grisebach and his colleagues[54]. Cell cultures of parsley produce the glycoside apiin (55) and the related flavonoid graveobioside B when exposed to high intensities of white light and detailed studies at the enzymic level[55–57] permitted a biosynthetic pathway (*Figure 6.4*) to be elaborated for the forma-

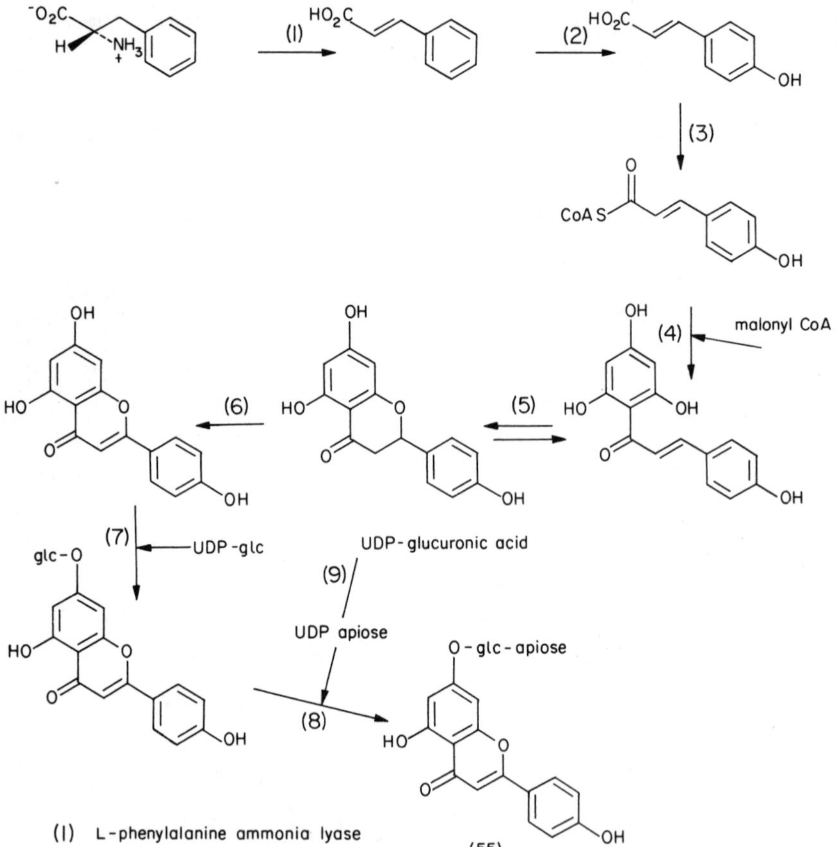

(1)   L-phenylalanine ammonia lyase
(2)   cinnamate – 4'– hydroxylase
(3)   p-coumarate CoA ligase
(5)   chalcone – flavanone isomerase
(7)   glucosyl transferase
(8)   apiosyl transferase

CoASH = coenzyme A

Figure 6.4. Biosynthesis of apiin in cell cultures of parsley (*Petroselinum hortense*)[54]

tion of apiin from L-phenylalanine. Methods for the determination and assay of seven of the enzymes (*1, 2, 3, 5, 7, 8* and *9*) in the pathway were investigated and it was shown that the initial stages of glycoside accumulation under illumination were accompanied by a dramatic increase in the associated enzyme activities. The responses to light of the first three enzymes in the pathway differed considerably from those of the remaining four (*5, 7, 8* and *9*) which were measured. Thus enzymes *1, 2* and *3* showed maximal activity 15 hours after the commencement of illumination, whilst enzymes *5, 7, 8* and *9* showed a peak of activity after 24 hours and then declined. It was tentatively suggested as an explanation of this behaviour that the biosynthesis of the flavone glycosides in cell cultures of parsley is associated with two or more differently regulated sequences of enzyme steps—those of group I (*1, 2* and *3*) produce substrates of the phenylpropanoid type whose metabolism is not restricted to flavonoid synthesis, and those of group II (*5, 7, 8* and *9*) whose formation is closely identified with flavonoid synthesis.

6.1.3 BIFLAVONYLS AND PROANTHOCYANIDINS

The biflavonyls and proanthocyanidins form two distinct groups of oligomeric flavonoid compounds whose structures are derived by combinations of the basic $C_6 . C_3 . C_6$ flavonoid unit. The biflavonyls[58], such as hinokiflavone (59), ginkgetin (57) and

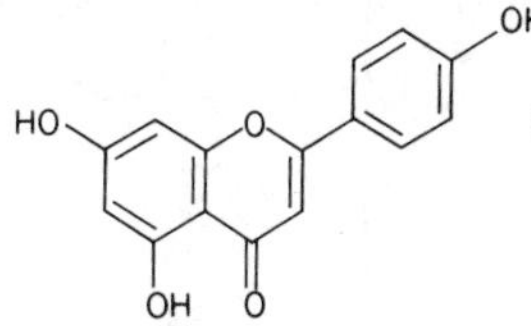

(56)

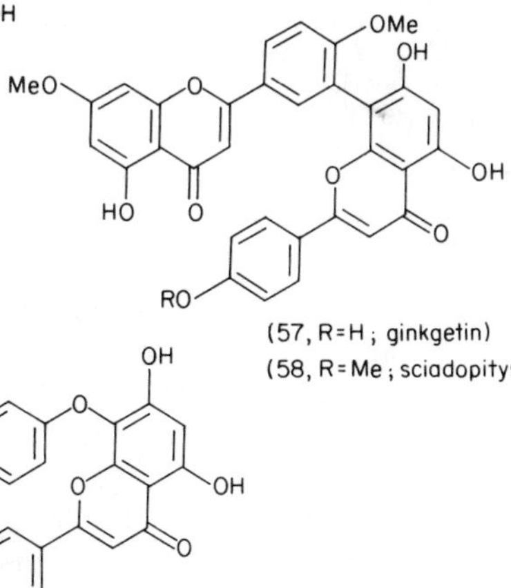

(57, R = H ; ginkgetin)
(58, R = Me ; sciadopitysin)

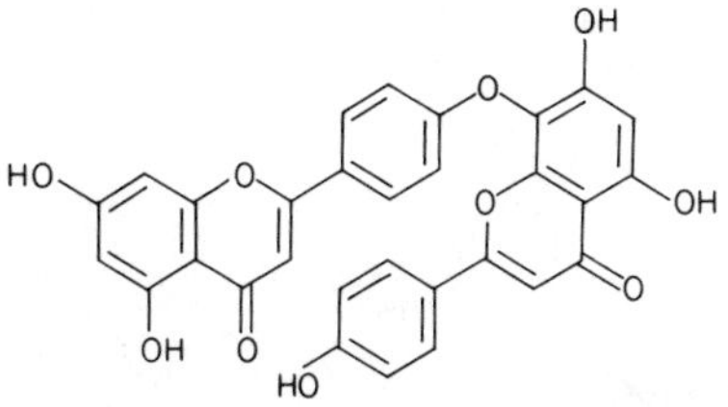

(59 , hinokiflavone)

sciadoptysin (58), are dimers composed of two flavonoid units at the flavone level of oxidation and they are widely distributed among the Gymnosperms. It has been suggested[58] that the biflavonyls are most probably synthesised in the plant tissue by oxidative coupling of a flavonoid precursor such as apigenin (56) but no definitive work has been presented to confirm or disprove this proposition.

The tissues of many plants with a woody habit of growth[59] are often sources of the second major group of oligomeric flavonoid compounds—the proanthocyanidins[60, 61, 62]. These compounds derive their names from their characteristic property of producing a typical deep red anthocyanidin pigment when treated with acid. The various dimeric proanthocyanidins which have been described[61-63] are almost invariably composed of two flavan-3-ol units ('catechins') either singly or doubly linked. The mode of linkage between the monomer units is quite different from that of the biflavonyls and typical structures are those of the widely distributed (−)-epicatechin (21) dimer procyanidin B2 (60) and the closely related but less frequently found proanthocyanidin A2 (61). Trimeric and tetrameric proanthocyanidins have also been described[62, 63] possessing similar structural characteristics. No biosynthetic work has yet been completed with this group of compounds

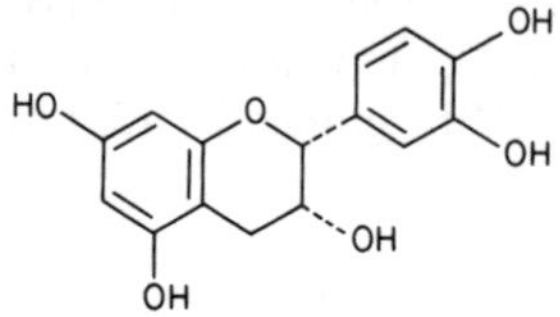

(21)

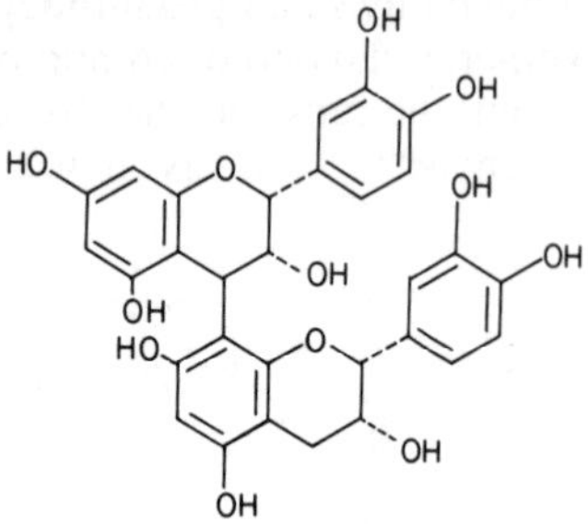

(60)

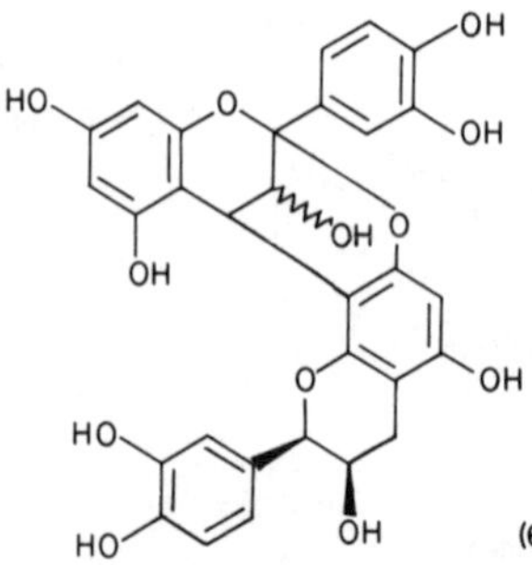

(61)

but two theories have been propounded to account for their biosynthesis. In the first of these[64], the inter-flavan linkage, it has been suggested, is formed by a simulated acid catalysed condensation between the flavan-3-ol and the corresponding flavan-3,4-diol. However, failure to locate flavan-3,4-diols in vegetative tissues has led to the alternative proposal that these compounds are formed by dehydrogenation of two, or more, molecules of the appropriate flavan-3-ol. Combinations of two or more flavonoid molecules at different levels of oxidation are, by comparison with the biflavonyls and proanthocyanidins, comparatively rare. A notable member of this group is xanthorrhone (64) isolated[64, 65] from Xanthorrhoea

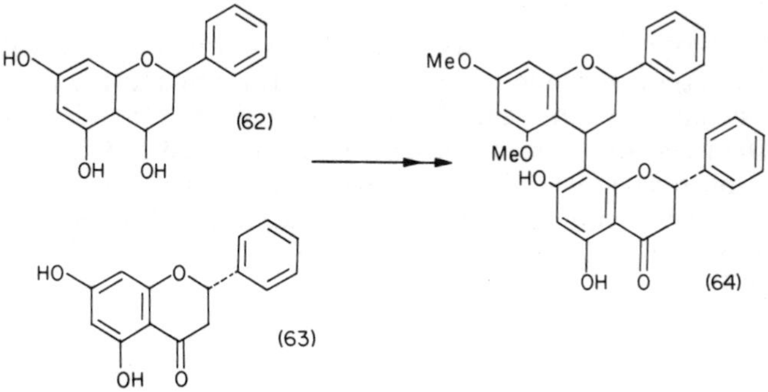

species. This is a dimer of the proanthocyanidin structural type composed of an 'upper' flavan and 'lower' flavanone unit. Birch and his collaborators suggested[65] that it was formed probably from pinocembrin (63), which is already present in the plant, and the flavan-4-ol (62) in a manner similar to that already proposed for the proanthocyanidins.

6.1.4 ISOFLAVONOIDS

The work of Grisebach and his associates[28-30, 34] has demonstrated conclusively that the biosynthesis of the isoflavone skeleton from precursors of the general flavonoid pathway involves a 1,2-aryl shift of the C ring and that this rearrangement takes place at or after the formation of the C_{15} chalcone intermediate. Thus investigations of isoflavone biosynthesis in *Trifolium pratense* and *Cicer arietinum* showed that L-phenylalanine, variously labelled with ^{14}C at C-2 and C-3 and in the carboxyl group (65), was incorporated into the isoflavone formononetin (24) as shown. The distribution of the radio-isotope was demonstrated by degradation. These results unequivocally demonstrate that a phenyl migration takes place in the course of the biosynthesis and experiments with chalcone glucosides[29, 34] indicate that the aryl rearrangement occurs after the formation of the $C_6 . C_3 . C_6$ intermediate.

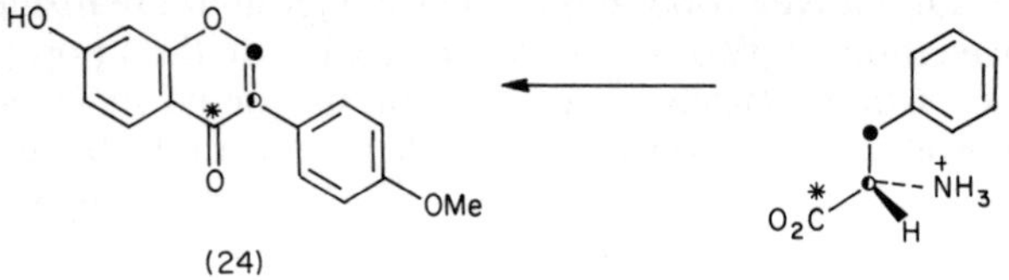

(24)   (65)

$\circ \bullet * = {}^{14}C$

A 1,2-aryl shift is similarly a notable feature of the biosynthesis of the carbon skeleton of both coumestrol (66) and the rotenoids (e.g. 67 and 68). Thus the ^{14}C labelled chalcone (48) administered to *Medicago sativa* gave coumestrol (66) labelled at C-2[66], and L-[^{14}C]-phenylalanine, labelled separately at the C-2 position and in the carboxyl group (65), fed to *Derris elliptica* and *Amorpha fructicosa* seeds gave respectively rotenone (67) and amorphigenin (68) labelled as shown.

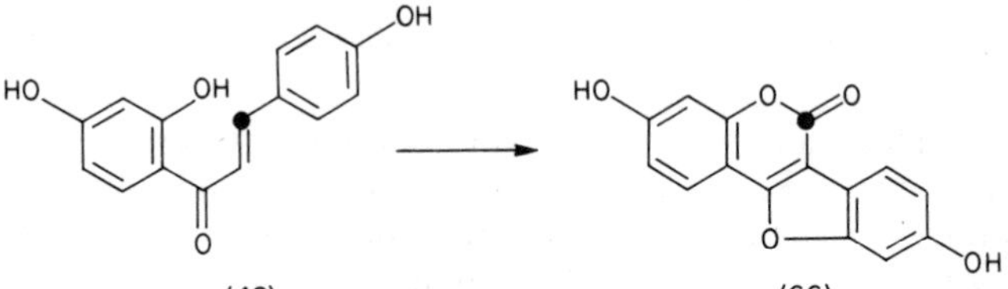

(48)   (66)

$\bullet = {}^{14}C$

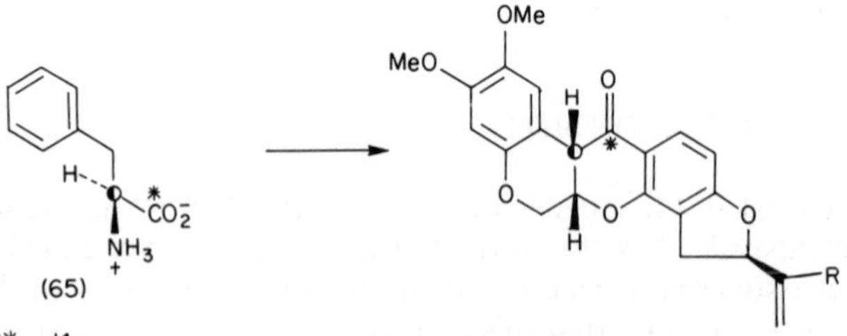

(65)

$\circ * = {}^{14}C$

(67, R = Me ; rotenone)
(68, R = CH$_2$OH ; amorphigenin)

Since the capacity to rearrange a flavonoid to an isoflavonoid structure is limited to essentially one phylogenetically related group of higher plants it is considered to be an isolated characteristic. It

has generally been assumed that there is only one such rearrangement leading from the biosynthetic pathway and that this gives a
unique isoflavonoid precursor from which the other isoflavonoids
are elaborated by secondary reactions (oxidation, reduction,
methylation and isoprenylation). The exact point at which this
rearrangement occurs has been the subject of some speculation but
some of the early proposals based on the involvement of a dihydroflavonol intermediate[25] have proved to be untenable[34, 68, 69].
Wong[70] has postulated, on the basis of competitive feeding experiments with a flavanone-chalcone pair, that isoflavones are derived
directly from the chalcone, and Ollis, Ormand and Sutherland[71]
have provided an elegant *in vitro* analogy for this conversion with
the demonstration that chalcones may be rearranged oxidatively
with thallic acetate to the isoflavone skeleton.

The nature of the first formed product on the biosynthetic
pathway has yet to be unequivocally established but present
evidence[26] favours the isoflavone as opposed to the isoflavanone.
A number of mechanisms have been proposed for the 1,2-aryl

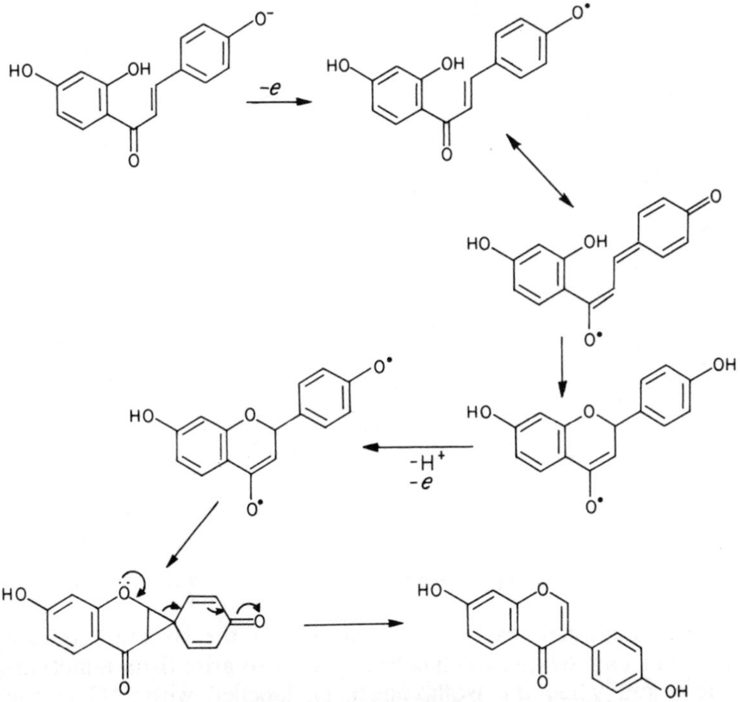

Figure 6.5. Suggested route for the 1,2-aryl migration of chalcones to isoflavones[72]

migration[25, 26] and the proposal of Pelter and his colleagues[72], which is shown in *Figure 6.5*, is an oxidative route which assigns a specific role to the 4-hydroxyl group of the migrating phenyl residue.

Direct experimental evidence for the various biosynthetic transformations which are necessary to establish the several classes of isoflavonoids has been obtained in just a few cases such as coumestrol (66)[72–74] and the rotenones (e.g. 67 and 68)[67, 75–77]. Thus from the results of comparative feeding experiments with labelled precursors in *Phaseolus aureus*, Grisebach and his associates[74] postulated that at least two routes to the coumestan coumestrol (66) existed. The situation was described as an example of a 'metabolic grid' in which the route *a* (*Figure 6.6*) was quantitatively of most significance. The formation of the five-membered ring was suggested, by chemical analogy[78], to occur by dehydration and cyclisation of the 2',4',7-trihydroxyiso-flavanone (69) and allylic oxidation of the product (70) would then give coumestrol (66).

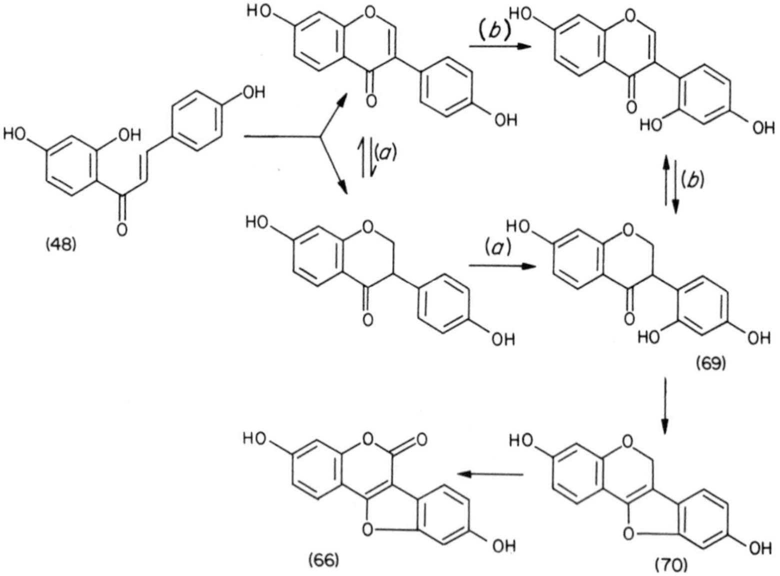

Figure 6.6. Probable biosynthetic route to coumestrol in *Phaseolus aureus*[74]

The 'additional' carbon atom at C_6 in the structure of the rotenoids (e.g. 67 and 68) has been shown to arise from L-methionine[75] and when the isoflavone (71), labelled with ^{14}C in the methoxyl group at the 2' position, was fed to germinating seedlings of *Amorpha fructicosa* an incorporation of 0.75 per cent was obtained into amorphigenin (68). Degradation of the rotenoid showed that all the radioactivity was localised at C-6 of the metabolite and hence a cyclisation process, analogous to the biological cyclisation of an ortho methoxyphenol to a methylenedioxy group[79], was formulated to elaborate the rotenoid skeleton from the isoflavone (71). Since 6-[$^3H_2$]-($\pm$)-9-demethylmunduserone (72) was also readily incorporated into amorphigenin (68), without randomisation of the radioactive label, it was inferred that prenylation to give the distinctive rotenoid structures, such as those of rotenone (67) and amorphigenin (68), occurs at a late stage of biosynthesis.

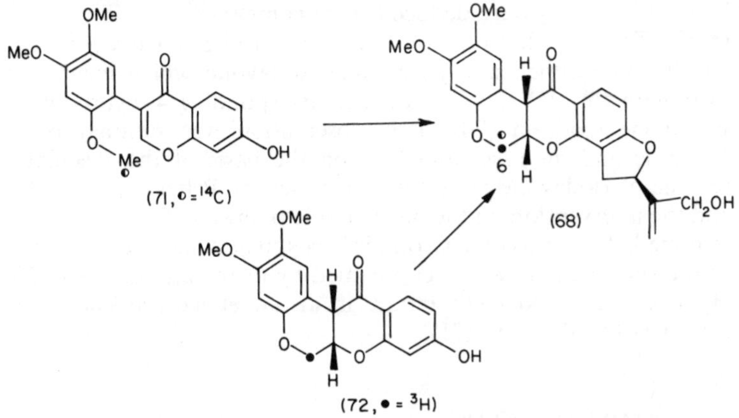

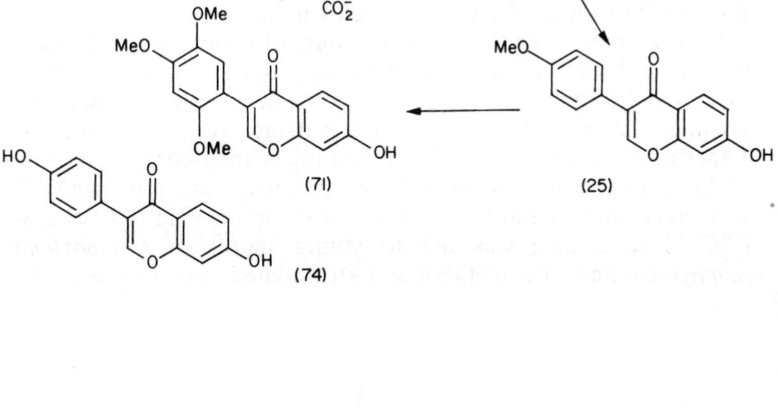

Earlier stages in the biosynthesis of the rotenoids to give the key intermediate (71) were deduced from comparative feeding experiments. Thus the chalcone (48) was shown to be an early precursor but the biosynthetically acceptable isoflavone which emerges is formononetin (25) rather than the corresponding 4′-hydroxylated isoflavone daidzein (74). This observation was rationalised by Crombie and his associates[76, 77] on the basis of the speculative scheme of isoflavone biogenesis proposed by Pelter[72] (*Figure 6.5*) and the intervention on the biosynthetic pathway of a spirodienone intermediate such as (73). Crombie[77] postulated that decomposition of the spirodienone was accompanied by methylation to give (25). Hydroxylation and methylation, in an undefined sequence, were then postulated to give (71).

6.1.5 NEOFLAVONOIDS

During the last decade a class of natural products—the neoflavonoids—which fall into a new structural pattern closely allied to both the flavonoids and the isoflavonoids has been discovered[10, 80]. The generic term neoflavonoid was proposed by Ollis and his collaborators to describe this group of natural phenolic compounds found in species of *Dalbergia* and *Macherium*, genera of the Leguminosae, and includes such substances as the dalbergiquinols (75), the dalbergiones (76), the neoflavenes (77) and the dalbergins (78). One member of the neoflavonoid class, the 4-phenylcoumarin group (e.g. 78 and 80) is also prolific amongst genera of the Guttiferae[81, 82] and it is with members of this group that successful biogenetic studies have been carried out[83–85].

In 1959, Seshadri[86] postulated that the 4-arylcoumarins might arise by an alternative linkage of a phenolic unit to a phenylpropanoid intermediate rather than a further aryl migration on the C_3 unit as is observed in the biosynthesis of the isoflavonoids. This proposal was subsequently elaborated and embodied by Ollis and Gottlieb[87, 88] in an elegant scheme of biosynthesis which they proposed for the neoflavonoids (*Figure 6.7*).

Strong support for the proposed route of formation of the neoflavonoids by a cinnamylation process, probably involving cinnamyl pyrophosphate, was provided by the discovery that cinnamyl phenols (79) are often found co-occurring with neoflavonoids. Experimental data which is also compatible with this proposed mode of biogenesis has been reported by Polonsky and Kunesch[83, 84] who have studied the biosynthesis of calophyllolide (80). D L-3-[^{14}C]-Phenylalanine was fed to young shoots of *Calophyllum inophyllum* and the metabolite (80) isolated and degraded by

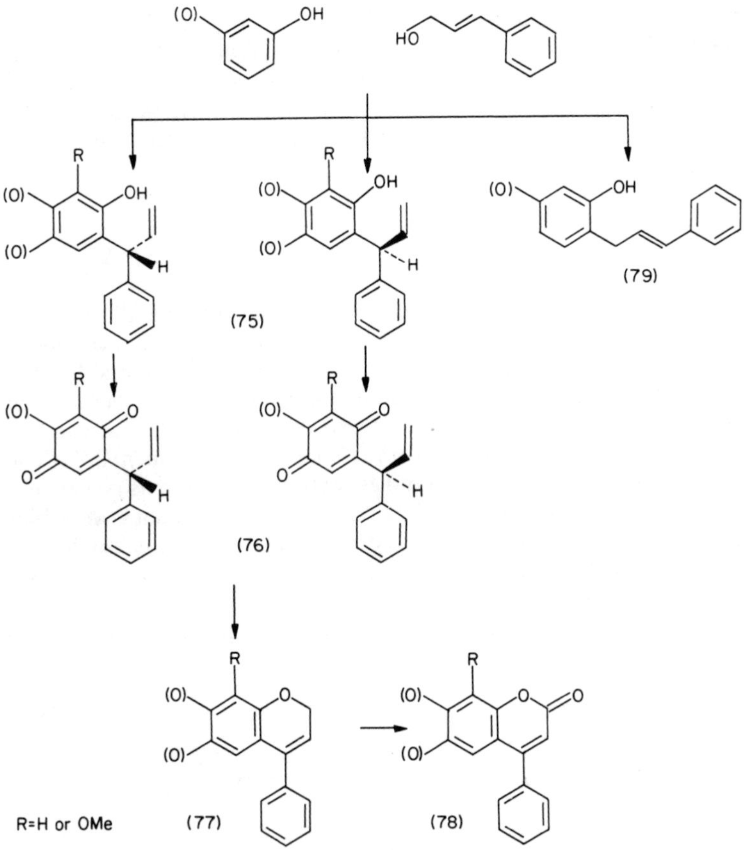

Figure 6.7. Probable routes for the biosynthesis of the neoflavonoids[87, 88]

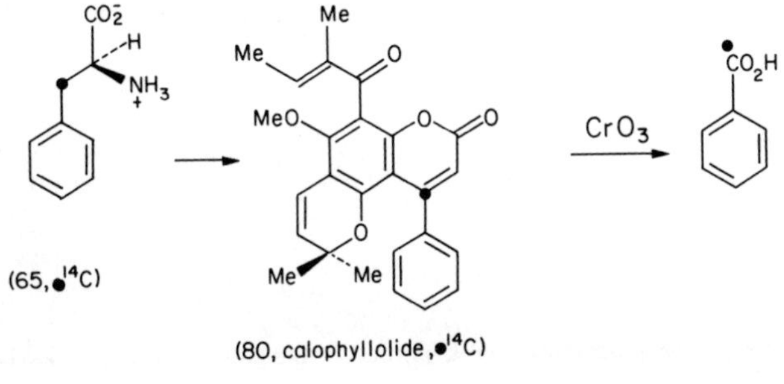

chromic acid oxidation to benzoic acid. More than 80 per cent of
the total activity incorporated into calophyllolide (80) was located
in the carboxyl group of benzoic acid and hence at C-4 in calophyllolide (80), and this observation shows that rearrangement of the
phenyl group of the $C_6 . C_3$ precursor does not occur during the
biosynthesis.

6.1.6 STILBENES

Although in number the stilbenes are a relatively small group of
natural phenols, they have nevertheless been found throughout
the plant kingdom from algae and liverworts to conifers[89]. They
range in structure from the parent hydrocarbon trans-stilbene which
has been isolated[90] from *Alnus firma* to polyphenols, such as the
pentahydroxystilbene (81) from *Vouacoupa macropetala*[91]. The
phenolic stilbenes are also found as methyl ethers and as glycosides.
Birch and Donovan originally suggested[1] that stilbenes were biosynthesised by a phenylpropanoid-polyacetate pathway, analogous
to that of the flavonoids, followed by an alternative mode of cyclisation of the poly-β-keto intermediate. Such a pathway is shown in
Figure 6.8 for pinosylvin (82) and experimental work with this and

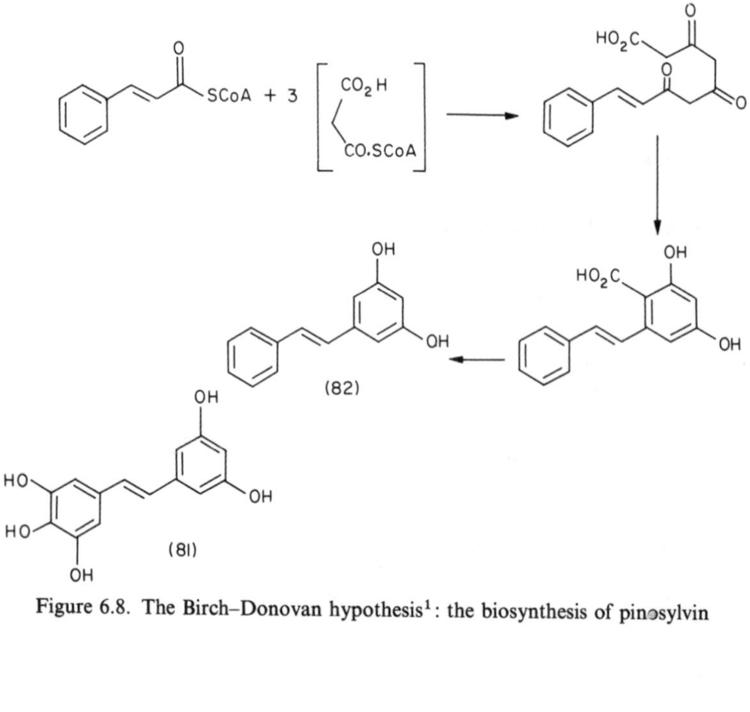

Figure 6.8. The Birch–Donovan hypothesis[1]: the biosynthesis of pinosylvin

other plant stilbenes has borne out the correctness of the original hypothesis[92-94].

Typical of the biosynthetic experiments in this field are those performed on oxyresveratrol (83) in *Morus alba* (*Figure 6.9*)[95]. Thus [14C]-carboxyl labelled acetate when it was administered gave labelling in only one of the aromatic rings (A) and [14C]-labelled cinnamic acid derivatives gave labelling of the alternative ring (B) and the vinyl linkage. These patterns of 14C isotope incorporation were revealed by standard chemical degradations (*Figure 6.9*).

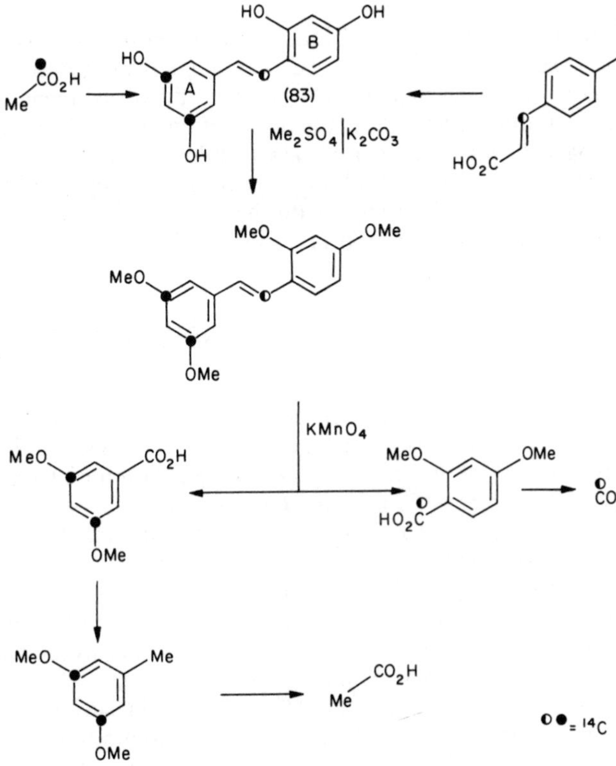

Figure 6.9. Biosynthesis and degradation of oxyresveratrol[95]

Further support for the Birch–Donovan scheme of biosynthesis for stilbenes has also been derived from similar studies of the C$_{15}$ stilbene carboxylic acids. These acids are postulated as intermediates in the biosynthesis of the stilbenes (*Figure 6.8*) and in support of this hypothesis hydrangeic acid (84) and the related isocoumarin

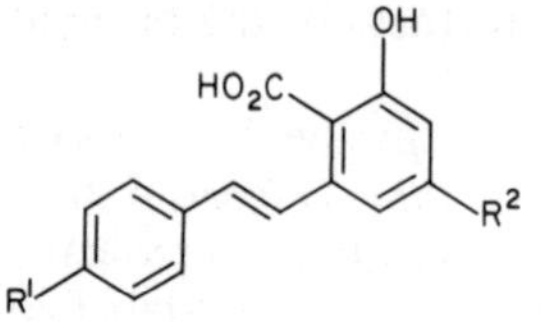

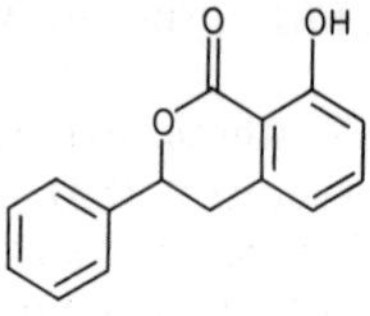

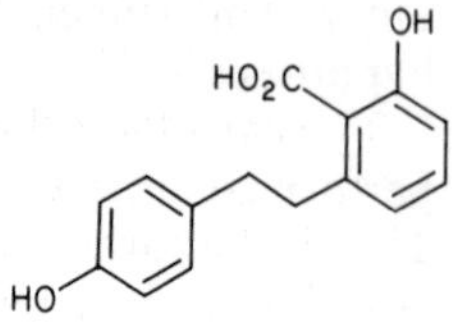

(84, R^1 = OH, R^2 = H;
hydrangeic acid)

(85; hydrangeol)

(87; lunularic acid)

(86, R^1 = H, R^2 = Oglc; gaylussacin)

hydrangeol (85) have been obtained from *Hydrangea macro-phylla*[96, 97], gaylussacin (86) has been isolated from *Gaylussacia baccata* and *Gaylussacia frondosa*[98], and the dihydrostilbene car-boxylic acid, lunularic acid (87), has been obtained from liverworts and algae[99]. The biosynthesis of these C_{15} stilbene carboxylic acids has been studied in some detail and conforms unexceptionally to the predicted pattern (*Figure 6.8*) with the carboxyl function being derived from the carboxyl group of acetate[99–103].

6.1.7   CURCUMIN AND THE 9-PHENYLPERINAPHTHENONE PIGMENTS

The unusual carbon skeleton of curcumin (88), which occurs in the genus *Curcuma* (Zingiberaceae) together with its mono- and bis-demethoxy derivatives has prompted some speculation on its mode of biogenesis[4]. Thomas[104] has pointed out that the carbon skeleton of curcumin is related to that of the 9-phenylperinaphthenone pigments and he suggested that these may be formed by the com-bination of one acetate and two shikimate derived C_6-C_3 units. Furthermore recent research has revealed a number of other plant constituents which have the same carbon skeleton as curcumin (88) and which have presumably similar biogenetic origins. Thus com-pounds such as yashabushiketol (89)[105] occur in several *Alnus* species whilst centrolobol (90)[106] and related compounds have been derived from *Centrolobium* species (Leguminosae). In addition, the unusual *meta* bridged biphenyl system found in myricanone (91) and myricanol from *Myrica nagi*[107, 108] and in asadanin from *Ostrya japonica*[109] likewise suggests a biosynthetic pathway closely related to that of curcumin and the 9-phenylperinaphthenone pigments.

  The experimental work which has subsequently been carried out in this area has demonstrated the validity of this biosynthetic hypothesis in relation to the 9-phenylperinaphthenone pigments

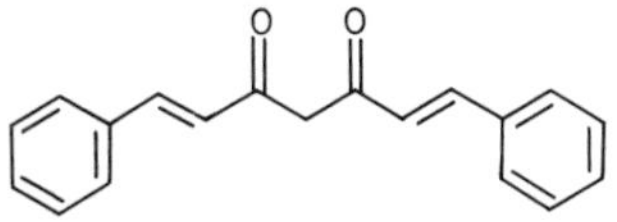

(88; curcumin)

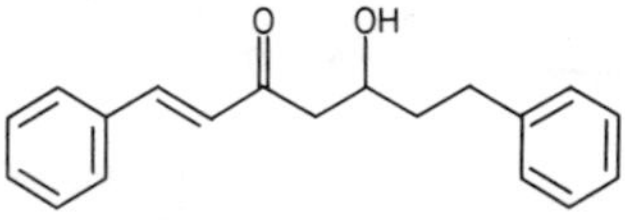

(89; yashabushiketol)

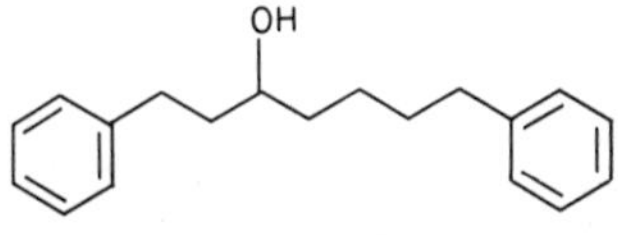

(90, centrolobol)

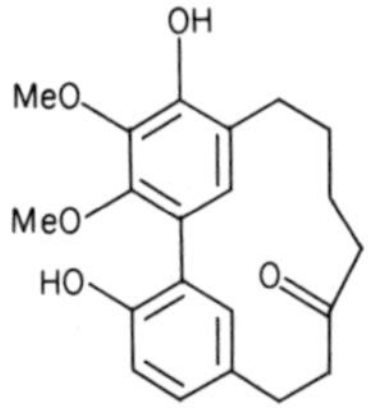

(91; myricanone)

(92 and 93)[110, 111] but has cast some doubt upon a similar biogenetic route to curcumin (88)[112]. The 9-phenylperinaphthenone pigments have so far only been found in plants of the family Haemodoraceae[113] and they differ in their mode of biogenesis from the related fully acetate-derived mould perinaphthenones such as atrovenetin (96) and herqueinone[114]. In a study of the biosynthesis of the aglycone (92) of the cellobioside haemocorin[115] in the rhizome of *Haemodorum corymbosum*, Thomas[110] established, using tracer techniques, that both L-phenylalanine and L-tyrosine were incorporated into the pigment and were specific precursors of different structural fragments within the molecule.

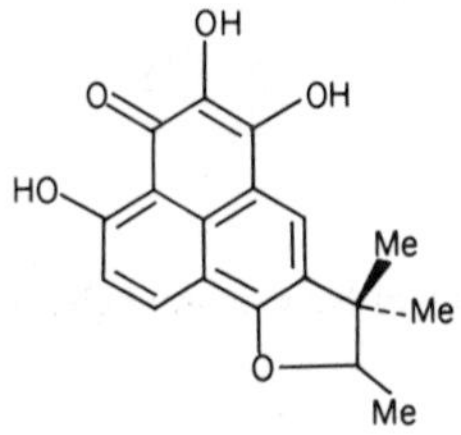

(96; atrovenetin)

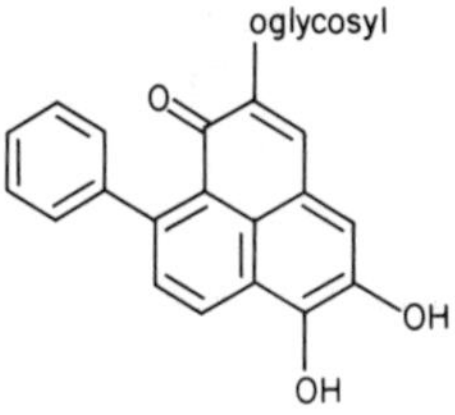

(93, lachnanoside)

A specific degradation was designed to isolate C-5 in haemocorin aglycone. Two tautomeric structures are possible for the pigment (92a and 92b) and methylation[115] gives a mixture of the corresponding methyl ethers (94a and 94b). Oxidation of the ether (94b) gives the anhydride (95) and C-5 is released as carbon dioxide. When 2-[^{14}C]-L-tyrosine was administered and the pigment degraded to the anhydride (95), this was found to contain less than 1 per cent of the original radioactivity of the pigment or the ether (94b). Hence

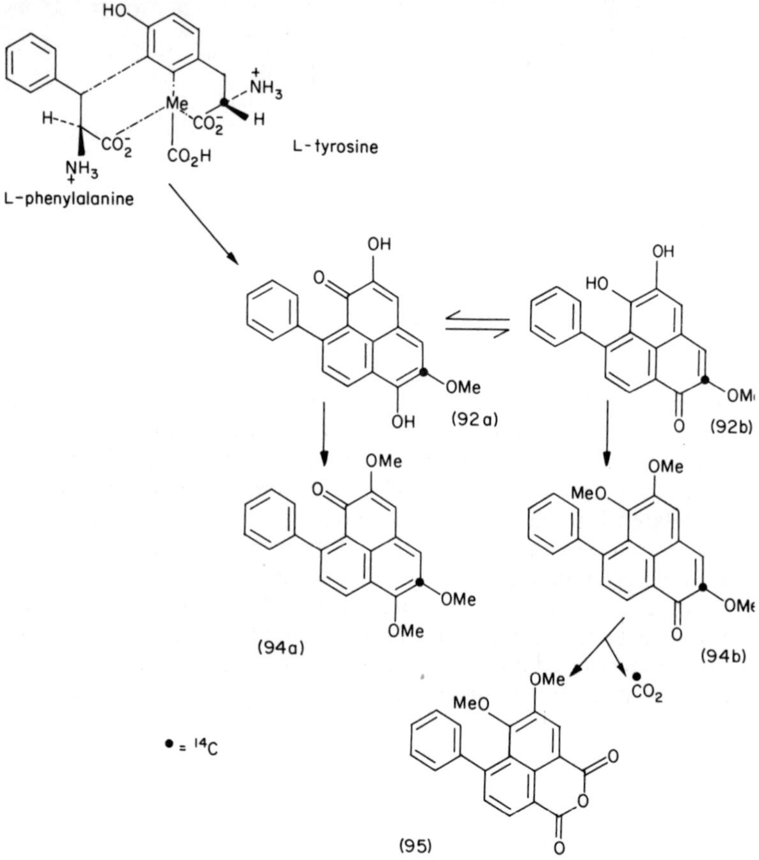

Figure 6.10. Biosynthesis and degradation of haemocorin aglycone[110]

the radio-isotope was located exclusively at C-5 in haemocorin aglycone (92) and his observation led to the proposal of the biosynthetic pathway shown in *Figure 6.10*. Similar experiments were carried out by Edwards, Schmitt and Weiss[111] on the biosynthesis of lachnanthoside (93) in the roots of the plant *Lachnanthes tinctoria*. These workers' results showed that here too L-phenylalanine, L-tyrosine and acetic acid were readily incorporated into the pigment and that the acetate carboxyl group was lost during biosynthesis. A similar biosynthetic pathway was formulated.

No further examples have yet been described in which two $C_6 . C_3$ units combine with one acetate fragment to form either an open chain compound or a complex ring system such as is found in the 9-phenylperinaphthenones.

6.2 MISCELLANEOUS PLANT PRODUCTS DERIVED FROM THE SHIKIMATE PATHWAY

6.2.1 PLANT QUINONES

Quinones form a large group of naturally occurring colouring matters with representatives in almost all of the phyla which have been examined. The biological significance of the various iso-prenoid quinones (ubiquinones, plastoquinones and menaquinones)

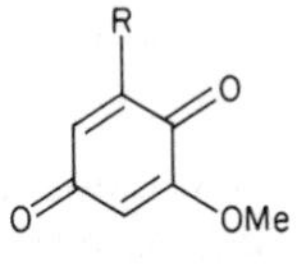

(97, R = H)
(98, R = OMe)

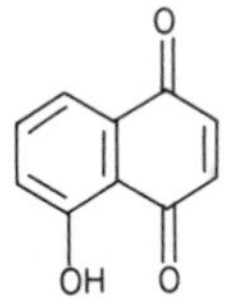

(99, juglone)

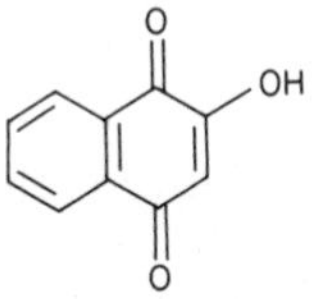

(100, lawsone)

has been discussed earlier but little is yet known about the function in living systems of the non-isoprenoid quinones. In the plant kingdom the largest number of these quinones occur in flowering plants and nearly half of these are found in one family, the Rubiaceae[116, 117]. They fall chemically into three major categories based on benzoquinone (97, 98), naphthoquinone (99, 100) and anthraquinone (101, 102) with a few additional minor classes such

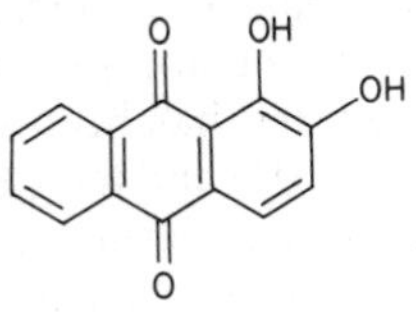

(101; alizarin)

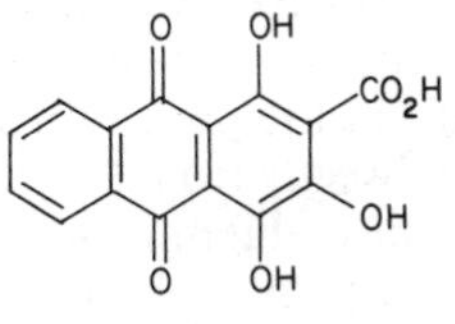

(102, pseudopurpurin)

as the phenanthraquinones and the extended quinones such as hypericin. In contrast to the situation with many fungal quinones, which are derived predominantly from acetate or malonate, the shikimate pathway appears to play a much more significant role in the biosynthesis of plant quinones and this contribution is conveniently discussed in terms of the three chemical classes defined above.

Benzoquinones

Although quinol (105) has been found in several plant species, notably within the Ericaceae as its β-D-glucoside arbutin (106) and its various acylated forms[118, 119], benzoquinone (104) itself has not been observed in the plant kingdom. Feeding experiments with ^{14}C labelled p-hydroxybenzoic acid showed an incorporation of up to 10 per cent of the radiocarbon into quinol and it was suggested that

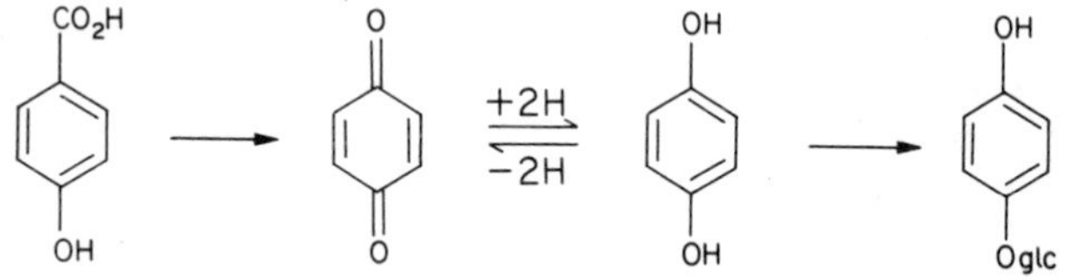

the latter is therefore formed by an oxidative decarboxylation of p-hydroxybenzoic acid (103)[120]. A similar mechanism, based on analogous evidence[121] has been suggested to explain the formation of methoxy-p-benzoquinone (97) and 2,6-dimethoxy-p-benzoquinone (98) in cereals from respectively vanillic acid and syringic acid. It should be similarly noted, however, that both of these latter quinones occur in the plant tissues as quinol glucosides and are not found as free quinones.

Naphthoquinones

Despite some uncertainties due to early observations in this area[122], the biosynthetic pathway to several plant naphthoquinones has been shown to involve (−)-shikimic acid and its metabolites. Thus the nucleus of both juglone (99) and lawsone (100) has been shown to be formed directly from (−)-shikimic acid (107) and a C_3 fragment derived from L-glutamic acid. The pattern of experimental results follows closely that obtained in the case of the bacterial mena-quinones (vitamin K_2).

Several groups have studied the biosynthesis of juglone (99) and lawsone (100)[123–128]. Degradation of radioactive juglone (99) obtained after feeding U-[^{14}C]-(−)-shikimic acid to *Impatiens balsamina* and *Juglans regia* showed that this precursor is readily incorporated into both quinone rings. The ring carbon atoms of shikimic acid provide one of the rings of the quinone and the carboxyl group is incorporated into one or both of the carbonyl

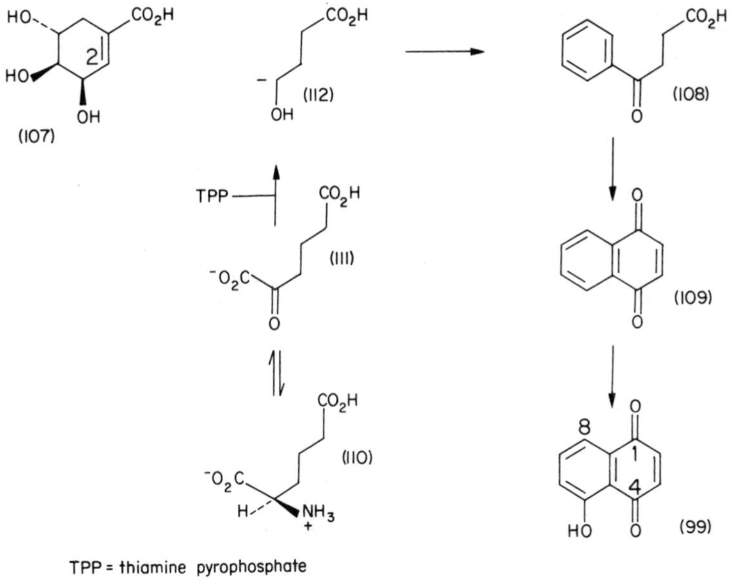

TPP = thiamine pyrophosphate

Figure 6.11. Biosynthesis of juglone[123–127]

groups[123–125]. In the case of juglone (99) it was also shown that the biosynthesis proceeds from (−)-shikimic acid through a symmetrical intermediate in which the two carbonyl groups are indistinguishable: thus [^{14}C]-1,4-naphthoquinone (109), labelled selectively in the quinone ring, was found to be an efficient precursor of the quinone (99). The origin of the remaining three carbon atoms of the naphthoquinone nucleus has, by analogy with the bacterial mena-quinones, been assumed to be C-2, 3 and 4 of L-glutamic acid (110) and the site of attachment has been demonstrated in a number of different ways[123–127] to be at C-2 of (−)-shikimic acid (107), Figure 6.11. Thus 1,6-[^{14}C$_2$]-(±)-shikimic acid fed to *Juglans regia* gave juglone which was degraded by the scheme shown in *Figure 6.12* to give, from C-8, bromopicrin (113) which contained one-quarter of the radioactivity incorporated into the quinone. Similarly 7-[^{14}C]-2[^{3}H]-(−)-shikimic acid gave juglone with almost complete loss of tritium thus indicating in confirmation of the previous result that the three carbon chain is attached at C-2 and that the hydrogen at C-2 of (−)-shikimic acid is eliminated during the biosynthesis[126]. In the case of the bacterial menaquinones, work with bacterial mutants led to the proposition that chorismic or *iso*chorismic acid is the intermediate to which the additional three carbon atoms derived from L-glutamic acid (110) become attached prior to formation of the second ring of the quinone nucleus. However, feeding experiments with G-[^{14}C]-chorismic acid showed[123, 124] that this was a poor precursor of juglone (99) and thus it is not possible, on the present evidence, to sustain the proposal that chorismic acid is also an intermediate in the formation of plant naphthoquinones such as juglone (99) and lawsone (100). [^{14}C]-o-Succinylbenzoic acid (108) was efficiently incorporated into both lawsone and juglone[127] and its formation must therefore be formulated at this stage as the result of a reaction between (−)-shikimic acid (107) and the fragment (112) derived from L-glutamic acid (110) via 2-keto glutaric acid (111) and thiamine pyrophosphate (Figure 6.11).

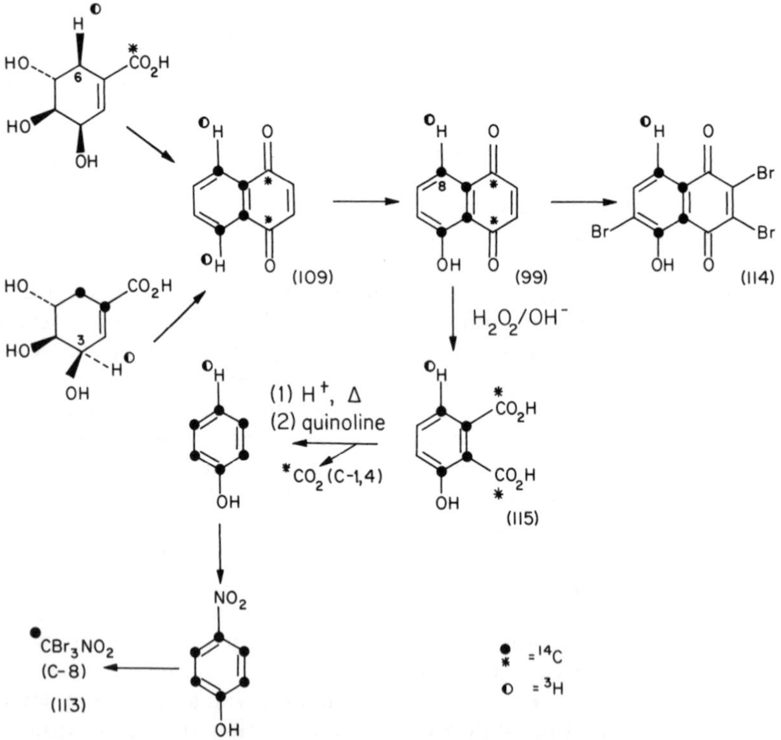

Figure 6.12. Degradation of juglone derived from isotopically labelled (−)-shikimic acid[123-127]

Novel observations have been made in an attempt to clarify the role of 1,4-naphthoquinone (109) as an intermediate in plant naphthoquinone biosynthesis and to investigate the hydroxylation reaction which finally yields juglone. Leduc, Dansette and Azerad[125] administered 3-[^{3}H]-(−)-shikimic acid to *Juglans regia* and obtained juglone (99) which was degraded to the bromo derivative (114) and 3-hydroxyphthalic acid (115). The ratio of the tritium retention in (114) and (115) was unity and this was interpreted[127] in terms of a hydroxylation reaction at the 5-position of 1,4-naphthoquinone which proceeds without an NIH shift but with complete displacement of the hydrogen atom at this position. Zenk and Floss[126] later obtained analogous results which corroborate this explanation and also therefore the idea that 1,4-naphthoquinone (109) is indeed an intermediate in the biosynthesis of juglone (99). Thus 6(s)-6(^{3}H)-7-[^{14}C]-(−)-shikimic acid when fed to *Juglans regia* gave the quinone (99) in which approximately 50 per cent loss of tritium (as measured by the change in ^{3}H : ^{14}C ratio) had occurred. An identical explanation of this result is possible invoking the intermediacy of 1,4-naphthoquinone (109).

On the other hand, Grotzinger and Campbell have described[128] the results of experiments on the biosynthesis of the quinonoid pigment lawsone (100) which show that a symmetrical intermediate such as free 1,4-naphthoquinone (109) cannot be involved in its biosynthesis. Isotopically labelled lawsone (100) was obtained after feeding 3,4-[^{14}C$_2$]-L-glutamic acid (110) and 2-[^{14}C]-sodium acetate to excised stems of *Impatiens balsima*. The biosynthesised lawsone was degraded by treatment with iodine and base to give C-3 of the quinone as iodoform. Oxidation of lawsone (100) to give phthalic acid permitted the total activity at C-(2+3) to be determined and hence by difference the activity at C-2. The results showed that, as predicted, the distribution of radioactivity between C-2 and C-3 was equal when 3,4-[^{14}C$_2$]-L-glutamic acid (110) was the precursor. However, in experiments with 2-[^{14}C]-sodium acetate as the substrate by far the greatest proportion of the radioactivity was localised at C-2 in lawsone (100) and hence it was deduced that a symmetrical intermediate such as 1,4-naphthoquinone (109) is not involved in the biosynthesis of the pigment. The results also clearly demonstrated that the phenolic hydroxyl group found in lawsone (100) is attached to the carbon atom which was originally C-2 in the intermediate *o*-succinylbenzoic acid (108). Grotzinger and Campbell[128] suggested two alternative, and on paper equally plausible, pathways from *o*-succinylbenzoic acid (108) to the quinone lawsone (100). In the first of these, cyclisation of (108) gives 1,4-dihydroxy-2-naphthoic acid and oxidative decarboxylation then gives lawsone. In the second pathway, the intermediate (108) is oxidised to the α,γ-diketo acid and direct oxidative decarboxylation followed by cyclisation then yields lawsone (100). Juglone (99) and lawsone (100) are apparently very similar naphthoquinone pigments which differ only in the position of a phenolic hydroxyl group and which it might be assumed are therefore closely related biogenetically. However, these observations underline the possible dangers in biosynthetic work inherent in making such an extrapolation.

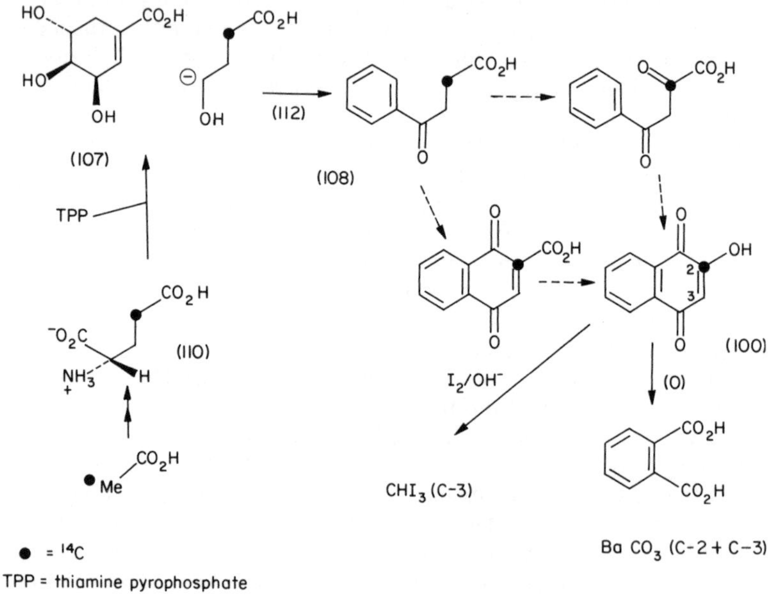

$\bullet$ = ^{14}C

TPP = thiamine pyrophosphate

Ba CO_3 (C-2 + C-3)

Chimaphilin (119) is a dimethylnaphthoquinone found in the Ericaceae and its biosynthesis has been shown to involve L-tyrosine and mevalonic acid[129] which accords neatly with the earlier hypothesis of Robinson[2] and Inouye[130]. Zenk and his collaborators[129] demonstrated that 2-[^{14}C]-L-tyrosine and U-[^{14}C]-tyrosine were efficiently incorporated into chimaphilin (119) in *Chimaphila umbellata*. Degradation showed that the former substrate was incorporated specifically into the quinone methyl group at C-2 and the latter into the quinonoid ring. Similarly, 2-[^{14}C]-mevalonic acid was also specifically incorporated into chimaphilin (119) and over 80 per cent of the ^{14}C label was located in the benzenoid methyl group at C-7.

Inouye suggested[130] that chimaphilin (119) might be formed by cyclisation of a glucoside bound 2-methyl-5-[-3-methylbut-2-enyl]-1,4-benzoquinone (118) and this hypothesis has been strengthened by the isolation of (118) from the genus *Pyrola*[131]. A plausible pathway for the biosynthesis of chimaphilin (119) which embodies

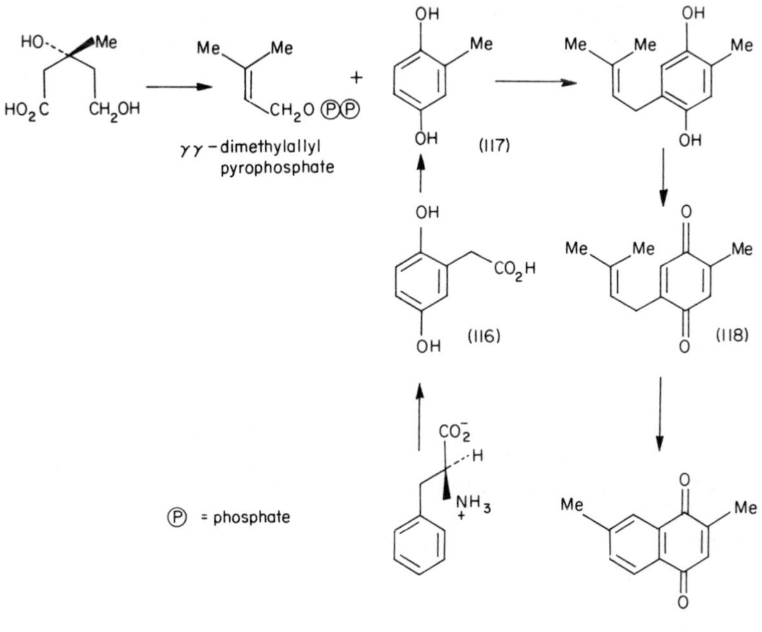

Figure 6.13. Biosynthesis of chimaphilin[129-131]

these ideas and the experimental data of Zenk and his group is shown in *Figure 6.13*. The quinone ring is clearly shikimate in origin and is presumed, on the present evidence, to be derived from L-tyrosine via homogentisic acid (116) and toluquinol (117). This route represents a second mode of synthesis of plant naphthoquinone metabolites from derivatives of the shikimate pathway.

Anthraquinones

Many anthraquinones found in higher plants, such as chrysophanol (120) and emodin (121), are of a polyketide origin[132] and they are often structurally distinguishable by the existence of substituents in the two benzenoid rings of the anthraquinone nucleus. Characteristically the anthraquinones found in the Verbenaceae (teak), Bignoniaceae (*Tabebuia* species) and Rubiaceae (*Galium* species) *almost* invariably carry substituents in only one of the benzenoid rings and, in contrast to those of a fungal origin, they are frequently devoid of a carbon chain[133, 134]. There is strong circumstantial and some experimental evidence to support the proposal[135] that these anthraquinones originate from a naphthalenic precursor to which a branched C_5 chain is attached and which then undergoes cyclisation and oxidation.

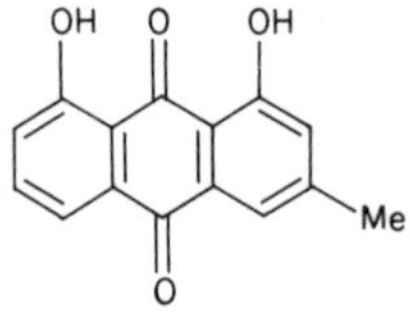

(120; chrysophanol) 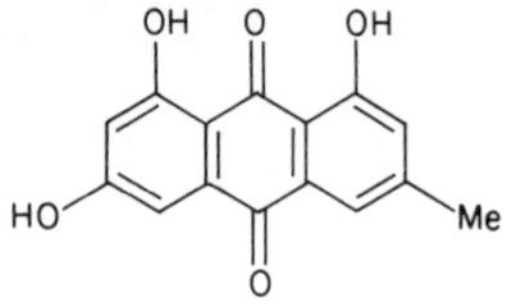   (121; emodin)

In an examination of the heartwood constituents of *Tabebuia avellandae* Burnett and Thomson[131] demonstrated the rather rare co-occurrence of naphthaquinones and anthraquinones. These C_{15} quinones had either a C_{10}-C_5 (naphthoquinone, e.g. lapachol, 122) or C_{14}-C_1 (anthraquinone, e.g. 123) structure and Burnett and Thomson[133, 134] suggested that the implication of this observation is that there exists a close biogenetic relationship between the two groups and in particular that one ring of the anthraquinone pigments, as in the analogous case of the naphthoquinone chimaphilin (119), is derived from mevalonic acid via $\gamma\gamma$-dimethylallylpyrophosphate.

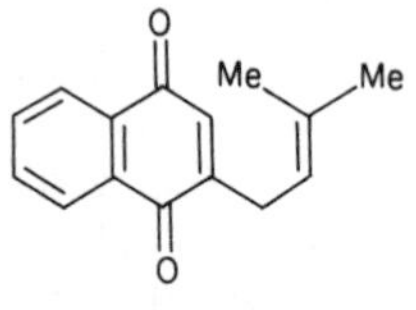

(122; lapachol) 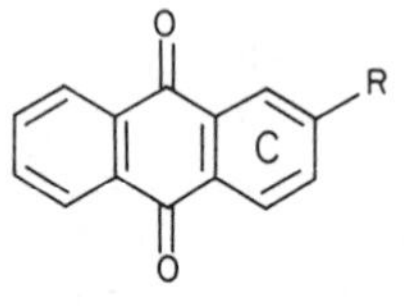   (123; R = CHO, CH_2OAc, CH_2OH, CH_3, CO_2H)

Evidence to support this proposal has been obtained by Burnett and Thomson[133, 134] in work with the madder plant (*Rubia tinctorum*). 2-[^{14}C]-Mevalonic acid was administered to this plant and was incorporated into a series of anthraquinone metabolites such as alizarin (101), purpurin (124) and pseudopurpurin (102). Degradation of the quinones to phthalic acid showed that all the radioactivity was confined to ring C. Decarboxylation of pseudo-purpurin (102) gave purpurin (124) containing one half of the total radioactivity and this is in agreement with the hypothesis outlined above namely that ring C of these quinones is derived by cyclisation of a precursor such as lapachol (122). Burnett and Thomson's[133, 134] observations imply that the two methyl groups in the precursor (122) become equivalent during the biosynthetic process and they differ from those of Zenk and Leistner[136] who demonstrated in a similar type of experiment that the carboxyl group of pseudopurpurin (102) was derived almost exclusively from only one of the two methyl groups of the $\gamma\gamma$-dimethylallyl pyrophosphate (126) precursor. This result is, however, quite analogous to the observations of Zenk and his associates[128, 129] in the case of the naphthoquinone chimaphilin (119) where the C-7 methyl group was shown to be derived predominantly from C-2 of the mevalonic acid intermediate (*Figure 6.13*).

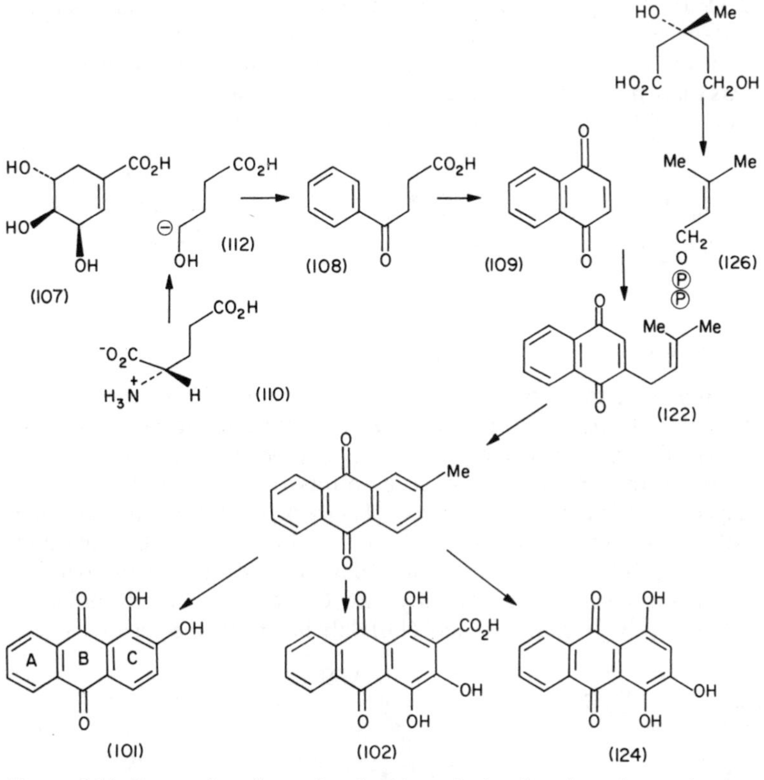

Figure 6.14. Proposed pathway for the biosynthesis of anthraquinones in the Rubiaceae

Investigations of the biosynthesis of alizarin (101) in *Rubia tinctora*[137] have revealed some of the details of the biosynthetic origins of rings A and B in the anthraquinone nucleus. Zenk and Leistner[137] have shown that 1,6-$[^{14}C_2]$-($\pm$)-shikimic acid is incorporated directly into ring A of alizarin (101) and pseudopurpurin (102), by degradation of the labelled metabolites to phthalic acid which retained all the activity of the quinonoid pigment. It was initially surmised that the biosynthetic pathway is thereafter very similar to that recognised for the hydroxynaphthoquinone juglone (99); namely that a C_3 fragment (from L-glutamic acid, 110) is attached to (−)-shikimic acid (107) at C-2 and that this intermediate then leads through o-succinylbenzoic acid (108) to 1,4-naphthoquinone (109), *Figure 6.14*. Evidence for this hypothesis was the finding that both 9,10-[$^{14}C_2$]-1,4-naphthoquinone (109) and 2,3,9,10-[$^{14}C_4$]-1,4-naphthoquinone (109) were specifically incorporated (0.63 per cent and 0.25 per cent respectively) into rings A and B of alizarin (101). The specific manner in which the precursor was incorporated into the quinone was demonstrated[137] by oxidation to phthalic acid (127) and decarboxylation.

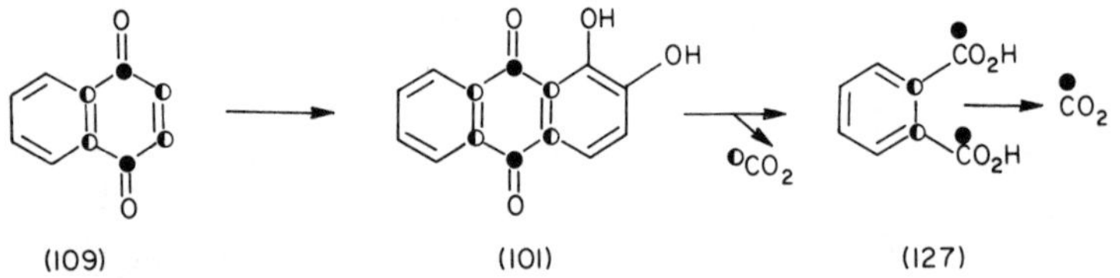

(109)   (101)   (127)

$O\ \bullet\ =^{14}C$

However, the more recent work of Leistner[138] on the biosynthesis of alizarin (101) in *Rubia tinctora*, although it has confirmed the position of o-succinylbenzoic acid (108) and L-glutamic acid (110) as intermediates, has cast doubts on the validity of some aspects of the proposed pathway (*Figure 6.14*) and in particular the role of 1,4-naphthoquinone (109). It was thus noted by Leistner that using 7-[^{14}C]-(−)-shikimic acid (107) as substrate the radioactive label was incorporated directly into C-9 of the alizarin molecule whilst 2-[^{14}C]-L-glutamic acid (110) labelled C-10 specifically. These

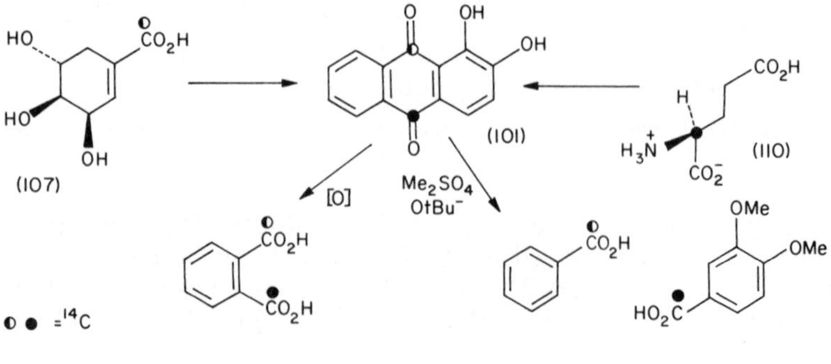

(107)   (101)   (110)

$O\ \bullet\ =^{14}C$

observations clearly suggest that there is *no* symmetrical intermediate such as 1,4-naphthoquinone (109) on the biosynthetic pathway to alizarin and that, although under appropriate conditions this substrate may be incorporated into the quinone[137], it is nevertheless not a true intermediate. Clearly, further more detailed investigations are required to establish the nature of the intermediates after *o*-succinyl benzoic acid (108) along the pathway. In addition evidence is still necessary to determine the stage at which prenylation of the naphthenoid precursor takes place and also to elucidate the sequence of events leading from the prenylquinone (122) to the anthraquinone structure.

6.2.2 BETALAIN PIGMENTS

The betalains are a distinctive group of red–violet (betacyanins) and yellow (betaxanthins) pigments which only occur, as far as is known, in flowering plants belonging to ten families usually allied in the order Centrospermae. They appear to replace in these plants the more widespread and common plant pigments the anthocyanins. Significant advances were only made in the chemistry of the betalains after the preparation in 1957 of the first crystalline samples of the pigment betanin[139, 140] and recent structural work[141–143] has revealed the true chemical nature of these pigments (128–131). The expressions betacyanin and betaxanthin used to describe these pigments reflect the long held, but erroneous, belief that there was a direct chemical relationship between them and the anthocyanins and flavonoids.

Experiments on the biosynthesis of the betalains have centred around the biogenesis of betanidin and the hypothesis that it is assembled from two molecules of L-dopa (3′,4′-dihydroxyphenylalanine)[140, 144, 145]. This idea was tentatively formulated[140, 146, 147]

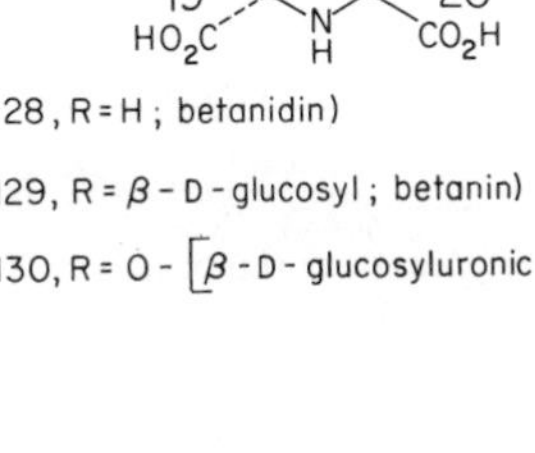

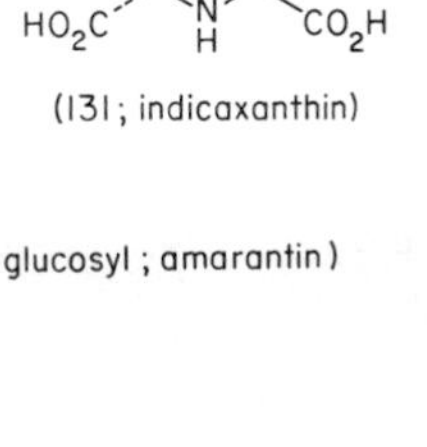

(128, R = H ; betanidin)

(131; indicaxanthin)

(129, R = β - D - glucosyl ; betanin)

(130, R = O - [β - D - glucosyluronic acid] - β - D - glucosyl ; amarantin)

in terms of the biogenetic scheme for the betalains shown in *Figure 6.15*. The dihydropyridine fragment of the pigments, betalamic acid (133), was suggested as being derived by oxidative ring cleavage of L-dopa (132) followed by ring closure on to the amino nitrogen. Condensation of betalamic acid (133) with L-proline would then afford a typical betaxanthin (e.g. 131) or condensation with L-cyclodopa (134), formed by oxidative cyclisation of L-dopa, would give betanidin (128).

Several groups of workers[149–152] have provided evidence to support these suggestions and have shown ^{14}C labelled dopa to be an excellent precursor of betanidin and its glycosides betanin and amarantin in several plant species, although in most cases no further chemical degradation of the pigments was undertaken in order to ascertain the positions of the labelled atoms. D L-1-[^{14}C]-

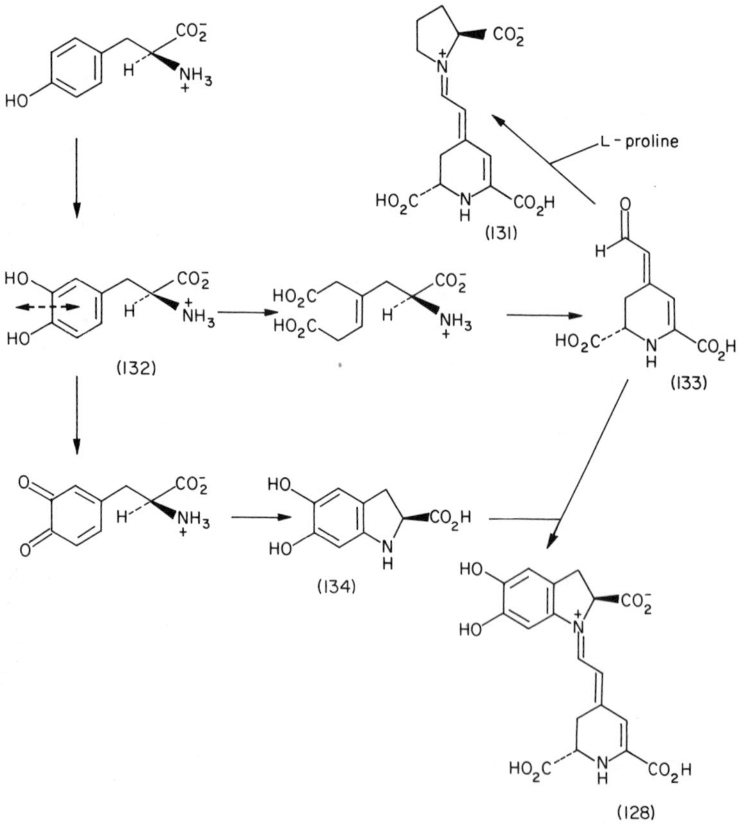

Figure 6.15. Proposed pathway of biosynthesis of the betalains[140, 146, 147]

Dopa and D L-2-[^{14}C]-dopa were both readily incorporated ($\sim$ 5 per cent) into the glucoside betanin (129) in the ripening fruits of *Opuntia decumbens* and *Opuntia bergeriana*[153]. Studies with beet seedlings afforded similar results but with much lower incorporations of radioactivity. The betanin (129) formed from D L-1-[^{14}C]-dopa contained 95 per cent of its radioactivity in the three carboxyl groups. Base catalysed exchange of the radioactive betanin (129) and L-proline gave[154] cyclodopa (134), isolated as its triacetylmethyl ester, and indicaxanthin (131) which contained 90 per cent of the total radioactivity of the original betalain. It was concluded from these observations that dopa is an efficient precursor of the betalamic acid fragment (133) of betanidin but not of the cyclodopa unit (134) in *Opuntia* species.

Schulte and his collaborators similarly demonstrated[146] that both 1-[^{14}C]-L-tyrosine and 2-[^{14}C]-L-tyrosine but not 1-[^{14}C]-L-phenylalanine were readily incorporated into betanin (129) in *Beta vulgaris* and *Kochia scoparia*. Utilising ^{15}N labelled L-tyrosine these workers showed that the intact amino acid (including the amino nitrogen) was incorporated into both parts of the pigment. Chemical degradation of betanin (129), isotopically labelled after feeding 1-[^{14}C]-L-tyrosine and 2-[^{14}C]-L-tyrosine, showed in contrast to the results of Dreiding's group[154] that the betalamic acid (133) portion of betanidin (128) contained only one-third of the total radioactivity of the pigment. These differences of experimental observations may result from the use of quite different experimental conditions or they may reflect the different patterns of metabolism operative in these plants.

It is assumed in the biogenesis of the betalains (*Figure 6.15*) that the dihydropyridine structural fragment (133) is generated by cleavage and recyclisation of L-dopa (132). Dreiding and his associates[155–158] have now provided definitive evidence on this point and they have delineated more precisely the chemical pathway which is followed. Monodecarboxylation of betanidin[150], isolated after the administration of 1-[^{14}C]-D L-dopa to *Opuntia decumbens*, gave the dicarboxylic acid (135) by loss of C-19 as carbon dioxide and concomitant migration of the C-17 to C-18 double bond to C-14 to C-15. Purification of the derived dicarboxylic acid (135) showed it to contain only 14 per cent of the total activity of the betanidin (128) and, taken in conjunction with the earlier results of Dreiding's group[153], this showed that almost all the radioactivity of the betalamic acid (133) was lost as carbon dioxide. Hence it was concluded that the carboxyl group of L-dopa becomes the α-amino carboxyl group of betalamic acid (133) and the betalains (*Figure 6.15*).

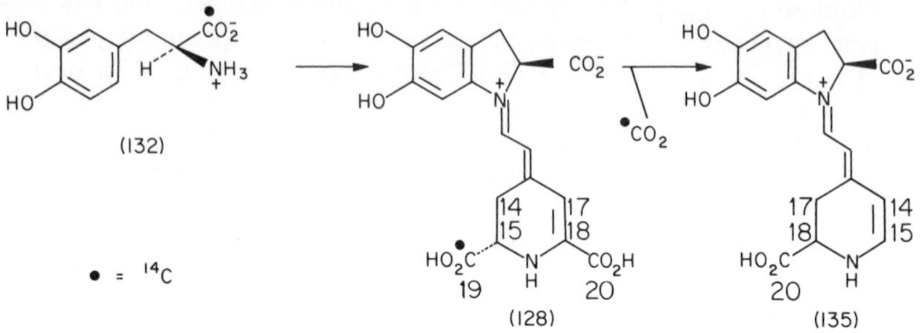

In principle, L-dopa can form the betalamic acid carbon skeleton (133) by initial cleavage of the aromatic ring either at the bond *between* the two phenolic hydroxyl groups (*Figure 6.15*) or at a bond *adjacent* to the catechol grouping. Both modes of cleavage have been characterised as methods of breakdown of the aromatic ring system in several micro-organisms[157–159]. In the formation of betalamic acid (133) from L-dopa in young cactus fruit (*Opuntia decumbens*), Dreiding and Fischer[156] have shown that cleavage occurs *adjacent* to the catechol grouping, between a hydroxylated and non-hydroxylated carbon atom. 3′,5′-[^{3}H$_2$]-L-Tyrosine (136) was incorporated into betanin (129, 0.03–0.08 per cent) in *Opuntia decumbens* and base exchange of the pigment with L-proline gave[154] indicaxanthin (131) and cyclodopa (134)—isolated as its triacetylmethyl ester. The ratio of the derived radioactivities in the two fragments (131) and (134) ranged between 7 and 16 to 1 and indicated that the biosynthesised betalamic acid (133) contained the majority of the tritium originally incorporated into the betanin. Dreiding and Fischer rationalised this result in the following way (*Figure 6.16*): the 3′,5′-[^{3}H$_2$]-L-tyrosine is first converted to 5′-[^{3}H]-L-dopa (132) and since no NIH shift would be expected to occur in this transformation[160], the L-dopa would contain statistically half the tritium of the precursor. Cleavage of the aromatic nucleus of L-dopa *adjacent* to the phenolic hydroxyl group (pathway *a*, *Figure 6.16*) would then lead to the formation of betalamic acid (133) with the tritium localised at the aldehydic centre and hence at C-11 of betanin (129). Fischer and Dreiding also argued that the alternative mode of fission of the aromatic nucleus, *between* the phenolic hydroxyl groups (pathway *b*, *Figure 6.16*), would lead to complete loss of tritium in the subsequent conversion to betalamic acid (133).

Results in agreement with this mode of cleavage of the L-dopa molecule have also been obtained by Impellizzeri and Piattelli[161]

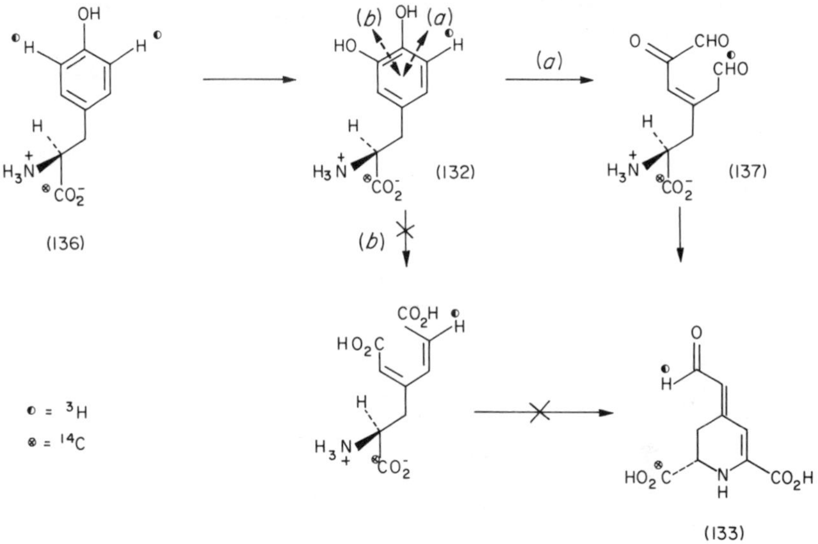

Figure 6.16. Formation of betalamic acid from 1-[^{14}C]-3′,5′-[^{3}H$_2$]-L-tyrosine in
Opuntia decumbens[156, 161]

in a study of the biosynthesis of indicaxanthin (131) in *Opuntia ficus-indica*. Doubly labelled 1-[^{14}C]-3′,5′-[^{3}H$_2$]-L-tyrosine was administered to the mature fruit of *Opuntia ficus-indica* immediately after collection and indicaxanthin (131) isolated after 24 hours. The results of experiments carried out with different ^{3}H:^{14}C ratios in the precursor, all gave indicaxanthin with a ^{3}H:^{14}C ratio approximately half that of the originally administered L-tyrosine. All the radioactivity in the indicaxanthin (131) was assumed to reside in the betalamic acid fragment (133) of the pigment and proof of this assumption was obtained in the case of the ^{14}C label by degradation to inactive L-proline and ^{14}C-labelled pyridine-2,4,6-tricarboxylic acid (derived from the betalamic acid). The change of ^{3}H:^{14}C ratio of the precursor amino acid on incorporation into indicaxanthin is exactly that predicted by the route of formation of the dihydropyridine (133) shown in *Figure 6.16*, pathway *a*, and this result therefore confirms that of Dreiding and Fischer[156] obtained with the pigment betanin. These results leave as yet unanswered the questions whether the cyclisation of the ring cleavage product (137) to the dihydropyridine (133) takes place before or after the condensation with L-cyclodopa (134), (or its glucoside), or L-proline to give the betalain pigment, and whether the ring closure is spontaneous or is enzymically controlled.

6.2.3   GLUCOSINOLATES AND CYANOGENIC GLYCOSIDES

Two small but distinctive groups of natural products found in higher plants which show a clear and definite relationship to their biosynthetic precursors, the α-L-amino acids, are the cyanogenic glycosides (138)[162, 163] and the mustard oil glucosides or glucosinolates (139)[163, 164, 165].

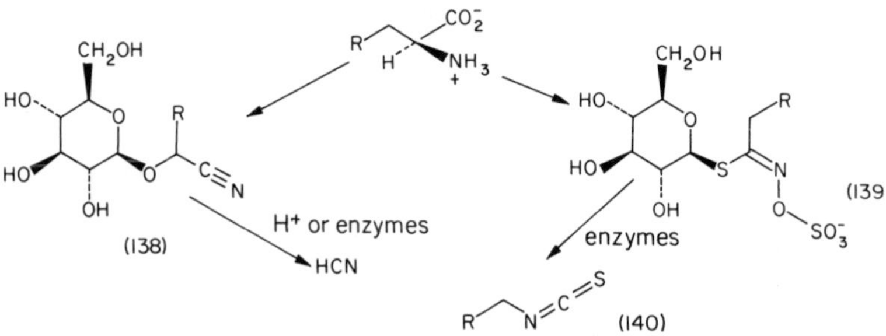

A characteristic feature of the cyanogenic glycosides is their ability to release hydrocyanic acid by hydrolysis and this process is often initiated by enzymes when the tissues which contain them are damaged by mechanical or other means. One of the principal groups of cyanogenic glycosides are those related to the amino acids L-phenylalanine and L-tyrosine in which the aglycone is derived from 2-hydroxy-2-phenylacetonitrile and its derivatives—the so-called amygdalin group[162]. Typical members are amygdalin itself (141), discovered in 1830 by Robiquet and Boutron-Chalard, and the diastereoisomeric pairs prunasin (142)[166] and sambunigrin (143)[167], and taxiphyllin (114)[168] and dhurrin (145)[169]. A further interesting member of this class is zierin (146)[170] which is formally related to the uncommon amino acid *m*-tyrosine.

The aglycones of the cyanogenic glycosides are derived biosynthetically from the closely related L-amino acids and this has been amply substantiated by numerous experiments for those of the amygdalin class[162, 163, 171–173]. It follows from this work that the α-amino acid carboxyl group is lost at some stage during the biosynthetic process but that the remaining $C_6 . C_2 . N$ skeleton remains intact. Thus it was observed that L-tyrosine doubly labelled with ^{14}C and ^{15}N was readily transformed to dhurrin (145), the cyanogenic glycoside of *Sorghum vulgare*, with a comparable but slightly less efficient incorporation of ^{15}N than ^{14}C into the product[174]. Analogous results have also been obtained with linamarin (147) in flax[175] and it has been concluded therefore that

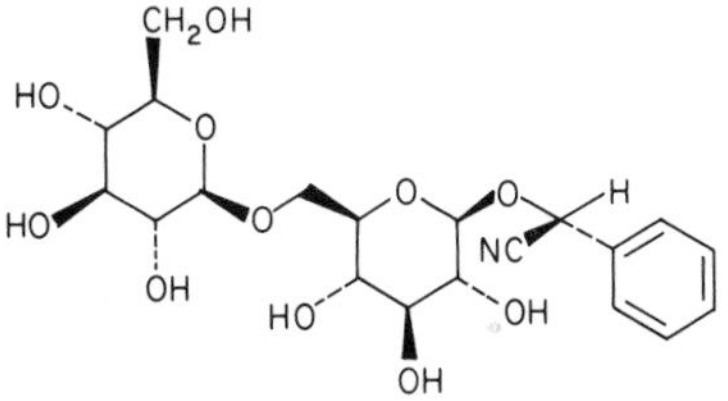

(141; amygdalin)

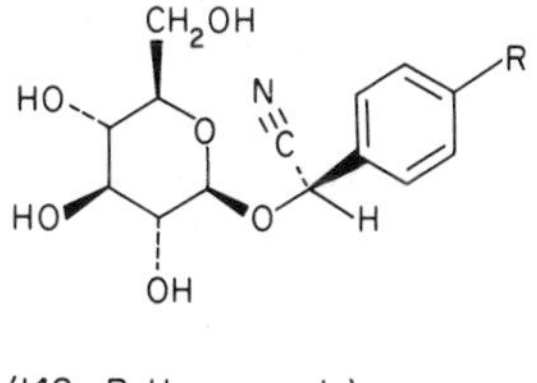

(142 , R=H ; prunasin)
(144, R=OH; taxiphyllin)

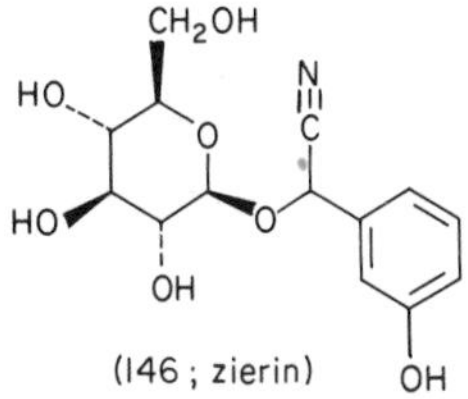

(146 ; zierin)

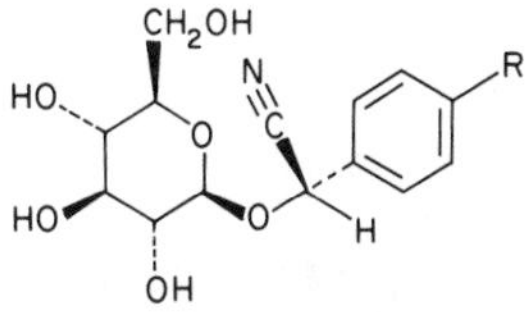

(143 , R=H; sambunigrin)
(145 , R = OH ; dhurrin)

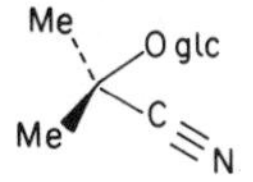

(147; linamarin)

the nitrogen of the nitrile function is directly derived biosynthetically from the α-amino group of the amino acid.

A great deal of the subsequent effort in this field has been devoted to the problem of establishing the nature of the remaining nitrogen containing intermediates on the biosynthetic pathway to the cyanogenic glycosides. Conn and his collaborators[176, 177] have for example shown that the corresponding ^{14}C labelled α-oximino acid, aldehyde oxime (148), nitrile (149) and α-hydroxynitrile (150) derived from L-phenylalanine are readily incorporated into the glucoside prunasin (142) in the leaves of cherry laurel (*Prunus laurocerasus*). On the basis of these observations, they proposed a pathway of biosynthesis (*Figure 6.17*) for the formation of prunasin (142) from L-phenylalanine. Subsequently, further evidence was derived by Tapper, Zilg and Conn[178] to show that an alternative pathway to mandelonitrile (150) exists through the 2-hydroxyaldoxime (151).

Analogous results were obtained in a study of the biosynthesis of linamarin (147) in flax[176, 177] and from this plant an enzyme has been isolated which specifically catalyses the last step of glucose transfer[179, 180] in the biosynthetic pathway.

Mustard oil glucosides (139) are found primarily in the plant family Cruciferae and they readily undergo enzymic hydrolysis

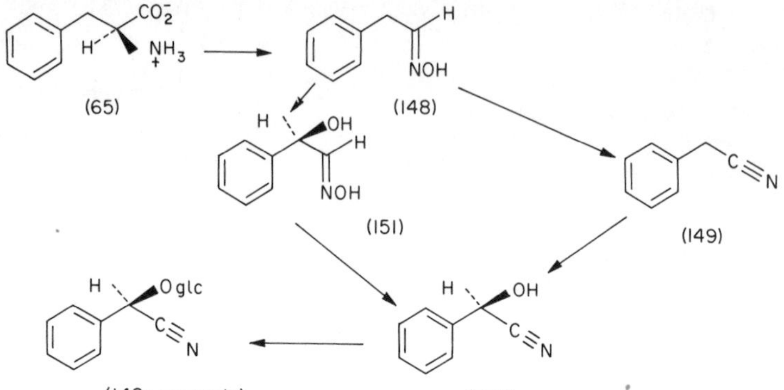

Figure 6.17. Biosynthesis of prunasin in *Prunus laurocerasus*[176-178]

followed by an intramolecular Lossen type rearrangement to give the corresponding isothiocyanates (140), D-glucose and sulphate ion. Most isothiocyanates are pungent smelling substances with a sharp taste and hence they are readily recognisable in plant materials. Like the cyanogenic glycosides the side chains of many of the natural mustard oil glucosides, or glucosinolates, are identical with the common α-L-amino acids and this observation led to an early postulate[181, 182] of a biogenetic relationship. Glucosinolates which bear a formal structural relationship to L-phenylalanine (152)[165, 183], L-tyrosine (153)[165, 183], L-dopa (154)[184] and L-tryptophan (155)[185] have been found in plants as have two compounds which appear to be derived from the unusual amino acid *m*-tyrosine (156 and 157)[186].

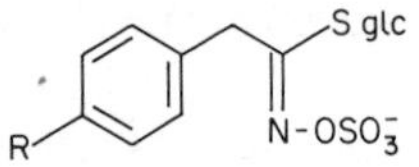

(152, R=H; glucotropaeolin)
(153, R=OH and OMe)

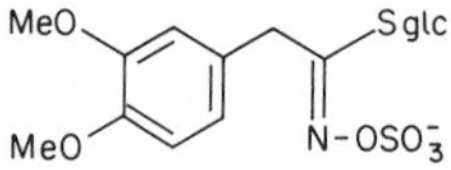

(154)

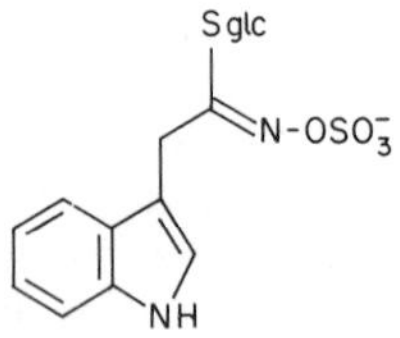

(155)

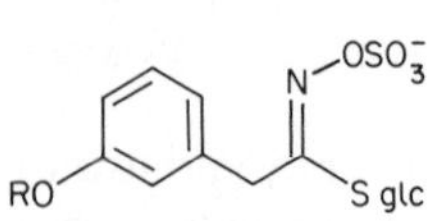

(156, R=H)
(157, R=Me)

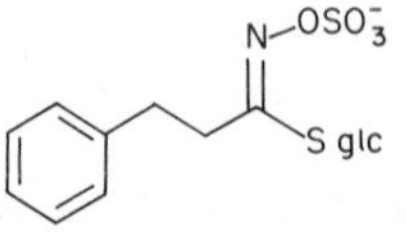

(158, gluconasturtiin)

Certain glucosinolates are, however, less clearly derived from the natural amino acids and of particular interest from the viewpoint of aromatic amino acid metabolism in higher plants are the 2-arylethyl glucosinolates such as gluconasturtiin (158).

The biosynthetic pathways to the glucosinolates, on the basis of the experimental data now available, show a striking resemblance to those of the corresponding cyanogenic glycosides. The evidence that the corresponding L-amino acids are direct precursors of the mustard oil glucosides has been obtained from orthodox experiments with isotopically labelled precursors[187, 188]. Thus $1-[^{14}C]$-D L-phenylalanine and $2-[^{14}C]$-D L-phenylalanine fed to the garden nasturtium (*Tropaeolium majus*) gave glucotropaeolin (152). Degradation to give carbon dioxide and benzylamine showed that the $C_6 . C_2$ carbon skeleton of glucotropaeolin (152) was derived directly from L-phenylalanine without rearrangement. Experiments with doubly labelled L-phenylalanine (^{14}C, ^{15}N) also demonstrated that the nitrogen atom of the amino acid was preserved in the biosynthetic transformation to the glucosinolate, and as in the biosynthesis of the cyanogenic glycosides, the corresponding aldoxime (148) serves as a highly efficient precursor of the natural product[189, 190].

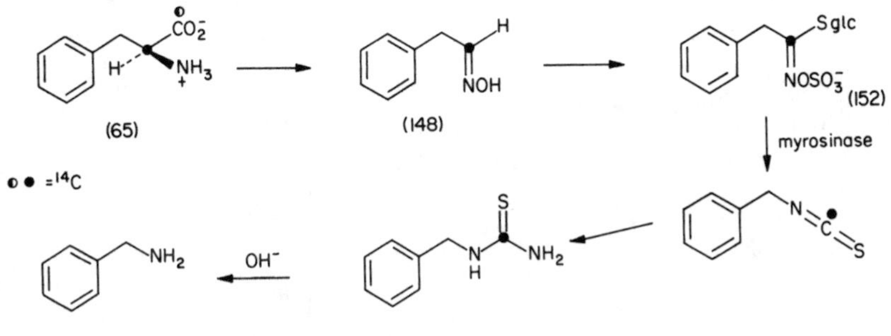

Substantial progress has been made by Underhill and his associates[191-194] in the elucidation of the intermediate stages of biosynthesis between the aldoxime (148) and glucotropaeolin (152) in *Tropaeolum majus*. Feeding experiments[191, 192] have established phenylacetothiohydroximate (162) as a precursor of benzyl glucosinolate (152) and this compound has also been shown to be present in *Tropaeolum majus*. Subsequently, Underhill[193] isolated an enzyme preparation from the same plant which catalyses the formation of desulphobenzylglucosinolate (163)—the penultimate step in the pathway—by transglycosylation of the thiohydroximate (162) with UDPG. Ettlinger and Kjaer[164] first noted that thiohydroximic acids such as (162) may be formed by a base catalysed reaction between thiols and the corresponding primary nitro compound (160) in its *aci*-tautomeric form (159) and Underhill[194] has confirmed this hypothesis at the biological level by isolation and feeding experiments. Thus 1-nitro-2-phenylethane (160) was demonstrated to be present in extracts of *Tropaeolum majus*. Feeding experiments also showed that the nitro compound (160), labelled with ^{14}C at position 1, was incorporated (dilution 130) into the benzylglucosinolate (152) and trapping experiments showed that it could also be derived biosynthetically from the aldoxime (148)[194]. These latter experiments not only confirmed the earlier suggestion of Gottleib and Magalhaes[195] that 1-nitro-2-phenylethane (160), which also occurs in two species within the family Lauraceae and in the fruits of *Dennettia tripeta*[196, 197], is derived from L-phenyla-lanine, but also allowed a detailed biogenetic scheme (*Figure 6.18*) to be proposed by Underhill and his group[194] for benzylglucosino-late (152). The origin of the sulphur atom of the thiohydroximate (162) remains a subject for speculation but Wetter and Chisholm[198] have shown that the sulphur atom of the amino acid L-cysteine is a very efficient source. Underhill proposed in his biogenetic scheme

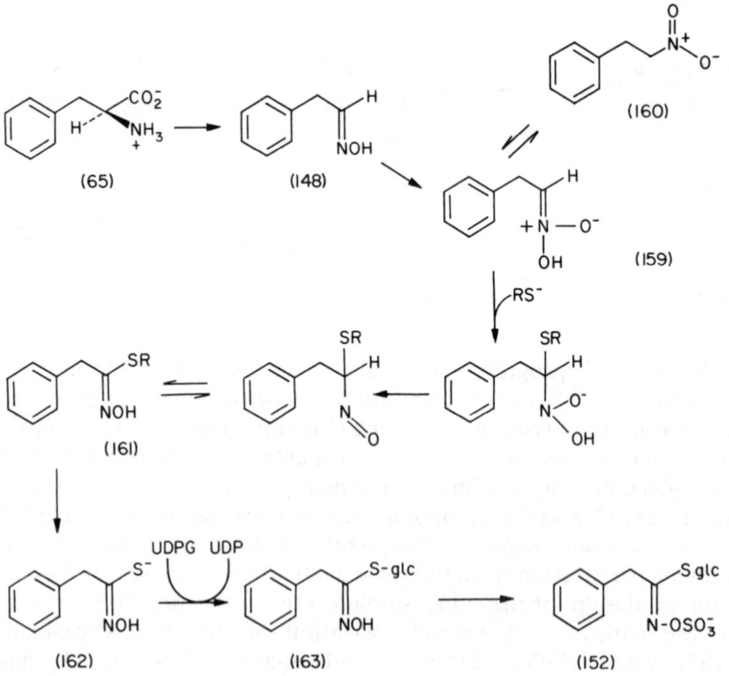

Figure 6.18. Proposed mechanism for the biosynthesis of benzylglucosinolate (glucotropaeolin) in *Tropaeolum majus*

that the substituted thiohydroximic acid (161, R = CH_2.CH_2.CH-(NH_3^+).CO_2H) was first formed from L-cysteine and the *aci*-nitro compound (159) and was then transformed to (162) by the intervention of a C-S lyase such as cystathionase[199, 200] or related enzymes.

A notable feature of the chemistry and biochemistry of the glucosinolates is the occurrence of homologues of the aglycones each differing by a single methylene group[201]. Knowledge of the biosynthesis of glucosinolates of this type, with side chains longer than those of the common α-L-amino acids, stems principally from the work of Underhill[188, 202–205] with 2-phenylethylglucosinolate (158, gluconasturtiin) in water cress (*Nasturtium officinalis*) and *Rorippa nasturtium-aquaticum*. Underhill[202] showed that both 2-[^{14}C] and 3-[^{14}C]-L-phenylalanine and 2-[^{14}C]-sodium acetate were readily and specifically incorporated, with loss of their carboxyl groups, into the aglycone of the thioglucoside nasturtiin (158). Degradation of the natural product showed that the methyl group of acetate provided the thiohydroximate carbon and the remaining C_6.C_2 fragment was furnished directly by the amino acid. However, neither of the substrates (acetate or L-phenylalanine) was as efficient a precursor of the aglycone of gluconasturtiin (158) as L-γ-phenylbutyrine (166, 2-amino-4-phenylbutyric acid) which when labelled with ^{14}C at C-2 or C-3 was utilised with very high efficiency (up to 40 per cent) as a specific precursor. Doubly labelled (2-^{14}C, ^{15}N) L-γ-phenylbutyrine (^{14}C:^{15}N = 1.82) was incorporated into gluconasturtiin (158) without appreciable change in the ^{14}C:^{15}N ratio and this observation demonstrated that the amino nitrogen was preserved in the transformation to the thioglucoside.

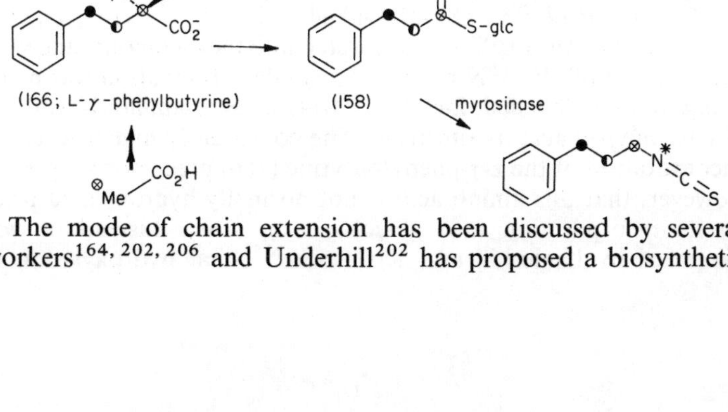

(166; L-γ-phenylbutyrine)   (158)   myrosinase

The mode of chain extension has been discussed by several workers[164, 202, 206] and Underhill[202] has proposed a biosynthetic pathway (*Figure 6.19*) analogous to the well authenticated pathways for the conversion of L-valine to L-leucine and L-aspartic acid to L-glutamic acid. In this scheme the aromatic amino acid is transformed to phenylpyruvic acid (164) by transamination and the latter then undergoes chain extension by condensation with acetate. In support of this pathway, Underhill has shown[202] that appropriately labelled [^{14}C]-3-phenylpropionaldehyde oxime (167) and 3-benzylmalic acid (165) were incorporated specifically and without randomisation of the label into gluconasturtiin (158).

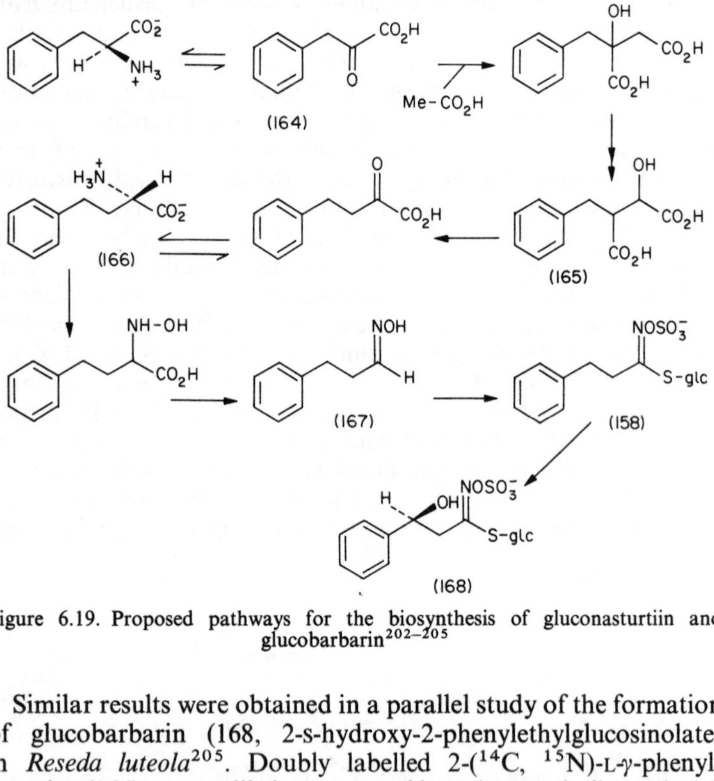

Figure 6.19. Proposed pathways for the biosynthesis of gluconasturtiin and glucobarbarin[202-205]

Similar results were obtained in a parallel study of the formation of glucobarbarin (168, 2-s-hydroxy-2-phenylethylglucosinolate) in *Reseda luteola*[205]. Doubly labelled 2-(^{14}C, ^{15}N)-L-γ-phenylbutyrine (166) was readily incorporated into the metabolite and with little change of ^{14}C:^{15}N ratio and a similar observation was made using doubly labelled 2-(^{14}C, ^{15}N)-(2s, 4s)-2-amino-4-hydroxyphenyl butyric acid as substrate. The consistently higher levels of incorporation of the L-γ-phenylbutyrine (166) precursor suggested, however, that this amino acid is not normally hydroxylated prior to its incorporation into glucobarbarin (168). Instead, it was proposed by Underhill and Kirkwood[205], that hydroxylation of gluconasturtiin (158) occurs as the final step in the biosynthesis of glucobarbarin (168) and in support of this proposal it was noted that G-[^{14}C]-2-phenylethylglucosinolate (158) was an efficient precursor of its hydroxylated analogue (168).

6.2.4 INDOLE ACETIC ACID

Plant hormones are substances which regulate a particular aspect of plant growth and which are produced by the plant itself. Among the well-established plant hormones, the group known as auxins and containing compounds chemically related to indole acetic acid (169) have been widely studied[207]. They have characteristic physiological properties which serve to define them and the most fundamental of these is their capacity to promote cell elongation. Specific bioassays are based on an evaluation of this property[208]. The mode of action of indole acetic acid is still not clear, but one possibility which has some support[209] is that the auxin causes the scission of the cross links between the cellulose microfibrils in plant cell walls. The breakage is most probably enzyme controlled and it

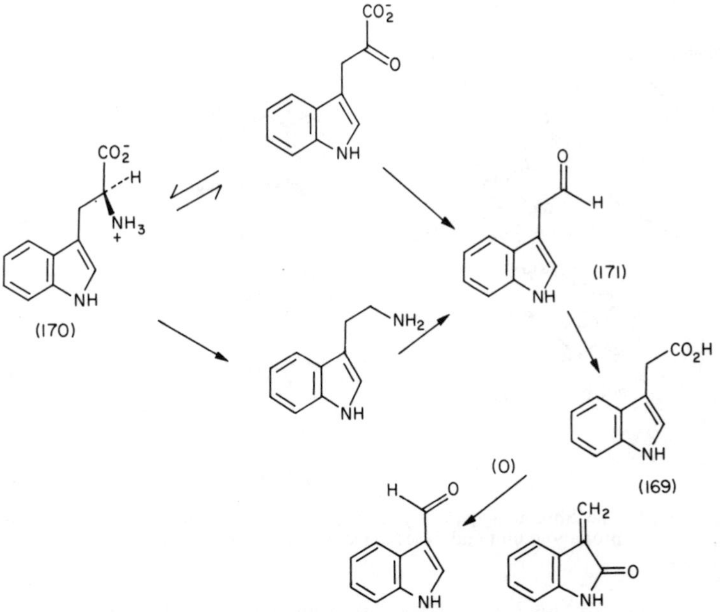

Figure 6.20. Biosynthesis and catabolism of indole acetic acid

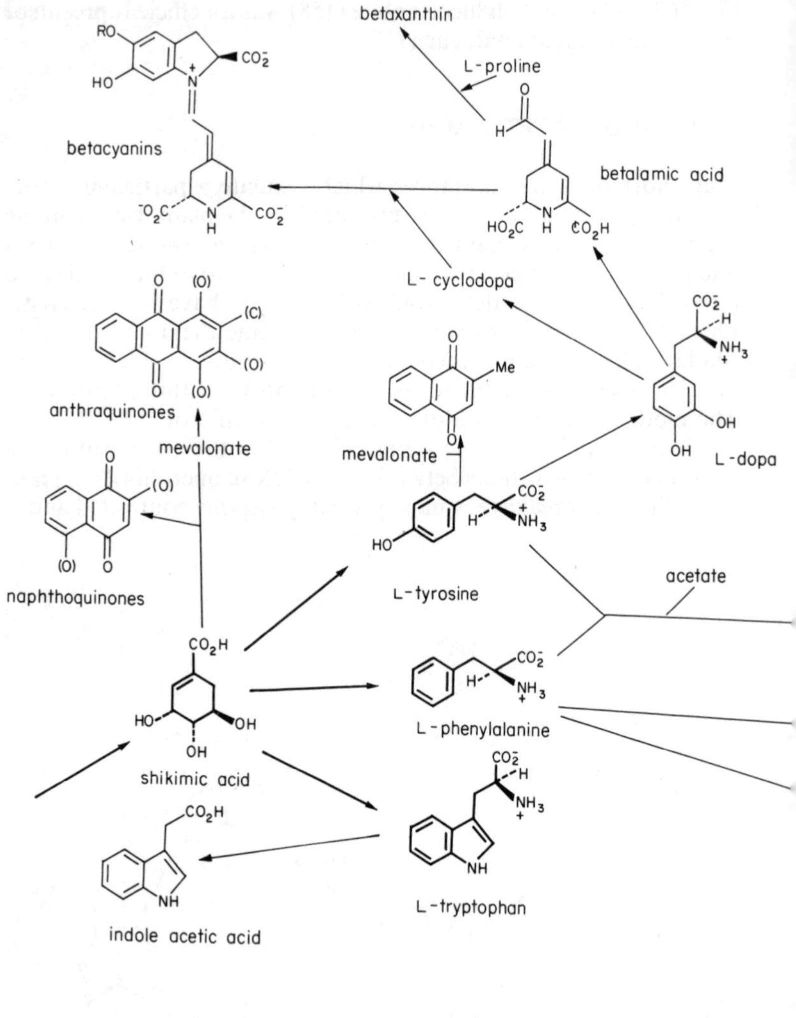

Figure 6.21. The Shikimate pathway in higher plants: extension of the phenylpropanoid unit and miscellaneous metabolites: summary

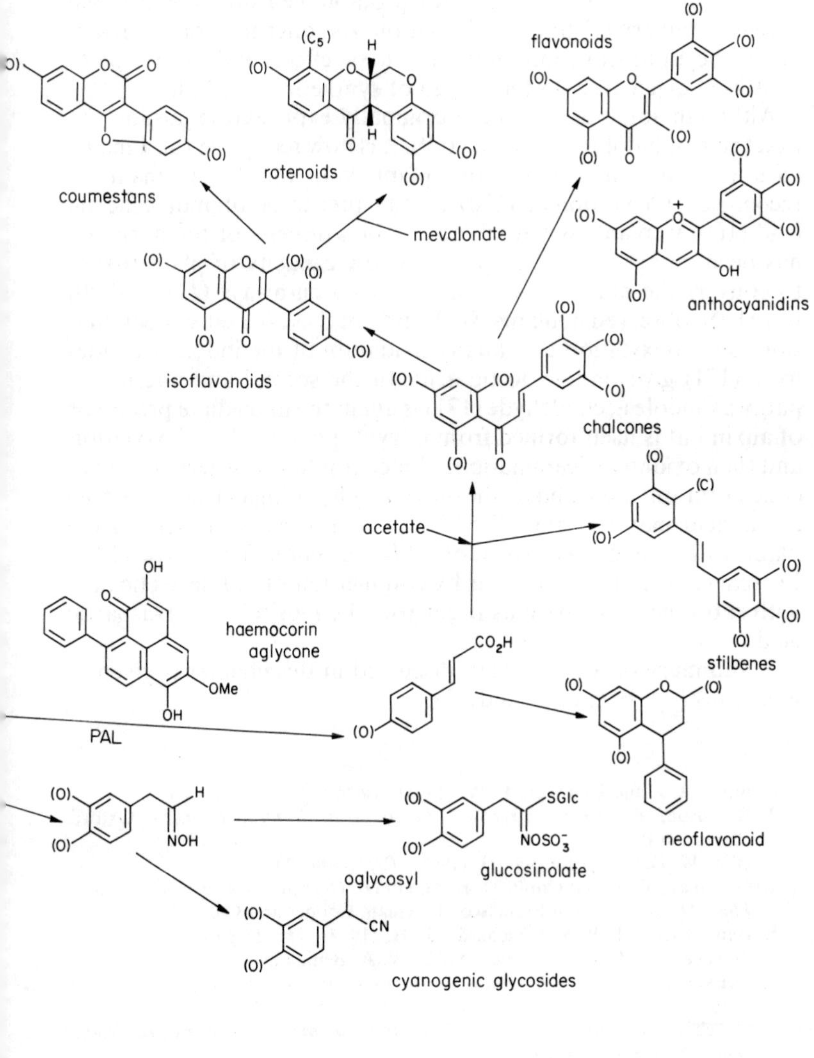

coumestans

rotenoids

mevalonate

flavonoids

anthocyanidins

isoflavonoids

chalcones

acetate

haemocorin
aglycone

stilbenes

PAL

neoflavonoid

glucosinolate

oglycosyl

cyanogenic glycosides

has been suggested that the synthesis or activation of these enzymes may be regulated by auxin. At low concentrations (10^{-5} to 10^{-8}M) auxins may also promote cell division, cambial activity, fruit growth and root initiation. As concentrations increase, auxins begin to show inhibitory and toxic effects and this property is utilised extensively in the design of synthetic weed killers.

Although some doubts were originally expressed regarding the exact role of indole acetic acid (169), it is now recognised as a major natural auxin throughout the plant kingdom[209]. Thimann[210] recognised L-tryptophan (170) as the precursor of indole acetic acid (169) in plants but work on the biosynthesis of the hormone has been hampered by the generally low capacity of plant tissues to convert the amino acid to auxin. Two pathways (*Figure 6.20*) have been observed in plants. In the first of these oxidative deamination, decarboxylation and finally oxidation of the indole acetaldehyde (171) gives indole acetic acid. In the second and alternative pathway indole acetaldehyde (171) is again the immediate precursor of auxin but is itself formed from L-tryptophan by decarboxylation and then oxidative deamination. Under conditions where the auxin concentration rises above the normal physiological level, indole acetic acid may be catabolised by phenoloxidases, present in the plant, to give methylene oxindole (172) or indole-3-aldehyde (173). Detoxification may also occur by conjugation of indole acetic acid with substances as varied as D-glucose, the inositols and L-aspartic acid.

A summary of the pathways discussed in this chapter is given in *Figure 6.21* on pages 292 and 293.

REFERENCES

1. Birch, A. J. and Donovan, F. W. (1953). *Austral. J. Chem.*, **6**, 360
2. Robinson, R. (1955). *Structural Relations of Natural Products*. Oxford, Clarendon Press
3. Ollis, W. D. and Grisebach, H. (1961). *Experentia*, **11**, 1
4. Geissman, T. A. and Crout. D. H. G. (1969). *Organic Chemistry of Secondary Plant Metabolism*. San Francisco; Freeman, Cooper and Co.
5. Hendrickson, J. B. and Richards, J. H. (1964). *The Biogenesis of Steroids, Terpenes and Acetogenins*. New York; W. A. Benjamin
6. Geissman, T. A. (1964) *Biogenesis of Natural Compounds*. Edited by P. Bernfeld. Oxford; Pergamon
7. Grisebach, H. (1967). *Biosynthetic Patterns in Micro-organisms and Higher plants*. New York; Wiley
8. Braga deOliveira, A., Gottlieb, O. R., Ollis, W. D. and Rizzini, C. T. (1971). *Phytochemistry*, **10**, 1863
9. Gottlieb, O. R. and Ollis, W. D. (1968). *Chem. Commun.*, 1396
10. Ollis, W. D. (1966). *Experentia*, **22**, 777
11. Carpenter, I., Locksley, H. D. and Scheinmann, F. (1969). *Phytochemistry*, **8**, 2013

Higher Plants: Extension of the Phenylpropanoid Unit 295

12. Aneja, R., Mukerjee, S. K. and Seshadri, T. R. (1958). *Tetrahedron*, **4**, 256
13. Ollis, W. D. and Sutherland, I. O. (1961). *Chemistry of Natural Phenolic Compounds*, p. 74. Edited by W. D. Ollis. Oxford; Pergamon
14. Pelter, A. and Hansel, R. (1968). *Tetrahedron Lett.*, 2911
15. Wagner, H., Seligman, O., Horhammer, L., Seitz, M. and Sonnenbichler, J. (1971). *Tetrahedron Lett.*, 1895
16. Grisebach, H. (1957). *Z. Naturforsch.*, **12B**, 227
17. Grisebach, H. (1958). *Z. Naturforsch.*, **13B**, 335
18. Underhill, W. E., Watkin, J. E. and Neish, A. C. (1957). *Canad. J. Biochem. Physiol.*, **35**, 219, 229, 230
19. Geissman, T. A. and Swain, T. (1957). *Chem. and Ind.*, 984
20. Shibata, S. and Yamazaki, M. (1958). *Pharm. Bull.* (Tokyo), **6**, 42
21. Zaprometov, M. N. (1962). *Biokhimiya*, **27**, 366
22. Comte, G., Ville, A., Zwinglestein, G., Favre-Bonvin, J. and Mentzer, C. (1960). *Bull. Soc. Chim. Biol.*, **42**, 1079
23. Hutchinson, A., Taper, C. D. and Towers, G. H. N. (1959). *Canad. J. Biochem. Physiol.*, **37**, 901
24. Grisebach, H. and Barz, W. (1969). *Naturwiss.*, **56**, 538
25. Grisebach, H. (1965). *Chemistry and Biochemistry of Plant Pigments*, p. 279. Edited by T. W. Goodwin. London and New York; Academic Press
26. Grisebach, H. (1968). *Recent Advances in Phytochemistry*, p. 379, Vol. I. Edited by T. J. Mabry, R. E. Alston and V. C. Runeckles. New York; Appleton-Century-Crofts
27. Grisebach, H. and Patschke, L. (1961). *Z. Naturforsch.*, **16B**, 645
28. Grisebach, H. and Patschke, L. (1960). *Chem. Ber.*, **93**, 2326
29. Grisebach, H. and Brandner, G. (1961). *Z. Naturforsch.*, **16B**, 2
30. Grisebach, H. and Brandner, G. (1962). *Experentia*, **18**, 400
31. Grisebach, H. and Patschke, L. (1965). *Z. Naturforsch.*, **20B**, 399
32. Grisebach, H. and Billhuber, W. (1967). *Z. Naturforsch.*, **22B**, 746
33. Grisebach, H. and Patschke, L. (1962). *Z. Naturforsch.*, **17B**, 857
34. Patschke, L., Grisebach, H. and Barz, W. (1964). *Z. Naturforsch.*, **19B**, 1110
35. Hess, D. (1964). *Planta*, **60**, 568
36. Hess, D. (1966). *Z. Pflanzenphysiol.*, **55**, 374
37. Patschke, L. and Grisebach, H. (1965). *Z. Naturforsch.*, **20B**, 1039
38. Vaughan, P. F. T., Butt, V. S., Grisebach, H. and Schill, L. (1969). *Phytochemistry*, **8**, 1373
39. Roberts, R. J. and Vaughan, P. F. T. (1971). *Phytochemistry*, **11**, 2649
40. Patschke, L., Barz, W. and Grisebach, H. (1966). *Z. Naturforsch.*, **21B**, 45
41. Ramakrishnan, V. T. and Kagan, J. (1970). *J. Org. Chem.*, **35**, 2898, 2901
42. Grisebach, H., Barz, W., Hahlbrock, K., Kellner, S. and Patschke, L. (1966). *Proc. Fed. Eur. Biochem. Soc.*, **3**, 25
43. Walton, E. and Butt, V. S. (1971). *Phytochemistry*, **10**, 295
44. Clark-Lewis, J. W. (1962). *J. Pure Appl. Chem.*, **12**, 96
45. Arakawa, H. and Nakazaki, M. (1960). *Annalen.*, **636**, 111
46. Moustafa, E. and Wong, E. (1967). *Phytochemistry*, **6**, 625
47. Hahlbrock, K., Wong, E., Schill, L. and Grisebach, H. (1970). *Phytochemistry*, **9**, 949
48. Grisebach, H., Patschke, L. and Barz, W. (1966). *Z. Naturforsch.*, **21B**, 201
49. Wong, E. (1968). *Phytochemistry*, **7**, 1751
50. Wong, E. and Grisebach, H. (1969). *Phytochemistry*, **8**, 1419
51. Barz, W., Patschke, L. and Grisebach, H. (1965). *Chem. Commun.*, 400
52. Fritsch, H., Hahlbrock, K. and Grisebach, H. (1971). *Z. Naturforsch.*, **26B**, 581
53. Grisebach, H. and Kellner, S. (1965). *Z. Naturforsch.*, **20B**, 446

296 Higher Plants: Extension of the Phenylpropanoid Unit

54. Hahlbrock, K., Ebel, J., Ortmann, R., Sutter, A., Wellmann, E. and Grisebach, H. (1971). *Biochem. Biophys. Acta*, **244**, 7
55. Hahlbrock, K. and Grisebach, H. (1970). *FEBS Lett.*, **11**, 62
56. Hahlbrock, K., Sutter, A., Wellmann, E., Ortmann, R. and Grisebach, H. (1971). *Phytochemistry.* **10**, 109
57. Ortmann, R., Sanderman, H. and Grisebach, H. (1970). *FEBS Lett.*, **7**, 164
58. Ollis, W. D. (1961). *Chemistry of Natural Phenolic Compounds*, p. 152. Edited by W. D. Ollis. Oxford; Pergamon
59. Bate-Smith, E. C. and Lerner, N. H. (1954). *Biochem. J., **58**, 126
60. Robinson, G. M. and Robinson, R. (1933). *Biochem. J.*, **27**, 206
61. Weinges, K., Kaltenhauser, W., Marx, H.-D., Nader, E., Perner, J. and Seiler, D. (1968). *Annalen*, **711**, 184
62. Thompson, R. S., Jacques, D., Haslam, E. and Tanner, R. J. N. (1972). *J. Chem. Soc.* (Perkin I), 1387
63. Roux, D. G. (1972). *Phytochemistry*, **11**, 1219
64. Birch, A. J., Dahl, C. J. and Pelter, A. (1967). *Tetrahedron Lett.*, 481
65. Birch, A. J., Dahl, C. J. and Pelter, A. (1969). *Austral. J. Chem.*, **22**, 423
66. Grisebach, H. and Barz, W. (1964). *Z. Naturforsch.*, **19B**, 569
67. Crombie, L. and Thomas, M. B. (1967). *J. Chem. Soc.* (C), 1796
68. Grisebach, H. and Barz, W. (1966). *Z. Naturforsch.*, **21B**, 47
69. Wong, E. (1965). *Biochem. Biophys. Acta*, **111**, 358
70. Wong, E. (1968). *Chem. Commun.*, 395
71. Ollis, W. D., Ormand, K. L. and Sutherland, I. O. (1968). *Chem. Commun.*, 1237
72. Pelter, A., Bradshaw, J. and Warren, R. F. (1971). *Phytochemistry*, **10**, 835
73. Dewick, P. M., Barz, W. and Grisebach, H. (1970). *Phytochemistry*, **9**, 775
74. Berlin, J., Dewick, P. M., Barz, W. and Grisebach, H. (1972). *Phytochemistry*, **11**, 1689
75. Crombie, L., Green, C. L. and Whiting, D. A. (1968). *J. Chem. Soc.*, (C), 3029
76. Dewick, P. M., Whiting, D. A. and Crombie, L. (1970). *Chem. Commun.*, 1469
77. Dewick, P. M., Whiting, D. A. and Combie, L. (1971). *Chem. Commun.*, 1182, 1183
78. Dewick, P. M., Barz, W. and Grisebach, H. (1969). *Chem. Commun.*, 466
79. Barton, D. H. R., Kirby. G. W. and Taylor, J. B. (1962). *Proc. Chem. Soc.*, 340
80. Ollis, W. D. (1968). *Recent Advances in Phytochemistry*, p. 329, Volume I. Edited by T. J. Mabry, R. E. Alston and V. C. Runeckles. New York; Appleton-Century-Crofts
81. Polonsky, J. (1957). *Bull. Soc. Chim.*, 1079
82. Crombie, L., Games, D. and McCormick, A. (1966). *Tetrahedron Lett.*, 145
83. Kunesch, G. and Polonsky, J. (1967). *Chem. Commun.*, 317
84. Kunesch, G. and Polonsky, J. (1969). *Phytochemistry*, **8**, 1221
85. Polonsky, J., Gautier, J., Cave, A. and Kunesch, G. (1972). *Experentia*, **28**, 759
86. Seshadri, T. R. (1959). *Tetrahedron*, **6**, 169
87. Ollis, W. D. and Gottlieb, O. R. (1968). *Chem. Commun.*, 1396
88. Braga deOliveira, A., Gottlieb, O. R., Ollis, W. D. and Rizzini, C. T. (1971). *Phytochemistry*, **10**, 1863
89. Billek, G. (1964). *Fortschritt. Chem. Org. Naturstoffe*. **22**, 115
90. Asakawa, Y. (1970). *Bull. Chem. Soc.* (Japan), **43**, 575
91. King, F. E., King, T. J., Godson, D. H. and Manning, L. C. (1956). *J. Chem. Soc.*, 4477
92. Billek, G. and Ziegler, W. (1962). *Monats. Chem.*, **93**, 1430
93. von Rudloff, E. and Jorgensen, E. (1963). *Phytochemistry*, **2**, 297
94. Hillis, W. E. and Hasegawa, M. (1962). *Chem. and Ind.*, 1330
95. Billek, G. and Schimpl, A. (1966). *Biosynthesis of Aromatic Compounds*, p. 37. Oxford; Pergamon

96. Asahina, Y. and Asano, J. (1929). *Chem. Ber.*, **62**, 171
97. Asahina, Y. and Asano, J. (1930). *Chem. Ber.*, **64**, 2059
98. Askari, A., Northen, L. R. and Shimizu, Y. (1972). *Lloydia*, **35**, 49
99. Pryce, R. J. (1971). *Phytochemistry*, **10**, 2679
100. Ibrahim, R. K. and Towers, G. H. N. (1960). *Can. J. Biochem. Physiol.*, **38**, 627
101. Ibrahim, R. K. and Towers, G. H. N. (1962). *Can. J. Biochem. Physiol.*, **40**, 449
102. Billek, G. and Kindl, H. (1961). *Monats. Chem.*, **92**, 493
103. Billek, G. and Kindl, H. (1962). *Monats. Chem.*, **93**, 814
104. Thomas, R. (1961). *Biochem. J.*, **78**, 807
105. Asakawa, Y., Genjida, F., Hayashi, S. and Matsuura, T. (1969). *Tetrahedron Lett.*, 3235
106. Aragao Craviero, A., Da Costa Prado, A., Gottlieb, O. R. and Welerson de Albuquerque, P. C. (1970). *Phytochemistry*, **8**, 1869
107. Campbell, R. V. M., Crombie, L., Tuck, B. and Whiting, D. A. (1970). *Chem. Commun.*, 1206
108. Begley, M. J. and Whiting, D. A. (1970). *Chem. Commun.*, 1207
109. Yasue, M. (1965). *Nippon Mokuzai Gakkaishi*, **11**, 146, 153
110. Thomas, R. (1971). *Chem. Commun.*, 739
111. Edwards, J. M., Schmitt, R. C. and Weiss, U. (1972). *Phytochemistry*, **11**, 1717
112. Roughley, P. J. and Whiting, D. A. (1971). *Tetrahedron Lett.*, 3741
113. Edwards, J. M., Churchill, J. A. and Weiss, U. (1970). *Phytochemistry*, **9**, 1563
114. Kriegler, A. B. and Thomas, R. (1971). *Chem. Commun.*, 738
115. Cooke, R. G. and Segal, W. (1958). *Austral. J. Chem.*, **11**, 230
116. Thomson, R. H. (1962). *Comparative Biochemistry*, p. 630. Vol. IIIA, Edited by M. Florkin and H. S. Mason. London and New York; Academic Press
117. Thomson, R. H. (1971). *Naturally Occurring Quinones*. London and New York; Academic Press
118. Naumann, M. O., Britton, G. and Haslam, E. (1964). *J. Chem. Soc.*, 5649
119. Britton, G. and Haslam, E. (1965). *J. Chem. Soc.*, 7312
120. Zenk, M. H. (1964). *Z. Naturforsch.*, **19B**, 856
121. Bolkart, K. H. and Zenk, M. H. (1968). *Z. Pflanzenphysiol.*, **58**, 439
122. Chen, D. and Bohm, B. A. (1966). *Can. J. Biochem.*, **44**, 1389
123. Leistner, E. and Zenk, M. H. (1967). *Z. Naturforsch.*, **22B**, 460
124. Leistner, E. and Zenk, M. H. (1968). *Z. Naturforsch.*, **23B** 259
125. Leduc, M. M., Dansette, P. M. and Azerad, R. G. (1970). *Eur. J. Biochem.*, **15**, 428
126. Scharf, K.-H., Zenk, M. H., Onderka, D. K., Carroll, M. and Floss, H. G. (1971). *Chem. Commun.*, 576
127. Dansette, P. and Azerad, R. G. (1970). *Biochem. Biophys. Res. Comm.*, **40**, 1090
128. Grotzinger, E. and Campbell, I. M. (1972). *Phytochemistry*, **11**, 675
129. Knobloch, M., Bolkart, K. H. and Zenk, M. H. (1968). *Naturwiss*, **55**, 444, 445
130. Inouye, H. (1956). *J. Pharm. Soc.* (Japan), **76**, 976
131. Burnett, A. R. and Thomson, R. H. (1968). *J. Chem. Soc.* (C), 857
132. Leistner, E. (1971). *Phytochemistry*, **10**, 3015
133. Thomson, R. H. and Burnett, A. R. (1967). *J. Chem. Soc.* (C), 2100
134. Thomson, R. H. and Burnett, A. R. (1968). *J. Chem. Soc.* (C), 850, 854, 2437
135. Sandermann, W. and Simatupang, M. H. (1967). *Naturwiss*, **54**, 118
136. Zenk, M. H. and Leistner, E. (1968). *Tetrahedron Lett.*, 1395
137. Zenk, M. H. and Leistner, E. (1968). *Tetrahedron Lett.*, 861
138. Leistner, E. (1973). *Phytochemistry*, **12**, 337
139. Wyler, H. and Dreiding, A. S. (1957). *Helv. Chim. Acta*, **40**, 191
140. Schmidt, O. Th. and Schonleben, W. (1957). *Z. Naturforsch.*, **12B**, 262

141. Dreiding, A. S. (1961). *Recent Developments in the Chemistry of Natural Phenolic Compounds*, p. 104. Edited by W. D. Ollis. Oxford; Pergamon
142. Mabry, T. J. (1966). *Comparative Phytochemistry*, p. 231. Edited by T. Swain. London; Academic Press
143. Dreiding, A. S. and Wyler, H. (1961). *Experentia*, **17**, 23
144. Wyler, H., Mabry, T. J. and Dreiding, A. S. (1963). *Helv. Chim. Acta.*, **46**, 1745
145. Wilcox, M. E., Wyler, H., Mabry, T. J. and Dreiding, A. S. (1965). *Helv. Chim. Acta*, **48**, 252
146. Liebisch, H.-W., Matschiner, B. and Schutte, H. R. (1969). *Z. Pflanzenphysiol.*, **61**, 269
147. Mabry, T. J. and Dreiding, A. S. (1968). *Recent Advances in Phytochemistry*, p. 155. Edited by T. J. Mabry, R. E. Alston and V. C. Runeckles. New York; Appleton-Century-Crofts
148. Thomas, R. (1965). *Biogenesis of Antibiotic Substances.* p. 155. Edited by Z. Vanek and Z. Hostalek. New York; Academic Press
149. Horhämmer, L., Wagner, H. and Fritzsche, W. (1964). *Biochem. Z.*, **339**, 398
150. Minale, L., Piatelli, M. and Nicolaus, R. A. (1965). *Phytochemistry*, **4**, 593
151. Kohler, K. H. (1965). *Naturwiss.*, **52**, 561
152. Garay, A. S. and Towers, G. H. N. (1966). *Can. J. Bot.*, **44**, 231
153. Miller, H. E., Rosler, H., Wohlpart, A., Wyler, H., Wilcox, M. E., Frohofer, H., Mabry, T. J. and Dreiding, A. S. (1968). *Helv. Chim. Acta*, **51**, 1470
154. Wyler, H., Mabry, T. J. and Dreiding, A. S. (1965). *Helv. Chim. Acta*, **48**, 361
155. Dunkelblum, E., Mitter, H. E. and Dreiding, A. S. (1972). *Helv. Chim. Acta*, **55**, 642
156. Fischer, N. and Dreiding, A. S. (1972). *Helv. Chim. Acta*, **55**, 649
157. Hayaishi, O. and Nozaki, M. (1969). *Science*, **164**, 389
158. Evans, W. C. (1956). *Ann. Rep. Chem. Soc.*, **13**, 279
159. Ribbons, D. W. (1965). *Ann. Rep. Chem. Soc.*, **22**, 445
160. Daly, J. W., Guroff, G., Udenfriend, S. and Witkop, B. (1967). *Arch. Biochem. Biophys.*, **122**, 218
161. Impellizzeri, G. and Piatelli, M. (1972). *Phytochemistry*, **11**, 2499
162. Eyjolfsson, R. (1970). *Fortschritt. Chem. Org. Naturstoffe*, **28**, 74
163. Butler, G. W. and Conn, E. E. (1969). *Perspectives in Phytochemistry*, p. 47. Edited by J. B. Harborne and T. Swain. London; Academic Press
164. Ettlinger, M. and Kjaer, A. (1968). *Recent Advances in Phytochemistry*, p. 89. Edited by T. J. Mabry, R. E. Alston and V. C. Runeckles. New York; Appleton-Century-Crofts
165. Kjaer, A. (1960). *Fortschritt. Chem. Org. Naturstoffe*, **18**, 122
166. Herissey, H. (1907). *J. Pharm. Chem.*, **26**, 194
167. Fischer, E. and Bergmann, M. (1917). *Chem. Ber.*, **50**, 1047
168. Towers, G. H. N., McInnes, A. G. and Neish, A. C. (1964). *Tetrahedron*, **20**, 71
169. Dunston, W. R. and Henry, T. A. (1902). *Roy. Soc. Trans.*, **199A**, 399
170. Finnemore, H. and Cooper, J. M. (1938). *J. Soc. Chem. Ind.*, **57**, 162
171. Koukol, J., Miljanich, P. and Conn, E. E. (1962). *J. Biol. Chem.*, **237**, 3223
172. Gander, J. E. (1962). *J. Biol. Chem.*, **237**, 3229
173. Reay, P. F. and Conn, E. E. (1970). *Phytochemistry*, **9**, 1825
174. Uribe, E. G. and Conn, E. E. (1966). *J. Biol. Chem.*, **241**, 92
175. Butler, G. W. and Conn, E. E. (1964). *J. Biol. Chem.*, **239**, 1674
176. Tapper, B. A., Conn, E. E. and Butler, G. W. (1967). *Arch. Biochem. Biophys.*, **119**, 593
177. Hahlbrock, K., Tapper, B. A., Butler, G. W. and Conn, E. E. (1968). *Arch. Biochem. Biophys.*, **125**, 1013
178. Tapper, B. A., Zilg, H. and Conn, E. E. (1972). *Phytochemistry*, **11**, 1047
179. Hahlbrock, K. and Conn, E. E. (1970). *J. Biol. Chem.*, **245**, 917

180. Hahlbrock, K. and Conn, E. E. (1971). *Phytochemistry*, **10**, 1019
181. Ettlinger, M. G. and Lundeen, A. J. (1956). *J. Amer. Chem. Soc.*, **78**, 4172
182. Kjaer, A. (1954). *Acta Chem. Scand.*, **8**, 110
183. Barothy, J. and Neukom, H. (1965). *Chem. and Ind.*, 308
184. Ettlinger, M. G., Kjaer, A., Thompson, C. P. and Wagmieres, H. (1966). *Acta Chem. Scand.*, **20**, 1778
185. Kutazek, M., Prochazka, Z. and Veres, K. (1962). *Nature*, **194**, 393
186. Friis, P. and Kjaer, A. (1963). *Acta Chem. Scand.*, **17**, 1515
187. Benn, M. H. (1962). *Chem. and Ind.*, 1907
188. Underhill, W. E., Chisholm, M. D. and Wetter, L. R. (1962). *Can. J. Biochem.*, **40**, 1505
189. Kindl, H. and Underhill, W. E. (1968). *Phytochemistry*, **7**, 5145
190. Tapper, B. A. and Butler, G. W. (1967). *Arch. Biochem. Biophys.*, **120**, 719
191. Underhill, W. E. and Wetter, L. R. (1969). *Plant Physiol.*, **44**, 584
192. Matsuo, M. and Underhill, W. E. (1969). *Biochem. Biophys. Res. Comm.*, **36**, 18
193. Matsuo, M. and Underhill, W. E. (1971). *Phytochemistry*, **10**, 2279
194. Matsuo, M., Kirkland, D. F. and Underhill, W. E. (1972). *Phytochemistry*, **11**, 697
195. Gottlieb, O. R. and Magalhaes, M. T. (1959). *J. Org. Chem.*, **24**, 2070
196. Kosolapoff, G. M. and Brown, A. D. (1969). *Chem. and Ind.*, 1272
197. Gottlieb, O. R., Magalhaes, M. T. and Mors, W. B. (1961). *Anais Acad. Brasil Cienc.*, **33**, 301
198. Wetter, L. R. and Chisholm, M. D. (1968). *Can. J. Biochem.*, **46**, 931
199. Binkley, F. (1950). *J. Biol. Chem.*, **186**, 287
200. Schwimmer, S. and Kjaer, A. (1960). *Biochim. Biophys. Acta*, **42**, 316
201. Chisholm, M. D. (1972). *Phytochemistry*, **11**, 197
202. Underhill, W. E. (1965). *Can. J. Biochem.*, **43**, 179, 189
203. Underhill, W. E. (1967). *Eur. J. Biochem.*, **2**, 61
204. Underhill, W. E. (1968). *Can. J. Biochem.*, **46**, 401
205. Underhill, W. E. and Kirkland, D. F. (1972). *Phytochemistry*, **11**, 1973
206. Wetter, L. R. and Chisholm, M. D. (1964). *Can. J. Biochem.*, **42**, 1033
207. Stowe, B. B. (1963). *Comprehensive Biochemistry*. p. 117. Volume II. Edited by M. Florkin and E. H. Stotz. Amsterdam; Elsevier
208. Tukey, H. B., Went, F. W., Muir, R. M. and van Overbeek, J. (1954). *Plant Physiol.*, **29**, 307
209. Galston, A. W. and Davies, P. J. (1970). *Control Mechanisms in Plant Development*, p. 56. New York; Prentice-Hall
210. Thimann, K. V. (1935). *J. Biol. Chem.*, **109**, 279

Appendix

ENZYMOLOGY OF THE PATHWAY AND ITS DERIVATIVES

A review on the enzyme L-tryptophan synthetase has been published[1] and O'Neil Hoch has recently described[2] her detailed investigations of the properties of the enzyme from *Bacillus subtilis*. In all organisms which have been studied as yet, L-tryptophan synthetase has been shown to be a tetrameric complex capable of performing both the A and B reactions without releasing indole. The enzyme in procaryotic organisms and plants is readily dissociable into two dissimilar units (the α and β_2 components) and the L-tryptophan synthetases which have been obtained from two Gram-negative bacteria, *Escherichia coli* and *Pseudomonas putida*, show many similarities (molecular weight, sub-unit structure and amino acid sequence adjacent to the cofactor binding site). There are, however, distinct differences in the two organisms in the organisation and regulation of the *trp* genes. Whereas the genes coding for the α and β chains are part of the *trp* operon in *Escherichia coli* for the *Pseudomonas* enzyme, they are separate from the other *trp* genes and are subject to repression by indole glycerol 3-phosphate.

L-Tryptophan synthetase in eucaryotic organisms, apart from that in *Chlorella ellipsoidea*[3], is, in contrast, not readily dissociable and the study of the enzyme from the Gram-positive bacterium *Bacillus subtilis* was undertaken from the comparative viewpoint since the life style of this organism is intermediate in character between bacteria with only a vegetative life cycle and the differentiating moulds and yeasts. Both α (molecular weight 26 000) and β_2 (molecular weight 82 000) sub-units were isolated. The β_2 component required the α sub-unit to attain maximal activity in the B reaction but successful complementation was also achieved with the α sub-units from the *Escherichia coli* and *Pseudomonas putida* enzymes to produce a level of activity 30 per cent of the homologous complementation. O'Neil Hoch concluded that the β_2 sub-unit from the *Bacillus subtilis* enzyme occupied an intermediate position in the comparative sense and appeared to be more closely related to the algal β_2 sub-units than those from the Gram-negative bacteria. The α sub-unit had an amino acid composition which was quite different from that of the *Pseudomonas putida* enzyme but only showed significant divergences in its content of L-serine and L-proline from the α sub-unit of the *Escherichia coli* enzyme. Unlike the β_2 sub-unit, the α sub-unit, which displayed an absolute requirement for the β_2 sub-unit in the A reaction, was not complemented by the β_2 components from *Escherichia coli* or *Pseudomonas putida*.

Further observations have been made by Arroyo-Begovich and DeMoss[4] on the components of the multi enzyme anthranilate synthetase complex from *Neurospora crassa*. From purification and characterisation studies, it was proposed that this complex has a molecular weight of 240 000 and is composed of six monomers of two distinct types which are specified by the *trp* 1 and *trp* 2 genes. The components in the *trp* 1 and *trp* 2 mutants of *Neurospora crassa*, which interact *in vitro* to form the active anthranilate synthetase complex, have now been purified and characterised. Component I (the *trp* 2 gene product, molecular weight $\sim$ 80 000) catalyses the anthranilate synthetase reaction only with ammonia as the amino donor. The undissociated form of component II (the *trp* 1 gene product, molecular weight $\sim$ 200 000) interacts with component I to produce a 1:1 enzyme complex with glutamine dependent anthranilate synthetase activity. Component II alone catalyses the PRA isomerase and IGP synthetase reactions. The anthranilate synthetase complex of *Neurospora crassa* is therefore organised in a structure analogous to that of the same enzyme isolated from several bacteria. It differs from these systems, however, in the nature of the further reactions of L-tryptophan biosynthesis which the second component of the complex catalyses and also in the fact that this component separates into two further sub-units (IIa and IIb) during purification. The 30 000 molecular weight sub-unit IIa interacts with component I to produce the glutamine linked anthranilate synthetase and sub-unit IIb (molecular weight 160 000) contains both the PRA isomerase and IGP synthetase activities.

The properties of the enzyme chorismate mutase have been similarly further examined in a variety of organisms and a remarkable diversity of properties has been demonstrated. In *Escherichia coli*, *Aerobacter aerogenes* and *Salmonella typhimurium*, two enzymes function in the utilisation of chorismate for the biosynthesis of L-phenylalanine and L-tyrosine but in a number of eucaryotic organisms there appears to exist only one chorismate mutase which does not carry prephenate dehydratase or prephenate dehydrogenase activities. This latter observation has been further extended by Lowe and Westlake[5] and Gorisch and Lingens[6, 7] with *Streptomyces* species. An interesting feature of the operation of the shikimate pathway in the *Streptomyces* strains which were examined by Lowe and Westlake[5] was the apparent absence of effective systems of control of the flow of metabolites. Gorisch and Lingens[7] similarly concluded that the level of L-tryptophan concentration acts as the only signal to increase or decrease the rate of synthesis of L-phenylalanine or L-tyrosine in *Streptomyces aureofaciens*. In contrast Woodin and Nishioka[8] have demonstrated that the fungi *Penicillium chrysogenum* and *Penicillium duponti* contain two electrophoretically distinct chorismate mutases both of which are inhibited by L-phenylalanine and L-tyrosine and activated by L-tryptophan. A third chorismate mutase was distinguished in *Penicillium duponti* whose activity was unaffected by L-phenylalanine or L-tyrosine. Three forms of the enzyme chorismate mutase were also detected and separated—DEAE cellulose and gel electrophoresis—from alfalfa plants[9]. The iso-enzymes (CM_1, CM_2 and CM_3) were distinguished by their differing molecular weights (46 000, 58 000 and 69 000) and their differing electrophoretic mobility. The bulk of the chorismate mutase activity was found as CM_1. Studies presently underway in this laboratory suggest that most plants, including a wide variety of monocotyledons and dicotyledons possess three chorismate mutase iso-enzymes. Only *Pisum sativum* and *Phaseolus vulgaris* of the dicotyledons and a few primitive plants such as fern and pine are limited to two chorismate mutase enzymes.

Whitt and DeMoss[10] have made observations on the molecular size of the active species of the enzyme tryptophanase from *Bacillus alvei*. They showed that the tetrameric form of the enzyme[11], although it readily dissociates into dimers, is responsible for both of the activities in the α, β-elimination and β-replacement reactions.

BIOSYNTHESIS

Baldwin, Snyder and Rapoport[12] in an important paper have directed attention at the symmetry (or otherwise) of the as yet unidentified naphthalenic intermediate in bacterial menaquinone biosynthesis. 7-[^{14}C]-Shikimic acid, prepared by the procedure of Grewe and Vangermain from 3,4,5-triacetoxycyclohexanone, was incorporated (0.25 per cent) into 9-dihydromenaquinone (1) in *Mycobacterium phlei*. Degradation gave C-1 and C-4 of the original menaquinone as the quinoxaline derivatives (2) and (3) respectively. The radioactive label administered was observed to reside almost exclusively at C-4 but not C-1 of the menaquinone. This result therefore excludes symmetrical compounds such as 1,4-naphthoquinone as precursors of the bacterial menaquinones and in this respect supports the observations of Grotzinger and Campbell on 2-hydroxy-1,4-naphthoquinone biosynthesis in plants.

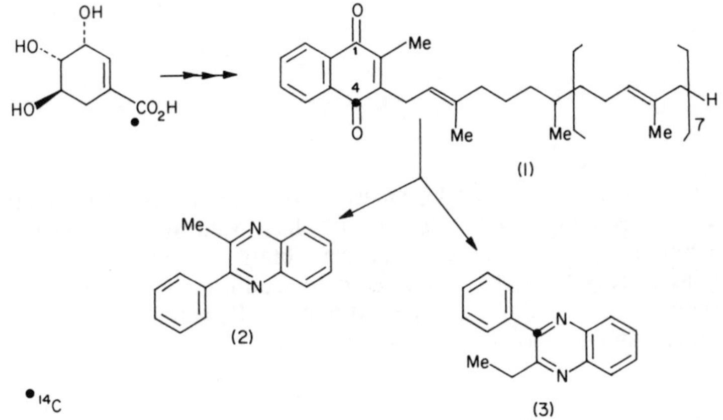

Measurement of the isotope effect following the NIH shift has been made using the hydroxylation of L-phenylalanine by *Pseudomonas* species[13]. Labelled ($4'$-^{3}H; carboxy-^{14}C and $4'$-^{3}H; $3,5$-^{2}H$_2$; carboxy-^{14}C) specimens of D L-phenylalanine were administered to the intact organism of a *Pseudomonas* strain and L-tyrosine was isolated after approximately 6 hours incubation. Degradation of the L-tyrosine showed that, in accord with earlier work, hydroxylation of $4'$-[^{3}H]-D L-phenylalanine by the intact organism proceeds with high (~ 95 per cent) migration and retention of the tritium. With the deuteriated precursor, a lower retention (~ 74 per cent) of tritium was observed. Ignoring secondary isotope effects and assuming the Swain relationship a k_H/k_D ratio of 10 ± 1 was calculated for the process. Kirby and his colleagues concluded, on the basis of this evidence and related work, that the hydroxylation of D L-phenylalanine by *Pseudomonas* species was best represented by the process ($4 \rightarrow 5$) below and that there is no indication that either of the last two steps are enzymically, and hence stereochemically, controlled.

Following its discovery by Erdtman dehydrododiisoeugenol has been considered as a model for lignin biosynthesis. Erdtman's view that one 'konnte die Moglichkeit ins Auge fassen, dass das

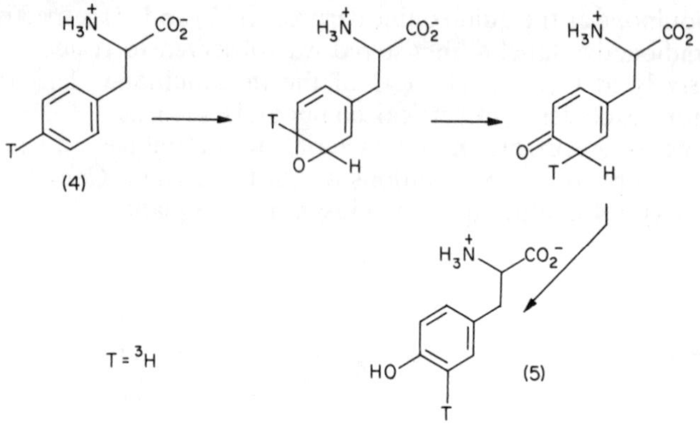

(4)

T = ³H

(5)

Lignin durch oxydative Polymerisation eines in der Seitenkette oxydierten Propylguaiakols ensteht' has been widely accepted. Further support for this idea has been obtained by Gottlieb and his collaborators[14] who have isolated dehydrodiisoeugenol (6) and the related methylenedioxy compound in optically active forms from a natural source, *Licaria aritu*, for the first time.

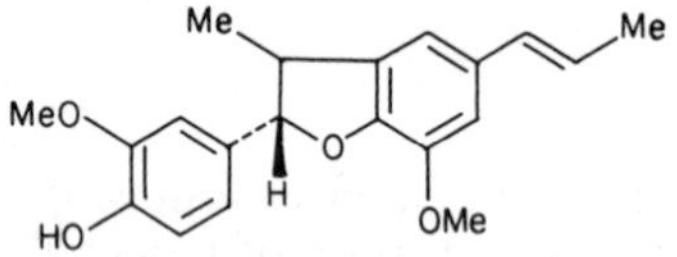

(6)

Further work on the mode of formation of the phenylpropanoid precursors of lignin in higher plants has also been published. Confirmatory observations on the stereochemical mode of action of tyrosine ammonia lyase have been made[15] and a review of the biochemical properties of phenylalanine ammonia lyase has been compiled by Towers and Camm[16]. Zenk and his group[17, 18], using a cell free system from cambial tissue of *Salix alba* and a cell free preparation from phytotron grown *Forsythia* species, have demonstrated for the first time the reduction of ferulate (7) to coniferyl alcohol (8) in a higher plant. The conversion is dependent on ATP, co-enzyme A and reduced pyridine nucleotides and Zenk has formulated the reaction sequence as shown below[18] (7 → 8).

Speculation still surrounds some of the aromatic hydroxylation processes which are believed to operate in higher plants. Griffith and Conn[19] have thus shown, contrary perhaps to expectation, that partially purified phenolase preparation from *Vicia faba* do *not* catalyse the conversion of L-tyrosine to L-dopa and they concluded that phenolase does not appear to play a role in L-dopa synthesis.

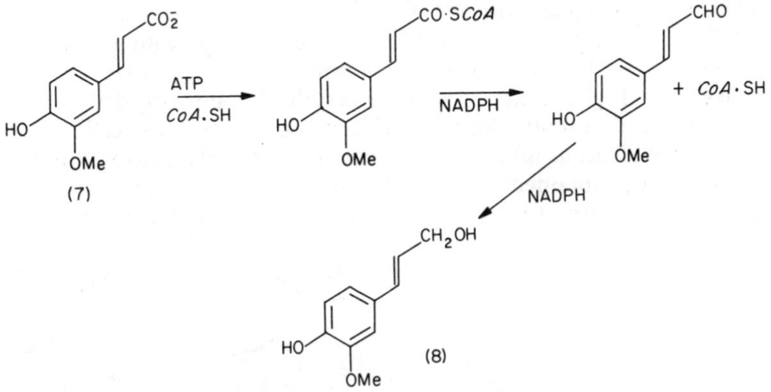

$$CO_2^- \xrightarrow[\;CoA.SH\;]{ATP} CO\cdot SCoA \xrightarrow[\;]{NADPH} CHO + CoA\cdot SH$$

(7)

$$\xrightarrow{NADPH} CH_2OH$$

(8)

Grisebach and Hahlbrock and their collaborators continue to successfully exploit the potential of the cell suspension culture technique to investigate the biosynthesis of flavonoids. Experiments have continued on the induction of the enzyme systems related to phenylpropanoid metabolism in cell suspension cultures of parsley. Hahlbrock and Wellmann[20] have shown that the three enzymes of group I (phenylalanine ammonia lyase, trans-cinnamate-4-hydroxylase and p-coumarate CoA ligase) may be induced in the absence of light by the transfer of aged parsley cells to fresh culture media. Sutter and Grisebach[21] have isolated a UDP-glucose:flavonol 3-O-glucosyltransferase from cell suspension cultures of parsley. The enzyme has a strict positional specificity and catalyses the 3-O-glucosylation of a number of flavonols. A p-coumarate; CoA ligase has also been isolated from cell suspension cultures of soya bean by Lindl, Kreuzaler and Hahlbrock[22] and a molecular weight of 55 000 was estimated for the enzyme. This observation complements the earlier work of Kreuzaler and Hahlbrock[23] who showed that extracts from cell suspension cultures of parsley were able to convert malonyl CoA and p-coumaroyl CoA to naringenin (5,7,4'-trihydroxy-flavanone). In related work Wallace and Grisebach[24] have noted the incorporation of 2-[^{14}C]-naringenin by C-glycosylation into C-glycosylflavones such as vitexin in Spirodela polyrhiza.

Miscellaneous

Sprinson and his co-workers[25] have produced syntheses of 2-R and 2-S-2-hydroxyquinic acid (9 and 10) and 2-R and 2-S-2 hydroxy-epiquinic acid (11 and 12) from (−)-shikimic acid via the bromo acid

(13) or its 1,5 or γ-lactone, *Figure A.1*. Treatment of (13) with baryta at 0° gave the epoxide (14) and oxirane cleavage with the same reagent at 100° gave (9) and (12) respectively. Cleavage of the epoxide with glacial acetic acid yielded the 1,4-anhydro derivative (15) and further treatment of its triacetyl derivative with acetic acid and saponification afforded the acid (10). 2R-2-Hydroxyepiquinic acid (11) was obtained by two independent routes and most simply by hydroxylation of methyl shikimate with osmium tetroxide.

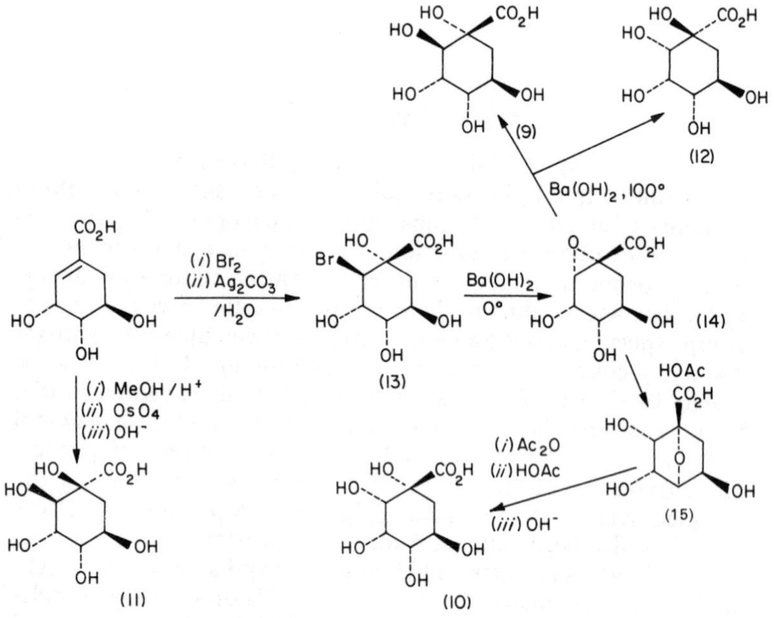

Figure A.1. 2-Hydroxyquinic and 2-hydroxyepiquinic acid[25]

Japanese workers[26] have described a novel microbiological synthesis of L-tyrosine and L-dopa from sodium pyruvate, ammonia and phenol or catechol respectively, using cells of *Erwinia herbicola*. The reaction is catalysed by tyrosine phenol lyase and is the reverse of the normal $\alpha\beta$-elimination reaction catalysed by this enzyme. The L-tyrosine or L-dopa synthesised by this enzymatic method precipitates from the reaction media during incubation. The maximum amount of L-tyrosine synthesised by this method was 60.5 g per litre and of L-dopa 58.5 g per litre. The process is a variation on the analogous procedure reported by the same school which used D L-serine in place of sodium pyruvate.

REFERENCES

1. Yanofsky, C. and Crawford, I. P. (1972). *The Enzymes*, p. 1. Vol. VII. Edited by P. Boyer. London and New York; Academic Press
2. O'Neil Hoch, S. (1973). *J. Biol. Chem.*, **248**, 2992, 2999
3. Sakuguchi, K. (1970). *Biochim. Biophys. Acta*, **220**, 580
4. Arroyo-Begovich, A. and DeMoss, J. A. (1973). *J. Biol. Chem.*, **248**, 126
5. Lowe, D. A. and Westlake, D. W. S. (1972). *Canad. J. Biochem.*, **50**, 1064
6. Gorisch, H. and Lingens, F. (1972). *Arch. Mikrobiol.*, **82**, 147
7. Gorisch, H. and Lingens, F. (1973). *J. Bacteriol.*, **114**, 645
8. Woodin, T. S. and Nishioka, L. (1973). *Biochim. Biophys. Acta*, **309**, 224
9. Woodin, T. S. and Nishioka, L. (1973). *Biochim. Biophys. Acta*, **309**, 211
10. Whitt, D. D. and DeMoss, R. D. (1973). *Biochim. Biophys. Acta*, **309**, 486
11. Hoch, S. O. and DeMoss, R. D. (1972). *J. Biol. Chem.*, **247**, 1750
12. Baldwin, R. M. Snyder, C. D. and Rapoport, H. (1973). *J. Amer. Chem. Soc.*, **95**, 276
13. Bowman, W. R., Gretton, W. R. and Kirby, G. W. (1973). *J. Chem. Soc.* (Perkin I), 218
14. Aiba, C. J., Campos Correa, R. G. and Gottlieb, O. R. (1973). *Phytochemistry*, **12**, 1163
15. Ellis, B. E., Zenk, M. H., Kirby, G. W., Michael, J. and Floss, H. G. (1973). *Phytochemistry*, **12**, 1057
16. Camm, E. L. and Towers, G. H. N. (1973). *Phytochemistry*, **12**, 961
17. Mansell, R. L., Stockigt, J. and Zenk, M. H. (1972). *Z. Pflanzenphysiol.*, **68**, 286
18. Gross, C. G., Stockigt, J., Mansell, R. L. and Zenk, M. H. (1973). *FEBS Lett.*, **31**, 283
19. Griffith, T. and Conn, E. E. (1973). *Phytochemistry*, **12**, 1651
20. Hahlbrock, K. and Wellmann, E. (1973). *Biochim. Biophys. Acta*, **304**, 702
21. Sutter, A. and Grisebach, H. (1973). *Biochim. Biophys. Acta*, **309**, 289
22. Lindl, T., Kreuzaler, F. and Hahlbrock, K. (1973). *Biochim. Biophys. Acta*, **302**, 457
23. Kreuzaler, F. and Hahlbrock, K. (1972). *FEBS Lett.*, **28**, 69
24. Wallace, J. W. and Grisebach, H. (1973). *Biochim. Biophys. Acta*, **304**, 837
25. Adlersberg, M., Bondinell, W. E. and Sprinson, D. B. (1973). *J. Amer. Chem. Soc.*, **95**, 887
26. Enei, H., Nakazawa, H., Okumura, S. and Yamada, H. (1973). *Agr. Biol. Chem.*, **37**, 725
27. Enei, H., Matsui, H., Nakazawa, H., Okumura, S. and Yamada, H. (1973). *Agr. Biol. Chem.*, **37**, 493

Index

309